T0302076

Geometry of the Phase Retrieval Problem

Recovering the phase of the Fourier transform is a ubiquitous problem in imaging applications from astronomy to nanoscale X-ray diffraction imaging. Despite the efforts of a multitude of scientists, from astronomers to mathematicians, there is as yet no satisfactory theoretical or algorithmic solution to this class of problems. Written for mathematicians, physicists, and engineers working in image analysis and reconstruction, this book introduces a conceptual, geometric framework for the analysis of these problems, leading to a deeper understanding of the essential, algorithmically independent, difficulty of their solutions. Using this framework, the book studies standard algorithms and a range of theoretical issues in phase retrieval and provides several new algorithms and approaches to these problems with the potential to improve the reconstructed images. The book is lavishly illustrated with the results of numerous numerical experiments that motivate the theoretical development and place it in the context of practical applications.

ALEXANDER H. BARNETT is Group Leader for Numerical Analysis at the Center for Computational Mathematics in the Flatiron Institute. He has published around 60 papers on partial differential equations, waves, fast algorithms, integral equations, neuroscience, imaging, signal processing, inverse problems, and physics, and received several research grants from the National Science Foundation.

CHARLES L. EPSTEIN is the Thomas A. Scott Professor Emeritus of Mathematics at the University of Pennsylvania, where he founded the graduate group in Applied Mathematics and Computational Science. He is currently a Senior Research Scientist at the Centre for Computational Mathematics at the Flatiron Institute. He has worked on a wide range of problems in pure and applied analysis and is the author of a widely used textbook *An Introduction to the Mathematics of Medical Imaging* (SIAM 2008). He shared the Bergman Prize in 2016 with Francois Treves and is a fellow of the American Association for the Advancement of Science and the American Mathematical Society.

LESLIE GREENGARD is Silver Professor of Mathematics and Computer Science at the Courant Institute, New York University and Director of the Center for Computational Mathematics at the Flatiron Institute. He is coinventor of several widely used fast algorithms and a member of the American Academy of Arts and Sciences, the National Academy of Sciences, and the National Academy of Engineering.

JEREMY MAGLAND is a Senior Data Scientist at the Flatiron Institute. He received his PhD in Mathematics from the University of Pennsylvania. Prior to joining the Flatiron Institute in 2015, he worked for about a decade as a research scientist in the Radiology Department of the Hospital of the University of Pennsylvania, where he developed software systems that dramatically streamlined experimental work on MRI scanners.

The *Cambridge Monographs on Applied and Computational Mathematics* series reflects the crucial role of mathematical and computational techniques in contemporary science. The series publishes expositions on all aspects of applicable and numerical mathematics, with an emphasis on new developments in this fast-moving area of research.

State-of-the-art methods and algorithms as well as modern mathematical descriptions of physical and mechanical ideas are presented in a manner suited to graduate research students and professionals alike. Sound pedagogical presentation is a prerequisite. It is intended that books in the series will serve to inform a new generation of researchers.

A complete list of books in the series can be found at
www.cambridge.org/mathematics.
Recent titles include the following:

21. Spectral methods for time-dependent problems, *Jan Hesthaven, Sigal Gottlieb & David Gottlieb*
22. The mathematical foundations of mixing, *Rob Sturman, Julio M. Ottino & Stephen Wiggins*
23. Curve and surface reconstruction, *Tamal K. Dey*
24. Learning theory, *Felipe Cucker & Ding Xuan Zhou*
25. Algebraic geometry and statistical learning theory, *Sumio Watanabe*
26. A practical guide to the invariant calculus, *Elizabeth Louise Mansfield*
27. Difference equations by differential equation methods, *Peter E. Hydon*
28. Multiscale methods for Fredholm integral equations, *Zhongying Chen, Charles A. Micchelli & Yuesheng Xu*
29. Partial differential equation methods for image inpainting, *Carola-Bibiane Schönlieb*
30. Volterra integral equations, *Hermann Brunner*
31. Symmetry, phase modulation and nonlinear waves, *Thomas J. Bridges*
32. Multivariate approximation, *V. Temlyakov*
33. Mathematical modelling of the human cardiovascular system, *Alfio Quarteroni, Luca Dede', Andrea Manzoni & Christian Vergara*
34. Numerical bifurcation analysis of maps, *Yuri A. Kuznetsov & Hil G.E. Meijer*
35. Operator-adapted wavelets, fast solvers, and numerical homogenization, *Houman Owhadi & Clint Scovel*
36. Spaces of measures and their applications to structured population models, *Christian Düll, Piotr Gwiazda, Anna Marciniak-Czochra & Jakub Skrzeczkowski*
37. Geometry of the Phase Retrieval Problem, *Alexander H. Barnett, Charles L. Epstein, Leslie Greengard & Jeremy Magland*

Geometry of the Phase Retrieval Problem

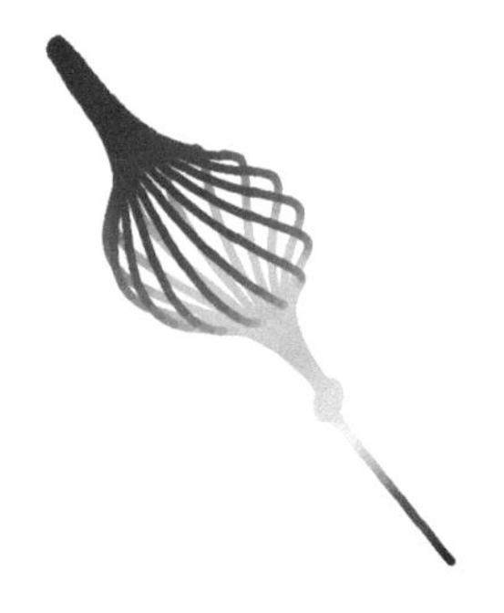

GRAVEYARD OF ALGORITHMS

ALEXANDER H. BARNETT
Flatiron Institute

CHARLES L. EPSTEIN
Flatiron Institute

LESLIE GREENGARD
Courant Institute

JEREMY MAGLAND
Flatiron Institute

CAMBRIDGE
UNIVERSITY PRESS

CAMBRIDGE
UNIVERSITY PRESS

University Printing House, Cambridge CB2 8BS, United Kingdom

One Liberty Plaza, 20th Floor, New York, NY 10006, USA

477 Williamstown Road, Port Melbourne, VIC 3207, Australia

314–321, 3rd Floor, Plot 3, Splendor Forum, Jasola District Centre,
New Delhi – 110025, India

103 Penang Road, #05–06/07, Visioncrest Commercial, Singapore 238467

Cambridge University Press is part of the University of Cambridge.

It furthers the University's mission by disseminating knowledge in the pursuit of
education, learning, and research at the highest international levels of excellence.

www.cambridge.org
Information on this title: www.cambridge.org/9781316518878
DOI: 10.1017/9781009003919

First published 2022

A catalogue record for this publication is available from the British Library.

Library of Congress Cataloging-in-Publication Data

Names: Barnett, Alex, 1972 December 7- author. | Epstein, Charles L.,
author. | Greengard, Leslie, author. | Magland, Jeremy, author.
Title: Geometry of the phase retrieval problem : graveyard of algorithms /
Alexander H. Barnett, Flatiron Institute, Charles L. Epstein, Flatiron
Institute, Leslie Greengard, Courant Institute, and Jeremy Magland, Flatiron Institute.
Description: New York, NY : Cambridge University Press, 2022. |
Series: Cambridge monographs on applied and computational |
Includes bibliographical references and index.
Identifiers: LCCN 2021044734 (print) | LCCN 2021044735 (ebook) |
ISBN 9781316518878 (hardback) | ISBN 9781009003919 (epub)
Subjects: LCSH: Geometry. | Algorithms.
Classification: LCC QC20.7.G44 B36 2022 (print) | LCC QC20.7.G44 (ebook) |
DDC 530.15/6–dc23/eng/20211117
LC record available at https://lccn.loc.gov/2021044734
LC ebook record available at https://lccn.loc.gov/2021044735

ISBN 978-1-316-51887-8 Hardback

Contents

Preface

Coherent diffraction imaging (CDI) is an experimental technique for determining the detailed structure of an object at the nanometer length scale. Coherent X-ray sources are used to illuminate the sample, and the scattered photons are captured in the far field on a detector array. In this regime, classical electromagnetic theory, in the Born approximation, shows that the measured intensity at each pixel in the detector is well approximated as the square of the modulus of the 3D Fourier transform, $|\widehat{\rho}(\boldsymbol{k})|^2$, of the X-ray scattering density, $\rho(\boldsymbol{x})$. Throughout this book we use the "mathematician's" convention for the $\boldsymbol{k}$ vector, in which exponentials take the form $e^{2\pi i\langle \boldsymbol{x},\boldsymbol{k}\rangle}$, rather than the "physicist's" convention, in which they are $e^{i\langle \boldsymbol{x},\boldsymbol{k}\rangle}$. With this convention the forward Fourier transform is defined by

$$\hat{\rho}(\boldsymbol{k}) = \mathscr{F}(\rho)(\boldsymbol{k}) \stackrel{d}{=} \int_{\mathbb{R}^d} \rho(\boldsymbol{x}) e^{-2\pi i\langle \boldsymbol{x},\boldsymbol{k}\rangle} d\boldsymbol{x}, \tag{1}$$

with inverse

$$\rho(\boldsymbol{x}) = \mathscr{F}^{-1}(\hat{\rho})(\boldsymbol{x}) \stackrel{d}{=} \int_{\mathbb{R}^d} \hat{\rho}(\boldsymbol{k}) e^{2\pi i\langle \boldsymbol{x},\boldsymbol{k}\rangle} d\boldsymbol{k}. \tag{2}$$

We assume that the illuminating light is a plane wave, $\boldsymbol{A}e^{2\pi i\langle \boldsymbol{x},\boldsymbol{p}_0\rangle}$, with wave vector $\boldsymbol{p}_0 = (0,0,p)$ in the above convention. If the detector array is oriented orthogonal to the illuminating light at a distance D downstream from the source, then the measured spatial frequency, $\boldsymbol{k}$, is related to the location of the measurement point $\boldsymbol{y} = (y_1, y_2, D)$ by the *Ewald Sphere* construction:

$$\boldsymbol{k} = p\left(\frac{\boldsymbol{y}}{\|\boldsymbol{y}\|} - (0,0,1)\right). \tag{3}$$

Here we assume that the experimental parameters are in the so-called Fraunhofer regime: D is much larger than R, the diameter of the support of ρ, and furthermore $\lambda D \gg R^2$, where $\lambda = 1/p$ is the wavelength. The maximum frequency component that is measured is determined by the frequency p, and the physical extent of the detector array. In particular, for small-angle scattering it is approximately p multiplied by the maximum scattering angle in radians.

In the small-angle limit, the Ewald sphere degenerates to a plane, and the 2D measurements lie on a 2D plane in $\boldsymbol{k}$-space, and thus, one has a 2D image reconstruction problem. Its solution is the x_3-axis projection of the 3D density function $\rho(\boldsymbol{x})$,

$$P[\rho](x_1,x_2) = \int_{-\infty}^{\infty} \rho(x_1,x_2,x_3)dx_3. \tag{4}$$

By rotating the object, in increments, about the x_1-axis, taking 2D measurements at each rotation angle, one may obtain samples of $|\rho(\boldsymbol{k})|^2$ thoughout a 3D volume. With this data one can then solve a phase retrieval problem to reconstruct the full 3D "image," $\rho(\boldsymbol{x})$. Since both $d = 2$ and $d = 3$ are of interest, d being the spatial dimension of the reconstruction, for much of the theory in this book, d is left general.

CDI is referred to as a "lensless" imaging modality, as there are no focusing optics involved. The full set of measurements is usually called a *diffraction pattern*. We assume that it is measured on (or interpolated to) a regular grid. Figures 1[a,b] show a synthetic 2D object of the sort used for numerical experiments in this book, and the diffraction pattern it would generate in the far field. The recent book, *X-Ray Microscopy* by Chris Jacobsen (Jacobsen 2019), is an excellent reference for the physics that underlie these imaging modalities and the techniques used to reconstruct images.

The measurement of $|\hat{\rho}(\boldsymbol{k})|^2$ allows for a direct reconstruction of a (band-limited) version of the autocorrelation function

$$\rho \star \rho(\boldsymbol{x}) = \int_{\mathbb{R}^3} \rho(\boldsymbol{y}+\boldsymbol{x})\overline{\rho(\boldsymbol{y})}d\boldsymbol{y}, \tag{5}$$

as $\mathscr{F}(\rho \star \rho)(\boldsymbol{k}) = |\widehat{\rho}(\boldsymbol{k})|^2$. An example of an autocorrelation function is shown in Figure 1[c]. To reconstruct ρ itself, the phase of $\widehat{\rho}(\boldsymbol{k})$ must be "retrieved." *In principle*, this is usually possible for a compactly supported object, provided that $|\widehat{\rho}(\boldsymbol{k})|^2$ is sampled on a sufficiently fine grid relative to the size of the support of $\rho(\boldsymbol{x})$. As a practical matter, this has proved very difficult to do. The book that follows discusses the reasons that underlie this difficulty, which are largely geometric, and considers approaches to circumventing these problems. The research described herein is an unusual combination of pure mathematics and computer experimentation, without which the pure mathematics would not have been possible to do.

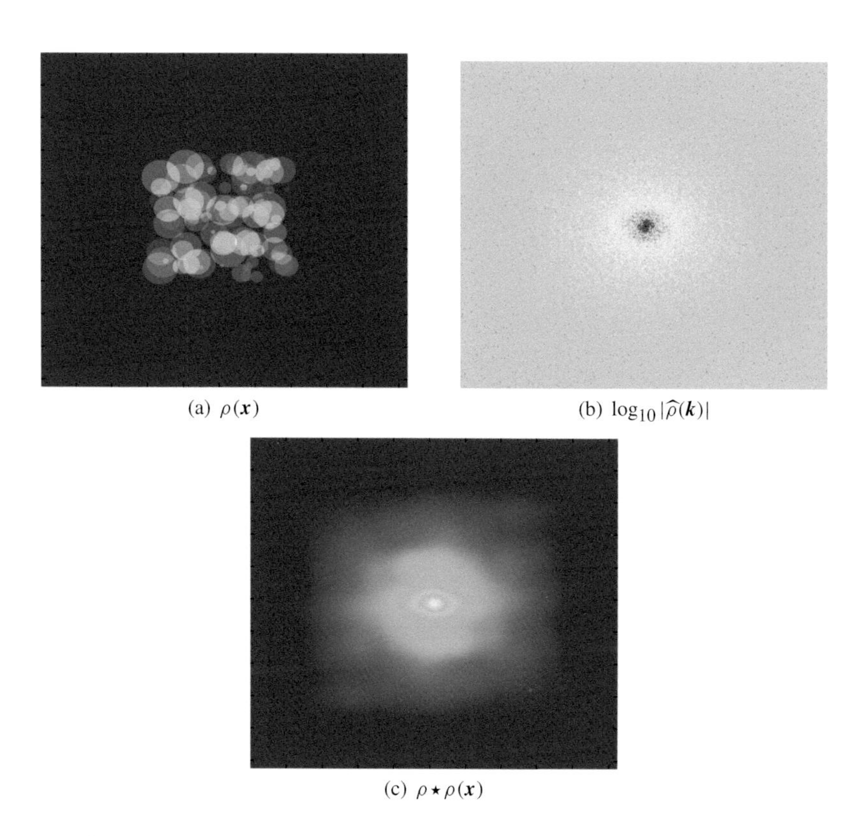

Figure 1 A synthetic object, the (false-colored) diffraction pattern it produces, and its autocorrelation function.

Acknowledgments

The work was largely carried out at, and supported by, the Flatiron Institute of the Simons Foundation, to which the authors are very grateful. Charles L. Epstein was also partially supported by the Mathematics Department at the University of Pennsylvania, and used their shared parallel computing facility (the GPC), for some of the computational experiments reported herein.

Charles L. Epstein would like to thank Leslie Greengard and the Flatiron Institute of the Simons Foundation for the supporting the very open-ended, five year exploration of the phase retrieval problem, and his other coauthors, Alex Barnett and Jeremy Magland, for joining him on this long journey.

We would also like to thank our colleagues at the Flatiron Institute, David Barmherzig, Michael Doppelt, Michael Eickenberg, and Marylou Gabrié, with whom we have explored many aspects of the phase retrieval problem. Our discussions led to unexpected insights into this problem, and many improvements to this book. Finally, we would like to thank Jim Fienup, who read the entire manuscript of the book, for sharing his enormous wealth of knowledge on the phase retrieval problem and its history. His comments led to many improvements in the text.

1
Introduction

In a wide range of applications, including crystallography, astronomy, coherent diffraction imaging (CDI), and ptychography, physical measurements are made that can be interpreted as samples of the magnitude of the Fourier transform, $\{|\widehat{\rho}(\boldsymbol{\xi})| : \boldsymbol{\xi} \in \Xi\}$, of an X-ray scattering density $\rho(\boldsymbol{x})$.[1] Here, Ξ denotes a finite discrete set of sample locations. Given this Fourier magnitude data, the problem of interest is the recovery of $\rho(\boldsymbol{x})$ itself or, more precisely, a bandlimited version of ρ sampled on a regular grid. This recovery entails first determining the phases of $\{\widehat{\rho}(\boldsymbol{\xi}) : \boldsymbol{\xi} \in \Xi\}$. Hence, this is called the *phase retrieval problem*. It is clear that the problem, as stated, is not solvable without some additional constraints on ρ, which we call *auxiliary information*. In CDI, it is typically known that $\rho(\boldsymbol{x})$ is supported in a compact set $S \subset \mathbb{R}^d$. As discussed in the Preface, in applications the space dimension d is typically 2 or 3, yet much of our analysis will apply to general d.

In this book, we study a discrete model of classical, phase retrieval, which, in the limit of high resolution, tends to the continuum problem. The discrete analysis brings many advantages mathematically (see Section 1.5) and allows clean numerical experiments (involving a single discrete Fourier transform, or DFT) that avoid complications of data interpolation or regridding. Yet the model preserves most of the interesting pathologies of the continuum case and is highly relevant because almost all numerical solution schemes ultimately rely on DFTs.

The model centers around a DFT of size $N_1 \times \cdots \times N_d$, i.e., a d-dimensional rectangular (cuboid) grid. More abstractly this grid will be denoted by $J \subset \mathbb{Z}^d$, a subset of the integer lattice, with vector indices $\boldsymbol{j} = (j_1, \ldots, j_d) \in J$. In the

[1] At the energies we consider in this book, the function $\rho(\boldsymbol{x})$ is connected to the refractive index, $n(\boldsymbol{x})$, of the material being imaged, by the relation $\rho(\boldsymbol{x}) = k^2(1 - n^2(\boldsymbol{x}))$. At X-ray energies $n(\boldsymbol{x}) = 1 - \delta(\boldsymbol{x}) - i\beta(\boldsymbol{x})$, with $|\delta(\boldsymbol{x}) + i\beta(\boldsymbol{x})| \ll 1$ (see Jacobsen, 2019, §3.3).

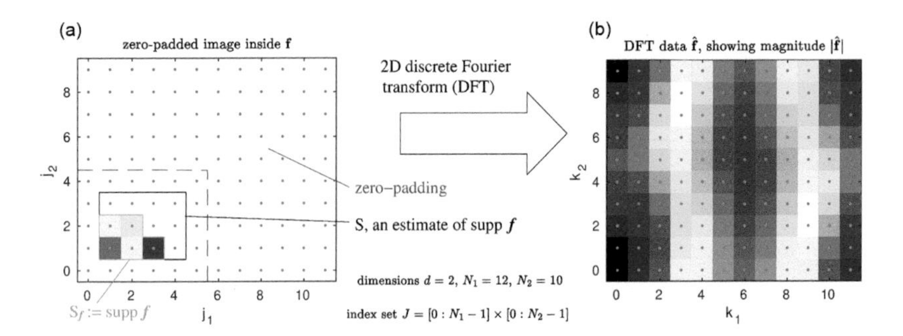

Figure 1.1 Setup for the discrete model analyzed in this book. A 2D example of DFT size 12×10 is shown, with index set J with $|J| = 120$. Part (a) is the image data $\boldsymbol{f} \in \mathbb{R}^J$, and part (b) is the resulting DFT data $\widehat{\boldsymbol{f}}$ which lives on the same index set J. The $\boldsymbol{k} = (0,0)$ index corresponds to zero spatial frequency. The constraint set S is a set containing the true support S_f, which, in this case, comprises only five pixels.

real case, the DFT (see (1.7)) takes an image vector $\boldsymbol{f} \in \mathbb{R}^J$, meaning a grid of real values living on the rectangle J, to its Fourier coefficient vector $\widehat{\boldsymbol{f}} \in \mathbb{R}^J$. The "measured data" is then the set of magnitudes $|\widehat{f}(\boldsymbol{k})|$ for $\boldsymbol{k} \in J$, where $\boldsymbol{k} = (k_1, \ldots, k_d)$ is a frequency index. This data comprises $N_1 \ldots N_d$ values, although in the real case, by symmetry, there are about half this number of independent values.

For reasons we will see shortly, for phase retrieval to stand a chance of success, the *unknown image* must live inside a rectangle at most *half* the size of J in each direction; the rest of the $\boldsymbol{f}$ vector contains zeros. For example, in the 2D case, the unknown image must live in a subrectangle of at most $N_1/2$ by $N_2/2$ pixels within J with the remaining pixel values zeros. An example is shown in Figure 1.1, where the nonzero pixels lie in the bottom-left region. In the physics literature this is often referred to as "oversampling" the Fourier coefficients of the image, but in fact is simply a matter of collecting Fourier data that contains information about support of the image within the field of view.

Unless stated, we allow J to be an arbitrary finite rectangular subset of a regular grid. The correspondence with the continuous problem it is easiest to describe in the "square" special case: $N_1 = \cdots = N_d = N$, where N sets the size of the $\boldsymbol{f}$ array. In this case, one may think of $\boldsymbol{f}$ as having elements

$$f_j = \rho\left(\frac{\boldsymbol{j}}{N}\right) \quad \text{for } \boldsymbol{j} \in J, \tag{1.1}$$

where $\boldsymbol{j}/N$ means the vector $(j_1/N, \ldots, j_d/N)$ and $\rho : \mathbb{R}^d \to \mathbb{R}$ is a continuous intensity function that vanishes outside of $[0, 1/2]^d$. This last condition ensures that the resulting $\boldsymbol{f}$ has support in a rectangular subset with sidelengths at most $N/2$. Then, fixing ρ and a nonnegative index vector $\boldsymbol{k}$, the DFT coefficient $\hat{f}_{\boldsymbol{k}}$ of $\boldsymbol{f}$ tends (disregarding a constant prefactor) as $N \to \infty$ to the Fourier transform $\hat{\rho}(\boldsymbol{k}) := \mathscr{F}(\rho)(\boldsymbol{k})$, using the definition (1). This is true because, when properly scaled, the DFT is a Riemann sum for the Fourier integral. This is discussed in greater detail in Section 1.5. In other words, the data in the discrete problem approximates the Fourier transform magnitudes for wavevectors Ξ on the integer lattice. (Note that the negative indices are to be found periodically "wrapped" into the other corners of the set J, as usual for the DFT, as shown in Figure 1.1.)

In the continuum case, the magnitude Fourier data is unchanged if $\rho(\boldsymbol{x})$ is replaced by a translate $\rho(\boldsymbol{x} - \boldsymbol{v})$, for a fixed vector $\boldsymbol{v}$, or by the inversion $\rho(-\boldsymbol{x})$, if ρ is real valued; for complex valued ρ, this operation is replaced by $\overline{\rho(-\boldsymbol{x})}$. These are called "trivial associates" of ρ. There is a similar phenomenon in the discrete problem. A more detailed discussion of the relationship between these two problems is given in Section 1.5.

We call the problem of reconstructing $\boldsymbol{f}$ from a knowledge of $|\widehat{\boldsymbol{f}}|$, plus some auxiliary information, the discrete, *classical*, phase retrieval problem, as the data are samples, $\{|\widehat{f}(\boldsymbol{k})|\}$, of the moduli of the Fourier transform itself, and what is sought are the phases of the complex numbers $\{\widehat{f}(\boldsymbol{k})\}$. For definiteness we define the *phase* of a nonzero complex number, z, to be the point on the unit circle in $\mathbb{C}$ given by

$$e^{i\theta} = \frac{z}{|z|}. \tag{1.2}$$

The phase of 0 is not defined, and the phase, as an S^1-valued function on $\mathbb{C} \setminus \{0\}$, cannot be extended continuously to $z = 0$. At times we are sloppy and refer to the *angle* θ as the phase; it is only defined up to the addition of multiples of 2π.

This problem is a member of a more general class of inverse problems where similar algorithms are used, generically referred to as "phase retrieval problems." Recently a great deal of effort has been directed toward the development and analysis of algorithms for solving these generalized problems (see Candès et al. 2015a, 2015b; Cahill et al. 2016; Alaifari et al. 2019). In this larger class of problems, the samples of the magnitude of the DFT data and information about its support are replaced with collection of absolute values of inner products

$$\mathfrak{M} = \{|\langle \boldsymbol{f}, \boldsymbol{m}_l \rangle| : l = 1, \ldots, L\}. \tag{1.3}$$

Here, $\boldsymbol{f} = (f(\boldsymbol{x}_1), \ldots, f(\boldsymbol{x}_K))^t$ is the unknown sample vector and $\{\boldsymbol{m}_l : l = 1, \ldots, L\}$ is a collection of "measurement vectors," which define a frame for an underlying Hilbert space, $\mathscr{H}$, of dimension $K < \infty$.

It has been shown that if $\mathscr{H}$ is complex and $L \geq 4K - 4$, or if $\mathscr{H}$ is real and $L > 2K$, then, for a generic set of measurement vectors, the measurements $\mathfrak{M}$ *uniquely* determine $\boldsymbol{f}$ (see Conca et al. 2015; Marchesini et al. 2016). We do not consider such generalized problems but restrict our attention to the classical phase retrieval problem, as its solution is required in many applications. While that solution is not unique, the nonuniqueness is (under broad conditions) a consequence of simple translation invariance or inversion symmetry. The well-posedness of the problem, however, is more subtle to analyze and there are many properties of the Fourier transform that lead to subtle geometric features in the problem, which can be both analyzed and probed numerically.

Remark 1.1 In current experimental practice, there has been a substantial shift of interest from CDI to ptychography (Rodenburg et al. 2007; Dierolf et al. 2010; Pfeiffer 2018; Thibault et al. 2008). Rather than using a single plane wave to scatter from the entire sample, one selectively illuminates a small subregion using either a mask or a more focused beam. This introduces a more complicated forward model but yields vastly more data – rendering the inverse problem much better posed and more easily solved. Recent work has achieved remarkably high resolution in this manner (see, for example, Guizar-Sicairos et al. 2014). Unfortunately, this technique requires rastering across the sample, longer acquisition times, and more complicated experimental protocols. Each of these leads to interesting new algorithmic challenges but for the sake of simplicity, we restrict our attention in this book to classical CDI itself. There are many important physical applications where one is *constrained* to solve the classical problem, especially when the sample is involved in a dynamic process and the global structure is desired over the smallest possible time window. Bragg CDI, which is used to study strain in crystals, is another inverse problem of the same general type as CDI (see Jacobsen, 2019, §10.3.8). We believe that the mathematical analysis presented here could also play a role in a deeper understanding of ptychography.

As in the continuum case, it is quite clear that magnitude DFT data alone does not suffice to determine any object, even up to translations and inversion, and therefore some auxiliary information is needed. Fixing the magnitudes of the DFT data determines a torus, which we denote by $\mathbb{A}$. This torus is typically very high dimensional; in the Fourier representation it is a Cartesian product of round circles, lying in complex lines, one corresponding to the unknown

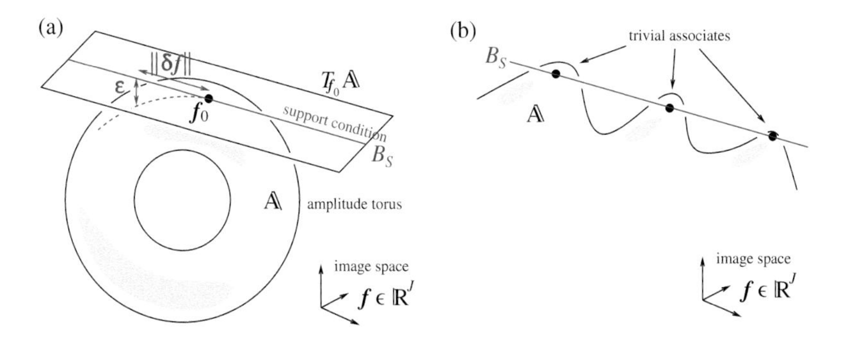

Figure 1.2 This figure illustrates (in a low dimensional cartoon) the full "image space" $\mathbb{R}^J$ in which $\boldsymbol{f}$ lives. In reality the ambient dimension, $|J|$, is thousands or more, so the 3D sketch misses many features. A valid image $\boldsymbol{f}_0$ must be at an intersection of the torus $\mathbb{A}$ (defined by given Fourier magnitude data) and the support condition B_S (the subset here shown as a line). Part (a) shows a nontransversal such intersection where $B_S \cap T_{\boldsymbol{f}_0}\mathbb{A} \neq \{\boldsymbol{f}_0\}$. The distance between B_S and $\mathbb{A}$ grows as $\epsilon \propto (\delta \boldsymbol{f})^2$. Part (b) shows a collection of trivial associates (separate intersection points between the two sets).

phase of each magnitude DFT value measured. As they are intrinsically flat, these tori do *not* metrically resemble the familiar "donut" embedded in $\mathbb{R}^3$, as shown in Figure 1.2. The simplest such flat torus is the set

$$\{(r_1 e^{i\theta_1}, r_2 e^{i\theta_2}) : \theta_1, \theta_2 \in [0, 2\pi)\} \text{ with } r_1, r_2 > 0,$$

sitting in $\mathbb{C}^2$.

Functions that satisfy the auxiliary information belong to a different subset, denoted by B. Thus, phase retrieval problems can be understood as the problem of searching for the intersections between $\mathbb{A}$ and B. This point of view, championed by Bauschke, Borwein, Combettes, Elser, Fienup, Luke, and many others is the basis for most of the results in this book (see Fienup 1987; Bauschke and Borwein 1993, 1996; Bauschke et al. 2002; Elser 2003; Elser et al. 2007; Gravel and Elser 2008; Borwein and Sims 2011; Borwein 2012). The set B is often taken to be the linear subspace of objects with support in a given set. If the set B is chosen well, then the set $\mathbb{A} \cap B$ consists of finitely many points. The major theoretical contributions of this book are concerned with the analysis of the geometry of the two subsets *near* to points in $\mathbb{A} \cap B$. This geometry, in particular the transversality properties of these intersections, has a decisive influence on the intrinsic difficulty of this problem, and the behavior of any practical algorithm for solving it.

In the DFT representation, the fact that the set $\mathbb{A}$ is a product of round circles, lying in coordinate planes, is clear; for a real image it is therefore a smooth, *nonlinear, nonconvex* subset of a Euclidean space, $\mathbb{R}^N$. We let $T\mathbb{A}$ denote the tangent bundle of $\mathbb{A}$. For $f \in \mathbb{A}$, the fiber of the tangent bundle at f, $T_f\mathbb{A}$, is naturally identified with the affine subspace of $\mathbb{R}^N$ that is the best linear approximation to $\mathbb{A}$ near to f, with a similar definition for $T_f B$. For a complex image, $\mathbb{A}$ is a subset of $\mathbb{C}^N$ and $T_f\mathbb{A}$ is the real affine subspace of $\mathbb{C}^N$ that is the best linear approximation to $\mathbb{A}$.

The natural linearization of the problem of finding points in $\mathbb{A} \cap B$, is to find points in $T_f\mathbb{A} \cap T_f B$. This linearization is a good model for the original problem *if the intersection between* $\mathbb{A}$ *and* B *is transversal*: that is, $T_f\mathbb{A} \cap T_f B$ consists only of the point f itself. If the intersection is not transversal, then $T_f\mathbb{A} \cap T_f B$ is positive dimensional, which vastly complicates the analysis and dramatically affects the behavior of algorithms for finding these intersections. This latter fact is already evident in 2-dimensional examples.

For illustration, consider locating the zero of the function $y = x^p$, for $p \in \mathbb{N}$. This can be understood as finding the intersection between the sets

$$A_p = \{(x,y) : y = x^p\} \text{ and } B = \{(x,y) : y = 0\}.$$

If $p > 1$, then $T_{(0,0)}A_p = B$ and therefore this intersection is nontransversal. Newton's algorithm, with a starting point $0 < x_0$, gives iterates

$$x_n = \left(\frac{p-1}{p}\right)^n x_0 \text{ if } 1 < p. \tag{1.4}$$

As is well-known, these converge *geometrically*, with higher-order contact producing a slower rate of convergence. This should be compared with the result of using Newton's algorithm to find the roots of $y = x^2 - 1$, starting with an $x_0 \in (1,2)$. In this case the intersection is transversal and the iterates, $x_{n+1} = \frac{1}{2}(x_n + x_n^{-1})$ satisfy

$$|x_n - 1| \leq \frac{|x_0 - 1|^{2^n}}{2^{2^n - 1}}, \tag{1.5}$$

and therefore, converge super-geometrically. Stated differently, higher-order contact between A_p and B, leads to a weaker attraction toward the fixed points of the iterated map.

Moreover, higher-order contact (nontransversal intersection) causes a more fundamental issue: it renders the *problem* of finding points in $A_p \cap B$ ill conditioned. This is a property independent of any algorithm used to attempt its solution. Intuitively, any small uncertainty in the specification of the set A_p leads to a strongly amplified error in the intersection point. In fact, this amplification factor can be arbitrarily large, as we now illustrate with an example where A_p is the graph of $y = x^p$. Supposing that an error shifts this set to the graph of $y = x^p - \epsilon$, for some small $\epsilon > 0$, then an intersection with B occurs (at least) at $x = \epsilon^{1/p}$, rather than the true value $x = 0$. In this example the error in the specification of the data has been amplified by factor $\epsilon^{(1-p)/p}$, which is clearly unbounded as $\epsilon \to 0$. This is known as an infinite *absolute* condition number: formally, the problem is infinitely sensitive to its input data.

In practice, intersection points do not usually lie at the origin; for a more generic variant of the above example one can take for A_p the graph of $y = (x - a)^p + b$, and for B the line $y = b$, for nonzero constants a and b. If B is perturbed to the line $y = b - \epsilon$, then the intersection (a, b) is replaced by $(a + \epsilon^{\frac{1}{p}}, b - \epsilon)$. In general, it is the *relative* condition number that is of interest. This is the largest possible ratio between the the *relative* error in output and the *relative* error in data that specifies the problem. For the second example this would be $(\epsilon^{\frac{1}{p}}/a)/(\epsilon/b) = b\epsilon^{(1-p)/p}/a$, which is again unbounded as $\epsilon \to 0$. This is further discussed in Appendix 1.B. Note that having an infinite relative condition number does not prevent the problem from being solved accurately; it merely implies some loss of digits. For instance, in the above example, if the problem "data" (b) is known to D digits of accuracy, then the Hölder continuity implies that the solution (x-value of an intersection point) can only be inferred, in principle, to around D/p digits of accuracy. This may be adequate for imaging purposes but is an important constraint on any solution algorithm.

Taking advantage of the special structure of the classical phase retrieval problem, we are able to analyze the tangent bundle to the torus and show that it usually meets the set B nontransversally. Regardless of the algorithm, this renders the problem of finding points in $\mathbb{A} \cap B$ somewhat ill posed. Most algorithms for finding these points involve iterating some map. The failure of transversality usually weakens the attraction toward the fixed-point sets of these maps, often leading to very complicated, nonconvergent dynamics. As we shall see, it also makes a definitive analysis of these maps on the fixed-point set very complicated to carry out in practice, and so one must resort to numerical experimentation. Finally, our analysis points to ways of understanding improved methods for *collecting data* that will allow for better image reconstruction.

In the remainder of this rather long introduction, we present the definitions used throughout the book, and briefly describe the main results. The book is divided into three parts. Part I contains theoretical results. Part II investigates the behavior of standard algorithms used in phase retrieval and Part III provides statistical analyses of the behavior of "hybrid iterative maps," as well as suggestions for more robust reconstruction methods.

1.1 Discrete, Phase Retrieval Problems

We now turn to a description of the finite model problem that is studied in this book. As noted above, the unknowns consist of finitely many samples of $\rho(\boldsymbol{x})$ for $\boldsymbol{x}$ lying on a uniformly spaced lattice, for example,

$$\left\{ f_{\boldsymbol{j}} = \rho\left(\frac{\boldsymbol{j}}{N}\right) : \boldsymbol{j} \in \mathbb{Z}^d \right\}, \tag{1.6}$$

where it is assumed that $\operatorname{supp} \rho$ is contained in a relatively compact subrectangle of $[0,1]^d$, with side-lengths less than $\frac{1}{2}$. The images we study in this book are often real valued, though many of our results, suitably modified, also hold for complex valued images. This is explained as required. Unless explicitly stated otherwise, real valued images can take both signs. We model the measurements as samples of the modulus of the DFT of this finite sequence. It is important to note that these measurements are not samples of the modulus of the continuous Fourier transform, $|\widehat{\rho}(\boldsymbol{\xi})|$, described above, but lead to a related problem that is easier to analyze. As noted above, when properly scaled, samples of the modulus of the DFT tend to uniformly spaced samples of $|\widehat{\rho}|$ in the limit $N \to \infty$.

For integers $n < N$, we define $[n : N] \stackrel{d}{=} \{n, n+1, \ldots, N-1, N\}$. A subset of $\mathbb{Z}^d$ of the form $[n_1 : N_1] \times \cdots \times [n_d : N_d]$ is called a *rectangular* subset. We call real or complex sequences indexed by such a set "images," or sometimes d-dimensional images. To define the discrete analogue of the classical phase retrieval problem we let $\boldsymbol{f}$ be an image $\{f_{\boldsymbol{j}} : \boldsymbol{j} \in J\}$, labeled by J a rectangular subset of $\mathbb{Z}^d$. We use the notation $\mathbb{R}^J$ to denote $\mathbb{R}^{|J|}$ with this indexing; respectively $\mathbb{C}^J$ to denote $\mathbb{C}^{|J|}$. Typically J takes the form $[0 : N]^d$, or $[-N : N]^d$. This approach to indexing images retains information about their underlying dimensions, and their organization in space.

Up to a rigid translation, any rectangular subset is equivalent, by translation, to a set of the following type: $J = [0 : N_1 - 1] \times \cdots \times [0 : N_d - 1]$. For an image $\boldsymbol{f} \in \mathbb{C}^J$, the DFT is defined, for $\boldsymbol{k} \in J$, to be

$$\mathscr{F}(\boldsymbol{f})_{\boldsymbol{k}} = \widehat{f}_{\boldsymbol{k}} \stackrel{d}{=} \sum_{j_1=0}^{N_1-1} \cdots \sum_{j_d=0}^{N_d-1} f_{\boldsymbol{j}} \prod_{l=1}^{d} \exp\left(-\frac{2\pi i j_l k_l}{N_l}\right). \tag{1.7}$$

The DFT could be defined for $\boldsymbol{k}$ in a different index set, as is sometimes done in the imaging literature; in this book we always take the domain of a d-dimensional image and its DFT to be the same rectangular subset of $\mathbb{Z}^d$. The formula for the DFT defines an extension of $\widehat{f}_{\boldsymbol{k}}$ from $\boldsymbol{k} \in J$ to all of $\mathbb{Z}^d$, as an image that is periodic in every index

$$\widehat{f}_{\boldsymbol{k}+(l_1 N_1, \ldots, l_d N_d)} = \widehat{f}_{\boldsymbol{k}} \quad \text{for all } (l_1, \ldots, l_d) \in \mathbb{Z}^d. \tag{1.8}$$

Definition 1.2 If $f \in \mathbb{C}^J$, with J equivalent to $[0 : N_1 - 1] \times \cdots \times [0 : N_d - 1]$, then f, extended to all of $\mathbb{Z}^d$ by setting

$$f_{j+(l_1 N_1, \ldots, l_d N_d)} = f_j \quad \text{for} j \in J \text{and all } (l_1, \ldots, l_d) \in \mathbb{Z}^d, \tag{1.9}$$

is said to be J*-periodically extended.* This agrees with the J-periodic extension of f given by inverse DFT.

The measurements in the discrete, classical, phase retrieval problem consist of a collection of nonnegative real numbers, which are interpreted as the magnitudes of the DFT of an unknown image f_0:

$$a_k = |\widehat{f_{0k}}| \quad \text{for } k \in J. \tag{1.10}$$

We denote the set of images with the given magnitude DFT data by $\mathbb{A}_a$:

$$\mathbb{A}_a = \{f \in \mathbb{R}^J : |\widehat{f_k}| = a_k \quad \text{for all } k \in J\} \tag{1.11}$$

and call this the *magnitude torus* defined by a. Geometrically, this is a real torus, i.e., a product of round circles, sitting in $\mathbb{C}^J$. We use $\widehat{\mathbb{A}}_a$ to denote its representation in the DFT domain

$$\widehat{\mathbb{A}}_a = \{(a_k e^{i\theta_k}) : \theta_k \in (-\pi, \pi] \quad \text{for all } k \in J\}. \tag{1.12}$$

As in the continuum case, there are two operations on a (periodic) image that leave the magnitude DFT data invariant.

Definition 1.3 Let f be an image indexed by $J \subset \mathbb{Z}^d$, which is extended periodically to $\mathbb{Z}^d$.

(i) The *inversion* of f is defined by

$$\check{f}_j \stackrel{d}{=} f_{-j}. \tag{1.13}$$

Note that if f is real, then $\mathscr{F}(\check{f})_k = \overline{\mathscr{F}(f)_k}$.

(ii) For $v \in \mathbb{Z}^d$ the *translation* of f by v, $f^{(v)}$, is defined by

$$f_j^{(v)} \stackrel{d}{=} f_{j-v}. \tag{1.14}$$

If $J = [1 : N]^d$, then

$$\mathscr{F}(f^{(v)})_k = e^{\frac{2\pi i k \cdot v}{N}} \mathscr{F}(f)_k. \tag{1.15}$$

(iii) If f is real valued, then $-f$ has the same magnitude DFT data. If f is complex valued, then the images $\{e^{i\theta} f : \theta \in [0, 2\pi)\}$ all have the same magnitude DFT data.

The set of images generated from $\boldsymbol{f}$ by these operations are called its *trivial associates.*

We let $S_f \subset J$ denote the *support of* $\boldsymbol{f}$

$$\boldsymbol{j} \in S_f \text{ if and only if } f_{\boldsymbol{j}} \neq 0. \tag{1.16}$$

We assume that we have an estimate for S_f, encoded as a subset $S \subset J$, such that $S_f \subset S$. The goal of the discrete, classical, phase retrieval problem, using support as the auxiliary information, is to find an image $\boldsymbol{f}_1$, indexed by J, with the given magnitude DFT data

$$\begin{aligned} |\widehat{f}_{1\boldsymbol{k}}| = a_{\boldsymbol{k}} \quad &\text{for all } \boldsymbol{k} \in J, \text{ and} \\ S_{f_1} \subset S&. \end{aligned} \tag{1.17}$$

That is, we search for points in $\mathbb{A}_{\boldsymbol{a}} \cap B_S$, where

$$B_S = \{\boldsymbol{f} \in \mathbb{R}^J : f_{\boldsymbol{j}} = 0 \text{ if } \boldsymbol{j} \notin S\}. \tag{1.18}$$

The solution to the phase retrieval problem is unique, *up to trivial associates*, if the set of points $\mathbb{A}_{\boldsymbol{a}} \cap B_S$ consist of trivial associates of one such point. From the perspective of applications, this sort of nonuniqueness does not pose any problems, and it is the strongest uniqueness statement that may be true.

Building on earlier work of Bruck and Sodin (1979), Hayes (1982) proved that, if the true support of $\boldsymbol{f}_0$, S_{f_0}, is contained in a rectangle $R \subset J$ with the length of each side of R at most half the length of the corresponding side of J, then "generically" the solution of the phase retrieval problem is unique up to trivial associates. The meaning of generic in this context is explained below. If S_{f_0} is contained in such a rectangle, then we say that $\boldsymbol{f}_0$ has *small support.*

The Z-transform of $\boldsymbol{f}$ is defined by

$$\boldsymbol{X}(z) = \sum_{j \in J} f_{\boldsymbol{j}} z^{-\boldsymbol{j}}; \tag{1.19}$$

using multi-index notation: $z = (z_1, \ldots, z_d)$ and $z^{\boldsymbol{j}} = z_1^{j_1} \ldots z_d^{j_d}$ is a monomial of degree $|\boldsymbol{j}| = j_1 + \cdots + j_d$. Note that the degree of a monomial can be either positive or negative. A polynomial is a finite sum of monomials of nonnegative degrees

$$p(z) = \sum_{j \in \mathscr{J}} a_{\boldsymbol{j}} z^{\boldsymbol{j}}, \text{ with } \mathscr{J} \text{ a finite set.} \tag{1.20}$$

Its degree is defined as the

$$\deg p \overset{d}{=} \max\{|\boldsymbol{j}| : \boldsymbol{j} \in \mathscr{J} \text{ and } a_{\boldsymbol{j}} \neq 0\}.$$

A polynomial, $p(z)$, is *reducible*, if there are polynomials $q_1(z)$, $q_2(z)$, both with positive degrees, so that $p(z) = q_1(z)q_2(z)$; if no such factorization exists, then the polynomial is *irreducible*.

The solution of the phase retrieval problem is unique, up to trivial associates, provided there is a monomial z^k such that $z^k X(z)$ is an irreducible polynomial. Irreducibility of polynomials in two or more variables is a generic property in that the set of generic polynomials of any fixed degree, labeled by their coefficients, is an open and dense set in the usual Euclidean topology. In fact, the reducible polynomials lie in a closed subset of considerably lower dimension. This remains true of data restricted to have support in a fixed subset $S \subset J$, as well as data restricted to be non-negative. These concepts from algebraic geometry are briefly explained in Appendix 1.A. The implications of these results, when working with finite precision arithmetic and noisy data, are still not well understood, as it is not known the extent to which the set of reducible polynomials in two or more variables becomes dense in the set of all polynomials as the degree tends to infinity.

Definition 1.4 We say that a subset $S \subset J$ is *adequate* for the phase retrieval problem if the support condition it defines, $\boldsymbol{f} \in B_S$, has the property that $\mathbb{A}_{\boldsymbol{a}} \cap B_S$ is finite.

Suppose the image $\boldsymbol{f}$ is defined, as in (1.6), as samples of a function ρ with compact support. If ρ is piecewise constant, then it is quite straightforward to define $S_{\boldsymbol{f}}$, and the border between points in the support and those not in the support is delineated by a large change in the value of $f_{\boldsymbol{j}}$. If ρ is even somewhat smooth, then the question of how to define and approximate its support, *in finite precision arithmetic*, is both complicated and potentially consequential. A smooth function goes to zero gradually and there is no sharp delineation between points in the support and points not in the support. We discuss this and the issue of "smoothness" for finitely sampled data in Section 1.4.

While a support constraint is the most common auxiliary condition, in some circumstances it is reasonable to assume that the unknown function is real and nonnegative. To make this precise, let us denote the closure of the positive orthant by

$$B_+ = \{\boldsymbol{f} : f_{\boldsymbol{j}} \geq 0 \quad \text{for all } \boldsymbol{j} \in J\}.$$

The inverse problem then consists of seeking points in $B_+ \cap \mathbb{A}_{\boldsymbol{a}}$. The solution to this problem is often referred to as a "positive" image, or one satisfying the "positivity constraint." This is a slight misnomer, since a strictly positive image in $B_+ \cap \mathbb{A}_{\boldsymbol{a}}$ cannot possibly be uniquely determined by its magnitude DFT data. If $(a_{\boldsymbol{k}} e^{i\theta_{\boldsymbol{k}}} : \boldsymbol{k} \in J)$ are the DFT coefficients of a strictly positive image, then there is an $\epsilon > 0$, so that, for any phases $\{e^{i\phi_k} : \boldsymbol{k} \in J\}$, with the symmetries required of the phases of the DFT of a real image, and $0 < \delta < \epsilon$, the point in $\mathbb{C}^J$ given by $(a_{\boldsymbol{k}} e^{i(\theta_k + \delta\phi_k)} : \boldsymbol{k} \in J)$ is the DFT of a strictly positive image. That is, the subset of $\mathbb{A}_{\boldsymbol{a}}$ corresponding to strictly positive images is an open set. Thus, for uniqueness up to trivial associates, some additional condition is needed such as some sort of loose support constraint – requiring that the image vanish at a subset of points. In this book we stick to the more precise "nonnegativity" constraint and avoid the usage of "positivity."

For a real image, the *autocorrelation image*, $\boldsymbol{f} \star \boldsymbol{f}$, is defined by

$$[f \star f]_j = \sum_{l \in J} f_l f_{l-j} = \sum_{l \in J} f_l f_{l+j}, \tag{1.21}$$

and for a complex image, by

$$[f \star f]_j = \sum_{l \in J} f_l \bar{f}_{l-j} = \sum_{l \in J} f_l \bar{f}_{l+j}. \tag{1.22}$$

All points on $\mathbb{A}_{\boldsymbol{a}}$ have the same autocorrelation image, as the DFT of $\boldsymbol{f} \star \boldsymbol{f}$ is

$$\widehat{\boldsymbol{f} \star \boldsymbol{f}}_{\boldsymbol{k}} = |\hat{f}_{\boldsymbol{k}}|^2 = a_{\boldsymbol{k}}^2. \tag{1.23}$$

In Chapter 4 we prove the statement: if there is a *nonnegative* image $\boldsymbol{f}_0 \in \mathbb{A}_{\boldsymbol{a}}$, and the support, $S_{f_0 \star f_0}$, of its autocorrelation image is sufficiently small, then the set of images in $\mathbb{A}_{\boldsymbol{a}} \cap B_+$ is finite, and generically consists of trivial associates. Note that, from (1.23) it is immediate that the condition on $S_{f \star f}$ can be checked from the magnitude DFT data. We know of no theorem in the literature that establishes a uniqueness result for the discrete, classical, phase retrieval problem when a sign constraint is the auxiliary information. Beinert (2017) considers the 1-dimensional case, showing that a sign constraint does *not* suffice to conclude generic uniqueness.

Finally, we note that the existence of trivial associates creates multiple basins of attraction for iterative algorithms. A priori, this might not seem to be a serious problem, as these basins are fairly well separated from one another; one might imagine that the competition between the different basins of attraction would quickly be settled in favor of one of them. Somewhat surprisingly, the mere existence of trivial associates often *forces* the intersections $\mathbb{A}_{\boldsymbol{a}} \cap B_S$ to be nontransversal. The failure of transversality leads to vastly more complicated dynamics for the maps used in iterative algorithms, and often prevents the iterates from converging.

1.2 Conditioning and Ill-Posedness of the Discrete, Classical, Phase Retrieval Problem

As noted above, if S is adequate, then the intersection $\mathbb{A}_{\boldsymbol{a}} \cap B_S$ is a discrete set of points. The number and distribution of points in this set depends on the accuracy of S as an estimate of the support of $\boldsymbol{f}$. If $\boldsymbol{f}_0 \in \mathbb{A}_{\boldsymbol{a}} \cap B_S$, then this intersection may fail to be transversal, that is $T_{\boldsymbol{f}_0}\mathbb{A}_{\boldsymbol{a}}$, the fiber of the tangent bundle to $\mathbb{A}_{\boldsymbol{a}}$ at $\boldsymbol{f}_0$, may intersect B_S in a positive dimensional subspace. As we shall see, the failure of these intersections to be transversal renders the phase retrieval problem ill posed and is a principal cause for the failure of existing algorithms to reliably converge. The phenomena of nontransversal intersections and trivial associates are illustrated in Figure 1.2. As noted above, metrically, magnitude tori are not like the donut depicted in Figure 1.2(a), as the metrics induced on them from their embeddings into $\mathbb{C}^J$ are intrinsically flat.

If $\boldsymbol{f} \in \mathbb{C}^J$ is an image, then we denote the measurement map, which takes $\boldsymbol{f}$ to its magnitude DFT data by

$$\mathscr{M}(\boldsymbol{f}) = \{|\widehat{f}_{\boldsymbol{k}}| : \boldsymbol{k} \in J\}. \tag{1.24}$$

If $\boldsymbol{a} \in \mathbb{R}_+^J$ is a nonnegative vector, then $\mathbb{A}_{\boldsymbol{a}} = \mathscr{M}^{-1}(\boldsymbol{a})$ is a real torus of dimension roughly $|J|/2$, if $\boldsymbol{f}$ is real, and $|J|$, if $\boldsymbol{f}$ is complex. The uniqueness results quoted above show that, if $S \subset J$ is sufficiently small, then the set $\mathbb{A}_{\boldsymbol{a}} \cap B_S$ is finite, and generically consists of trivial associates. In light of this, the restricted map, $\mathscr{M} \restriction_{B_S}$, is locally invertible on its range.

This means that, given an image $\boldsymbol{f}_0 \in B_S$ whose measured data is $\boldsymbol{a}_0 = \mathscr{M}(\boldsymbol{f}_0)$, there is a one-to-one map between points on B_S near to $\boldsymbol{f}_0$ and points on $\mathscr{M}(B_S)$ near to $\boldsymbol{a}_0$. In particular there is a local inverse function, which we denote by $\mathscr{M}^{-1}_{\boldsymbol{f}_0,S}$, which maps data points on the subset $\mathscr{M}(B_S)$ near to $\boldsymbol{a}_0$ back to their preimages $\boldsymbol{f} \in B_S$, which are near to $\boldsymbol{f}_0$. See Figure 1.3 to help visualize this. The local inverse map, if it were known in the neighborhood of the measured data, would constitute a local *solution operator* for the discrete, classical, phase retrieval problem.

In Chapter 3 we investigate the conditioning of this inverse and prove the following theorem:

Theorem 1.5 *Let $\boldsymbol{f}_0 \in \mathbb{A}_{\boldsymbol{a}} \cap B_S$; suppose that $\boldsymbol{a}$ is strictly positive. There exists a constant $0 < C$ so that, for $\boldsymbol{f}_1$ sufficiently near to $\boldsymbol{f}_0$ we have the estimate,*

$$C\|\boldsymbol{f}_0 - \boldsymbol{f}_1\| \leq \|\mathscr{M}(\boldsymbol{f}_0) - \mathscr{M}(\boldsymbol{f}_1)\|, \tag{1.25}$$

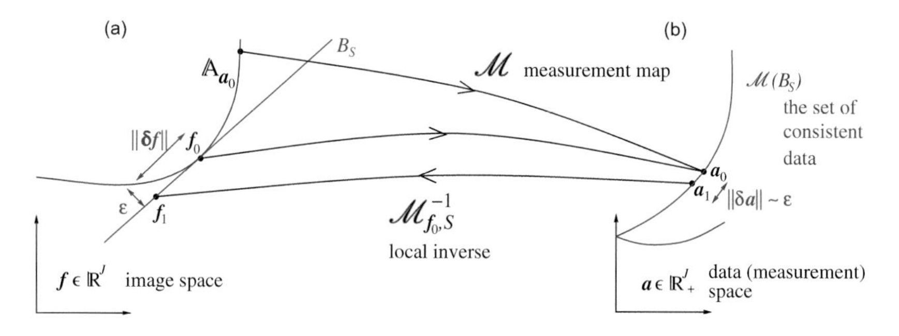

Figure 1.3 Sketch of the measurement map $\boldsymbol{a} = \mathcal{M}(\boldsymbol{f})$ from image space (a) to data or measurement space (b). This map, restricted to the images in B_S, is locally invertible on its range $\mathcal{M}(B_S)$. A nontransversal intersection is shown; it implies ill conditioning because two images exist whose separation $\|\delta \boldsymbol{f}\| = \|\boldsymbol{f}_0 - \boldsymbol{f}_1\|$ is large compared to their separation $\|\delta \boldsymbol{a}\|$ in measurement space.

if and only if the intersection of $\mathbb{A}_{\boldsymbol{a}}$ *with* B_S *is transversal at* $\boldsymbol{f}_0$.

The estimate in (1.25) is *necessary* for $\mathcal{M}$ to have a locally Lipschitz inverse. Near a nontransversal intersection, the inverse is only Hölder continuous in some directions, and therefore, in finite precision arithmetic, any approximate inverse can be expected to lose a fixed *fraction* of the digits available in the data. Whenever the intersection is nontransversal, the phase retrieval problem is ill posed, and any algorithm for solving it can be expected to reflect this sensitivity to the input data (see Appendix 1.B).

Recall that if $\boldsymbol{f}$ is an image indexed by J, extended periodically to $\mathbb{Z}^d$, then $\boldsymbol{f}^{(\boldsymbol{v})}$ is its translate by $\boldsymbol{v} \in \mathbb{Z}^d$, as defined in (1.14). The magnitude DFT data of $\boldsymbol{f}^{(\boldsymbol{v})}$ equals that of $\boldsymbol{f}$. The following theorem summarizes another, more surprising property, of these translates. It is proven in Section 2.1.

Theorem 1.6 *Let* $\boldsymbol{f}$ *be a real image parameterized by* J *whose magnitude DFT data defines the torus* $\mathbb{A}_{\boldsymbol{a}}$. *For each* $\boldsymbol{v} \in J$ *the differences*

$$\boldsymbol{f}^{(\boldsymbol{v})} - \boldsymbol{f}^{(-\boldsymbol{v})} \tag{1.26}$$

belong to fiber of the tangent bundle to $\mathbb{A}_{\boldsymbol{a}}$ *at* $\boldsymbol{f}$, *and the sums*

$$\boldsymbol{f}^{(\boldsymbol{v})} + \boldsymbol{f}^{(-\boldsymbol{v})} \tag{1.27}$$

belong to fiber of the normal bundle to $\mathbb{A}_{\boldsymbol{a}}$ *at* $\boldsymbol{f}$. *If* $\boldsymbol{f}$ *is complex valued, then*

$$\boldsymbol{f}^{(\boldsymbol{v})} - \boldsymbol{f}^{(-\boldsymbol{v})}, \text{ and } i(\boldsymbol{f}^{(\boldsymbol{v})} + \boldsymbol{f}^{(-\boldsymbol{v})}) \tag{1.28}$$

belong to fiber of the tangent bundle to $\mathbb{A}_{\boldsymbol{a}}$ *at* $\boldsymbol{f}$, *and*

$$\boldsymbol{f}^{(\boldsymbol{v})} + \boldsymbol{f}^{(-\boldsymbol{v})} \text{ and } i(\boldsymbol{f}^{(\boldsymbol{v})} - \boldsymbol{f}^{(-\boldsymbol{v})}) \tag{1.29}$$

belong to fiber of the normal bundle to $\mathbb{A}_{\boldsymbol{a}}$ at $\boldsymbol{f}$.

Remark 1.7 This is the deleterious echo of the existence of trivial associates mentioned above. For $\boldsymbol{f}_0 \in \mathbb{A}_{\boldsymbol{a}} \cap B_S$, this theorem allows one to compare the tangent bundle to $\mathbb{A}_{\boldsymbol{a}}$ at $\boldsymbol{f}_0$ to the linear subspace B_S. In particular, unless S is a very accurate estimate for $S_{\boldsymbol{f}_0}$, then

$$T_{\boldsymbol{f}_0}\mathbb{A}_{\boldsymbol{a}} \cap B_S \neq \{\boldsymbol{f}_0\}, \tag{1.30}$$

and this intersection is, therefore, not transversal.

Generically, the dimension of the intersection, $T_{\boldsymbol{f}_0}\mathbb{A}_{\boldsymbol{a}} \cap B_S$, depends only on the accuracy of S as an estimate of the support of $\boldsymbol{f}_0$, with $T_{\boldsymbol{f}_0}\mathbb{A}_{\boldsymbol{a}} \cap B_{S_{\boldsymbol{f}_0}} = \boldsymbol{f}_0$. This, in turn, explains the observation in the literature (see the discussion of Shrinkwrap in Chapman et al. 2006) that, for high contrast objects, a more precise estimate for the support sometimes leads to faster convergence to a more accurate limit.

In Section 2.2 we explore a nongeneric situation that nonetheless arises often in imaging applications. Let $\boldsymbol{a}_{\boldsymbol{g}}$ denote the magnitude DFT data defined by the image $\boldsymbol{g}$, which we assume has "small support." Suppose that the image $\boldsymbol{g}$ is real and inversion symmetric, that is $g_j = g_{-j}$, and has support $S_{\boldsymbol{g}}$. Because of this symmetry we can show that

$$\dim T_{\boldsymbol{g}}\mathbb{A}_{\boldsymbol{a}_{\boldsymbol{g}}} \cap B_{S_{\boldsymbol{g}}} \geq \frac{|S_{\boldsymbol{g}}|}{2}. \tag{1.31}$$

Thus, for an inversion symmetric, real image, even the intersection $\mathbb{A}_{\boldsymbol{a}_{\boldsymbol{g}}} \cap B_{S_{\boldsymbol{g}}}$ is nontransversal! Of course, the DFT coefficients of $\boldsymbol{g}$ are real, so the phase retrieval problem for $\boldsymbol{g}$ reduces to a much simpler sign retrieval problem. But that is not the end of the story. This problem is compounded by the fact that the failure of transversality is "inherited" by images obtained by convolving with $\boldsymbol{g}$. If $\boldsymbol{f} = \boldsymbol{g} * \boldsymbol{h}$, then it is also the case that

$$\dim T_{\boldsymbol{f}}\mathbb{A}_{\boldsymbol{a}_{\boldsymbol{f}}} \cap B_{S_{\boldsymbol{f}}} \geq \frac{|S_{\boldsymbol{g}}|}{2}. \tag{1.32}$$

This points to a danger in imaging applications, where it is natural to smooth out Gibbs' artifacts, or suppress noise by apodizing the measured data $|\widehat{\boldsymbol{h}}|$ by multiplying the magnitude DFT data by $|\widehat{\boldsymbol{g}}|$, with $\boldsymbol{g}$ a Gaussian. Mathematically this is equivalent to convolution of $\boldsymbol{g}$ with $\boldsymbol{h}$, in the image domain. The inversion symmetry of $\boldsymbol{g}$, and equation (1.32) show that, even if the intersection $\mathbb{A}_{\boldsymbol{a}_{\boldsymbol{h}}} \cap B_{S_{\boldsymbol{h}}}$ is transversal, the intersection $\mathbb{A}_{\boldsymbol{a}_{\boldsymbol{f}}} \cap B_{S_{\boldsymbol{f}}}$ will fail to be. The phase retrieval problem for $|\widehat{\boldsymbol{h}\boldsymbol{g}}|$ can, therefore, be expected to be much harder than that for $|\widehat{\boldsymbol{h}}|$ itself.

1.3 Algorithms for Finding Intersections of Sets

As noted above, the phase retrieval problem is a special case of the more general problem of finding the intersections of two subsets (Gravel and Elser 2008), which we denote by A and B, of a fixed Euclidean space, $\mathbb{R}^J$ or $\mathbb{C}^J$. In many imaging applications the ambient space has hundreds of thousands, millions, or even billions of dimensions, and the dimensions of the subsets are a substantial fraction of this dimension. One or both of these subsets may be nonlinear and/or nonconvex.

Most of the algorithms we examine for locating points in $A \cap B$ are defined as iterations of a map $F : \mathbb{R}^J \to \mathbb{R}^J$ (resp. $F : \mathbb{C}^J \to \mathbb{C}^J$). Algorithms defined by iterating a fixed map are essentially generalizations of "power methods" from linear algebra. Once they have found an attracting basin, their behavior depends on the linearization of F near to a fixed point. The simplest case is when the singular values of the linearization are all less than one. For algorithms typically used for phase retrieval this does *not* seem to be the rule. When iterative algorithms converge, the linearization at the limit point often turns out to be a nonnormal operator, with all of its (complex) eigenvalues less than 1 in modulus, but with many singular values larger than 1. This nonnormality leads to very complicated dynamics.

The maps most often used in algorithms to find points in $A \cap B$ are built from "nearest point maps." For an arbitrary subset $W \subset \mathbb{R}^N$, denote the point of W, nearest to f, by $P_W(f)$. Throughout this book, distance in Euclidean spaces is defined by the ℓ_2-norm

$$d(\boldsymbol{x}, \boldsymbol{y}) = \|\boldsymbol{x} - \boldsymbol{y}\|_2 \overset{d}{=} \left[\sum_{j=1}^{N} |x_j - y_j|^2 \right]^{\frac{1}{2}}. \tag{1.33}$$

For linear subspaces, these maps are orthogonal projections. If a set is nonconvex, then this map is not globally defined, but it is defined on the complement of positive codimensional subset.

The earliest approach to finding points in $A \cap B$ is called *alternating projection* and uses the iteration

$$f^{(n+1)} = P_A \circ P_B(f^{(n)}). \tag{1.34}$$

This algorithm was introduced in the context of Banach spaces by John von Neumann (1950), and in image reconstruction by Gerchberg and Saxton (1972), with an important simplification suggested by Fienup (1978). A reasonable measure of the error is the residual, $\|f^{(n)} - P_B(f^{(n)})\|_2$, which is the distance from $f^{(n)} \in A$ to the nearest point in B. The behavior of the distance between iterates $\{\|\boldsymbol{f}^{(n)} - \boldsymbol{f}^{(n+1)}\|_2\}$ indicates whether or not the iterates are converging.

Clearly, points in $A \cap B$ are fixed points of $P_A \circ P_B$. In the non-linear case, however, alternating projection can, in principle, have fixed points that are not related, in any simple way, to points in $A \cap B$. Indeed every local minimum of the function $d_{AB} : A \times B \longrightarrow [0, \infty)$, defined by

$$d_{AB}(\boldsymbol{f}_A, \boldsymbol{f}_B) = \|\boldsymbol{f}_A - \boldsymbol{f}_B\|_2 \tag{1.35}$$

produces a stable fixed point of the alternating projection map. In the context of phase retrieval, it is not obvious that such points exist, but, from numerical experiments, it appears that nonzero, local minima exist in great abundance. In these experiments, the iterates of alternating projection reliably *converge* to such local minima, which are usually at a considerable distance from points in $A \cap B$.

In earlier literature it is said that the alternating projection algorithm "stagnates," but, from our experiments, this terminology does not seem to be appropriate. In fact, stagnation is sometimes defined in the literature as a circumstance where the distance between successive iterates tends to zero. However, if the distance between iterates is summable, that is, if

$$\sum_{n=1}^{\infty} \|\boldsymbol{f}^{(n)} - \boldsymbol{f}^{(n+1)}\|_2 < \infty, \tag{1.36}$$

then, in fact, the sequence converges. In alternating projection we observe precisely this behavior, indicating that the sequence is converging, but to an incorrect solution. We introduce a different notion of stagnation here.

Definition 1.8 The iterates of an algorithm $\{\boldsymbol{f}^{(n)}\}$ *stagnate* if they remain in a *fixed* bounded set, but the distances between the successive iterates, $\{\|\boldsymbol{f}^{(n)} - \boldsymbol{f}^{(n+1)}\|_2\}$ do not converge to zero.

As such, stagnation and convergence are mutually exclusive, however, it is possible for the sequence $\{\|\boldsymbol{f}^{(n)} - \boldsymbol{f}^{(n+1)}\|_2\}$ to tend to 0, without $\{\boldsymbol{f}^{(n)}\}$ converging. This is not something we have observed in the algorithms for phase retrieval considered herein. Because of the failure of alternating projection, at present, phase retrieval is accomplished largely through the use of what we refer to as "hybrid iterative maps." A typical example is

$$D_{BA}(\boldsymbol{f}^{(n+1)}) = \boldsymbol{f}^{(n)} + P_B \circ R_A(\boldsymbol{f}^{(n)}) - P_A(\boldsymbol{f}^{(n)}), \tag{1.37}$$

where R_A is the "reflection across A" defined by

$$R_A(\boldsymbol{f}) = 2P_A(\boldsymbol{f}) - \boldsymbol{f}. \tag{1.38}$$

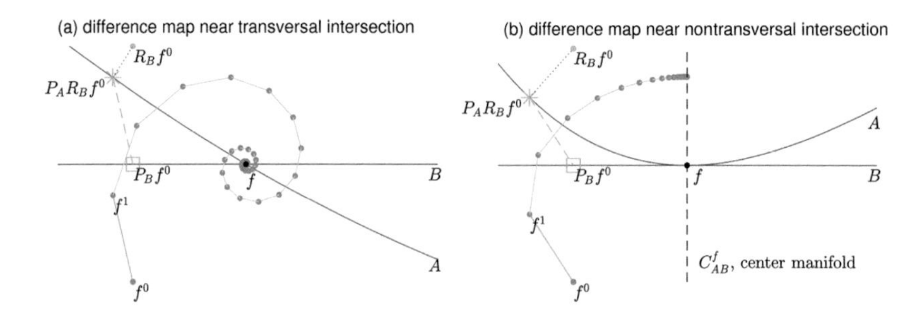

Figure 1.4 The hybrid iterative map $f^{n+1} = D_{AB}(f^n)$ defined by (1.37), in the setting where A (red curve) and B (blue line) are 1D manifolds in $\mathbb{R}^2$. (a) $f \in A \cap B$ is a transversal intersection; here the center manifold is the single point f. (b) Nontransversal case, with generic quadratic separation between the manifolds; here the center manifold is a vertical line. Each plot shows the iterates $f^0, f^1, \ldots$ (gray dots), and the construction of the update vector (green solid line) $f^1 - f^0$ as the difference between a projected reflection (green star) and a projection (green square). Note that in each plot the green solid and dashed lines are equal as displacement vectors.

A low dimensional example is shown in Figure 1.4.

This approach was originally introduced by Fienup (1978, 1982), where he defines the hybrid input-output (HIO) method. The map in (1.37) is a special case of an HIO provided $B = B_S$; for $B = B_+$ the map defined by the HIO method is quite different, and will be considered in Part II. For $B = B_S$, Fienup considers the more general family of maps

$$D^{\beta}_{BA}(f^{(n+1)}) = f^{(n)} + P_B \circ [(1+\beta)P_A(f^{(n)}) - f^{(n)}] - \beta P_A(f^{(n)}), \quad (1.39)$$

where $\beta \in (0, 1]$; we focus on the $\beta = 1$ case. See Bauschke et al. (2002) and Borwein and Sims (2011) for a discussion of the relation between alternating projection, HIO, and the algorithms due to Dykstra and Douglas-Peaceman-Rachford for convex set intersection and operator splitting in numerical analysis, respectively.

From (1.37) it is evident that fixed points of hybrid iterative maps always determine points in $A \cap B$:

$$\text{If } D_{BA}(f^*) = f^*, \text{ then } f^{**} = P_B \circ R_A(f^*) = P_A(f^*) \in A \cap B. \quad (1.40)$$

We call the set of points that satisfy (1.40) the *center manifold* defined by f^{**} and denote it by $\mathscr{C}^{f^{**}}_{BA}$. The "fixed bounded sets," appearing in the definition of stagnation, that arise for hybrid iterative maps are typically small neighborhoods of points on one or several center manifolds.

If such an algorithm converges to a point on $\mathscr{C}_{BA}^{f^{**}}$, then the limit determines a point $f^{**} \in A \cap B$. To obtain approximate reconstructions, $\{r^{(n)}\}$ from the iterates of an algorithm based on a hybrid map, one must project them onto either the A or B spaces:

$$r^{(n)} = P_A(f^{(n)}), \text{ or } \tilde{r}^{(n)} = P_B \circ R_A(f^{(n)}). \tag{1.41}$$

In this context, the residual

$$\|P_B \circ R_A(f^{(n)}) - P_A(f^{(n)})\|_2 = \|f^{(n)} - f^{(n+1)}\|_2 \tag{1.42}$$

is both a reasonable measure for the current error, and an indication of the extent to which the algorithm is converging. In the discrete, classical, phase retrieval problem, the set A is always a magnitude torus $\mathbb{A}_{a}$, whereas B is fixed by the choice of auxiliary information.

Definition 1.9 Let $B \subset \mathbb{R}^J$, or $B \subset \mathbb{C}^J$, be a subset specified by auxiliary information in the phase retrieval problem. This information, or the set itself, is said to be *adequate* if $\mathbb{A}_{a} \cap B$ is finite and, when non-empty, generically consists of trivial associates of a single point in this intersection.

If support is the auxiliary information being utilized, then B equals B_S, defined in (1.18). If the image is known to be nonnegative, which is often the case in astronomical applications and sometimes the case in X-ray diffraction imaging, then one can use the nonnegativity constraint. That is, we look for points in $\mathbb{A}_{a} \cap B_+$. For adequately sampled, compact objects, the intersections of interest lie on ∂B_+, which is a nonsmooth stratified space. Near such points B_+ is "more convex" than the linear subspace B_S, and hence, it is more likely that $\mathbb{A}_{a}$ meets ∂B_+ transversally. This suggests that algorithms using the nonnegativity constraint can be expected to work better than those employing the support constraint, an expectation that is borne out in experiments below.

For an image f, the ℓ_1-norm is given by

$$\|f\|_1 = \sum_{j \in J} |f_j|. \tag{1.43}$$

If $f \in B_+$, then evidently $\|f\|_1 = \widehat{f_0}$. It is a simple consequence of the triangle inequality that $\|f\|_1$ assumes its strict global minimum on $\mathbb{A}_{a}$ *only* at images of a single sign. The fact that the ℓ_1-norm assumes a local minimum, $r_1 = \widehat{f_0}$, on $\mathbb{A}_{a}$ at a nonnegative image suggests a different formulation of the phase retrieval problem: seek points in $\mathbb{A}_{a} \cap B_{r_1}^1$, where

$$B_{r_1}^1 = \{f \in \mathbb{R}^J : \|f\|_1 = r_1\}. \tag{1.44}$$

There is a very efficient implementation of the map $P_{B^1_{r_1}}$, which makes this approach practical to implement. In practice, algorithms using P_{B_+} and $P_{B^1_{r_1}}$ behave similarly.

The problem of searching for intersections of sets has two distinctly different phases: there is an initial "global" phase where the iteration searches for an "attracting basin" and a subsequent "local" phase where the iterates are confined to this basin but may or may not converge. The global phase is hard to analyze, especially in the very high-dimensional settings of interest in the phase retrieval problem. It has been observed experimentally that the iterates behave extremely chaotically at this stage: adding a machine-precision-sized perturbation to a random initial condition typically leads to an $O(1)$ difference in the trajectories after approximately one hundred iterates. So long as the iterates do not diverge to infinity, this chaotic behavior can be desirable as it allows the iterates to quickly sample a wide range of the possible starting configurations. This gives the iteration a better chance of finding a basin of attraction, which, in our experience, seems to occur reasonably quickly. This perspective was noted by Elser and colleagues (Elser 2003; Elser et al. 2007; Gravel and Elser 2008; Elser et al. 2018).

In our analysis, we concentrate on the behavior of maps once they have settled into an attracting basin, and demonstrate, in some cases, that sufficiently close to an intersection point, these maps are weakly contracting in directions transverse to the fixed-point set. Because the attraction is weak, the iterates often do not get pulled close enough to a single fixed-point set, but rather remain influenced by the weak attraction of several nearby fixed-point sets. Even when the iterates are started very close to a point in $\boldsymbol{f}_0 \in \mathbb{A}_{\boldsymbol{a}} \cap B$, if this intersection is nontransversal, then the iterates generally fail to converge to $\boldsymbol{f}_0$, but will stagnate, with the differences $\{\|\boldsymbol{f}^{(n)} - \boldsymbol{f}^{(n+1)}\|_2\}$ remaining proportional to the square of the distance to the nearest intersection point. A surprising finding is that, even after millions of iterations, these points are not observed to diverge. Rather they remain in a small neighborhood of the fixed-point sets defined by a small number of trivial associates. This topic is investigated in detail in Sections 7.2.1 and 7.3.1– 7.3.2.

Since hybrid iterative maps for solving the phase retrieval problem are essentially nonlinear power methods, the rates of convergence depend on the angles between $T_{\boldsymbol{f}_0}\mathbb{A}_{\boldsymbol{a}}$ and B_S. This is true even though the attracting fixed points lie on the center manifold, quite far from the intersection points themselves. The failure of transversality at points in $\mathbb{A}_{\boldsymbol{a}} \cap B$ complicates the analysis, rendering the question of convergence in these directions fundamentally nonlinear. It often seems to have the effect of producing directions in which the contraction rate toward the fixed-point set is effectively zero. Beyond this exact intersection, it is usually the case that there is a further subspace of $T_{f_0}\mathbb{A}_a$ that makes a very shallow angle with B_S. The dimension of this subspace is data dependent, increasing monotonely with the smoothness of the image, and can be quite large. For even moderately smooth images, these nearly tangent directions can lead to stagnation. For many types of images, even a very precise support condition does not lead to a well-conditioned reconstruction problem.

Beyond these essentially local issues, which arise near to exact intersection points, the function from $\mathbb{A}_a \times B \to [0,\infty)$,

$$d_{\mathbb{A}_a B}(f_{\mathbb{A}_a}, f_B) = \|f_{\mathbb{A}_a} - f_B\|_2, \tag{1.45}$$

has many critical points where $d_{\mathbb{A}_a B}$ is nonzero. It turns out that these nonminimal critical points can themselves define attracting basins for hybrid iterative map algorithms. This phenomenon was noted earlier by Seldin and Fienup (1990).

As suspected by practitioners, the analysis in this book suggests that there are fundamental obstacles to CDI and that progress on phase retrieval might require changes in the way the data itself is collected. In Section 7.4, we explore an alternative experimental protocol that would allow an entirely different type of reconstruction algorithm, which does not depend on auxiliary information. This approach is a form of "external holography," wherein a small, highly diffracting external object is placed very near to the object we are trying to image. With the magnitude DFT measurements of the combined objects and a knowledge of the DFT of the external object, the reconstruction is done directly using the 1-dimensional Hilbert transform. We call this the *Holographic Hilbert Transform* method. In many cases this method gives an accurate reconstruction. A variety of other forms of external holography have been considered in the literature (see, for example, Barmherzig et al. 2019a, 2019b; Maretzke and Hohage 2017; Jacobsen 2019; Guizar-Sicairos and Fienup 2007, 2008 and the references therein); there have also been previous attempts to use the Hilbert transform, without holography, to solve the phase retrieval problem, see Nakajima and Asakura (1985, 1986); Nakajima (1995).

1.4 Numerical Experiments

The results in this book are both illustrated with and motivated by numerical experiments. In this section, we describe how the test objects used in most of these experiments are generated. The images we use are obtained by sampling randomly generated sums of compactly supported radial functions. These functions are supported in a square with side length equal to, at most, half that of a larger enclosing square, with the larger square defining the field of view. In most cases, the disks in a given example all have the same degree of smoothness which we parameterize using the variable "k." The least smooth examples ($k = 0$) correspond to a sum of piecewise constant functions. In the CDI literature this corresponds to a "high contrast," or "hard" object. Larger values of k correspond to smoother, or "softer" objects.

More precisely, the radial functions we use are of the form

$$\rho_{k.\mathscr{G}}(\boldsymbol{x}) = \sum_{i=1}^{l} m_i \psi_k \left(\frac{\|\boldsymbol{x} - \boldsymbol{c}_i\|_2}{r_i} \right), \tag{1.46}$$

with $\chi_{[-1,1]}(t) = 1$, for $|t| \le 1$, and zero otherwise, we set

$$\psi_k(t) = (1 - t^2)^k \chi_{[-1,1]}(t), \tag{1.47}$$

The magnitudes, centers and radii, $\mathscr{G} = \{(m_i, \boldsymbol{c}_i, r_i) : i = 1, \ldots, l\}$, are randomly selected. As noted, when $k = 0$ these functions are piecewise constant. For integral $0 < k$, $\psi_k \in \mathscr{C}^{k-1}(\mathbb{R})$, with $\partial_t^{k-1}\psi_k$ a Lipschitz continuous function. For nonintegral $0 < k$, $\psi_k \in \mathscr{C}^{[|k|]}(\mathbb{R})$, with $\partial_t^{[|k|]}\psi_k$ a $(k - [|k|])$-Hölder continuous function. The support of $\rho_{k,\mathscr{G}}$ typically does not depend on $0 \le k$.

To construct an image indexed by $J \subset \mathbb{Z}^d$, we sample a function like $\rho_{k,\mathscr{G}}$ at a set of uniformly spaced points $\{\frac{\boldsymbol{j}}{N}, \boldsymbol{j} \in J\}$, where, for simplicity we use the same spacing in all directions. The image $\boldsymbol{f}$ is then defined by

$$f_{\boldsymbol{j}} = \rho_{k,\mathscr{G}}\left(\frac{\boldsymbol{j}}{N}\right) \quad \text{for } \boldsymbol{j} \in J; \tag{1.48}$$

$\boldsymbol{f} \in \mathbb{R}^J$ if the $\{m_i\}$ are real and $\boldsymbol{f} \in \mathbb{C}^J$ if the $\{m_i\}$ are complex. As indicated above, there are two features of an image that play a significant role in determining the difficulty of the phase retrieval problem:

(i) The image's support, S_f, and our knowledge of it.
(ii) The "smoothness" of the image.

While the meanings of both concepts are clear for $\mathbb{R}$- (or $\mathbb{C}$-) valued functions defined in a continuum, neither concept has an unambiguous meaning for discrete images in finite precision arithmetic. For this discussion, we let ϵ_{mach} denote the *machine epsilon* – the smallest positive number representable on a computer for which $1 + \epsilon_{\text{mach}}$ is distinguishable from 1 in floating point arithmetic. For double precision calculations, $\epsilon_{\text{mach}} \approx 10^{-16}$. For the discussion that follows we restrict ourselves to functions $\rho_{k,\mathscr{G}}$ and images f that are normalized so that $\|\rho_{k,\mathscr{G}}\|_\infty = 1$ and

$$\max\{|f_j| : \boldsymbol{j} \in J\} = 1. \tag{1.49}$$

The notion of support is simpler. If $k = 0$, then the set of values

$$\{|f_j| : \boldsymbol{j} \in J \text{ with } |f_j| \neq 0\}$$

has a minimum, which is usually many orders of magnitude larger than machine precision, and therefore the set

$$S_f^{\epsilon_{\text{mach}}} = \{\boldsymbol{j} \in J : |f_j| \geq \epsilon_{\text{mach}}\} \tag{1.50}$$

usually coincides with the set $S_f = \{\boldsymbol{j} \in J : |f_j| \neq 0\}$. If $k > 0$, then the function $\rho_{k,\mathscr{G}}$ goes to zero gradually. For larger values of k and N there may be many indices $\boldsymbol{j} \in J$ for which

$$0 < |f_j| \ll \epsilon_{\text{mach}}. \tag{1.51}$$

For any $\epsilon > 0$ we define the ϵ-support of $\boldsymbol{f}$ to be

$$S_f^{\epsilon} = \{\boldsymbol{j} \in J : |f_j| \geq \epsilon\}. \tag{1.52}$$

Note that if $\|\boldsymbol{f}\| \neq 1$, then

$$S_f^{\epsilon} = \{\boldsymbol{j} \in J : |f_j| \geq \epsilon \|\boldsymbol{f}\|\}. \tag{1.53}$$

Each of these sets provides some definition of the "support of f," however for the sake of consistency and mathematical correctness we *define* the support of f by

$$S_f \stackrel{d}{=} \{\boldsymbol{j} \in J : |f_j| \geq 0\}. \tag{1.54}$$

Remark 1.10 (Important Remark) For the reason indicated above, using the exact support of an image is often not desirable (or necessary) in numerical experiments. In this context, the support constraint is usually defined by S_f^{ϵ}, with $\epsilon = 10^{-14}$. To simplify the notation, **in numerical examples,** we often write S_f instead of the more accurate, but cumbersome $S_f^{10^{-14}}$

If we choose $\epsilon \gg \epsilon_{\text{mach}}$, then, depending on k and N, and the particular function $\rho_{k,\mathscr{G}}$, it may well happen that, at machine precision, the set $\mathbb{A}_{\boldsymbol{a}_f} \cap S_f^{\epsilon}$ is empty. Hence, even at machine precision, the phase retrieval with this data has no solution. In the sequel we show that the difficulty of the phase retrieval problem increases with the smoothness of the data, and that it is sometimes advantageous to use a set S_f^{ϵ}, with $\epsilon > \epsilon_{\text{mach}}$ to define the support constraint, even though it leads to a formally inconsistent problem.

Providing a meaning to the notion of "smoothness" for a sampled image is a bit more subtle. This is even a hard question within the restricted class of images defined by (1.46) and (1.48). In this context, one can phrase the question as: "given the finite image $\boldsymbol{f}$, can one determine the value of k used to define the function $\rho_{k,\mathscr{G}}$ that was sampled to define $\boldsymbol{f}$?" A somewhat less ambitious goal would be the introduction of quantities that would allow us to compare the smoothness of the functions used to define two different images indexed by the same set J.

Since our measurements take place in the Fourier domain it seems reasonable to consider how the smoothness of $\rho_{k,\mathscr{G}}$ is reflected in the behavior of $\widehat{\boldsymbol{f}}$. The answer to this question depends, to some extent, on the dimension d of the image. We restrict our attention here to the case of $d = 2$. The continuous Fourier transform of $\psi_k(\|\boldsymbol{x}\|_2/r)$, where $\boldsymbol{x} \in \mathbb{R}^2$ is given by

$$C_k r^2 \frac{J_{k+1}(\|r\boldsymbol{\xi}\|_2)}{\|r\boldsymbol{\xi}\|_2^{k+1}}, \tag{1.55}$$

where C_k is a constant and J_{k+1} is the J-Bessel function of degree $k+1$ (see Watson 1922). This function is oscillatory and satisfies the sharp estimate

$$\frac{|J_{k+1}(\|\boldsymbol{\xi}\|_2)|}{\|\boldsymbol{\xi}\|_2^{k+1}} \leq \frac{C_k'}{(1+\|\boldsymbol{\xi}\|_2)^{k+\frac{3}{2}}}; \tag{1.56}$$

the larger k is, the faster this function eventually decays.

We consider the notion of smoothness of sampled images for a simple set of examples. The DFT coefficients of the image

$$f_{k,r,N} = \left\{\psi_k\left(\frac{\boldsymbol{j}}{r(2N+1)}\right) : \boldsymbol{j} \in [-N : N]^2\right\}$$

give approximations to

$$\left\{C_k r^2 \frac{J_{k+1}(\|r\boldsymbol{k}\|_2)}{\|r\boldsymbol{k}\|_2^{k+1}} : \boldsymbol{k} \in [-N : N]^2\right\}. \tag{1.57}$$

From results in Epstein (2005) we know that these approximations are relatively accurate for $\boldsymbol{k} \in \left[-\frac{N}{2} : \frac{N}{2}\right]^2$, and have the correct order of magnitude over the entire range. The question of whether we can detect the underlying smoothness of $\psi_k(\|\boldsymbol{x}\|_2/r)$ is, therefore, equivalent to the question of whether we can detect the asymptotic behavior of the $r^2\widehat{\psi}_k(r\boldsymbol{\xi})$ from the DFT of the samples, $\{\widehat{f}_{k,r,N;\boldsymbol{k}} : \boldsymbol{k} \in [-N : N]^2\}$. This, in turn, depends on how large rN is, and how long it takes for the magnitude of $J_{k+1}(t)/t^{k+1}$ to reliably follow its asymptotic formula.

It is well-known that $J_k(t)$ settles into its asymptotics once it has passed its second negative minimum, which occurs at a value of $t \approx 11.94 + 1.32k$. For a fixed value of N, and $0 < r_{\min} \leq r$, it should be possible to correctly order the different values of k from sampled data for a range $k \in [0, k_{\max}(r_{\min}, N)]$. For example, it is reasonable to expect that we could discriminate the smoothness of a collection of examples of this type for which

$$11.94 + 1.32k < \frac{r_{\min} N}{2}. \tag{1.58}$$

Qualitatively, this analysis carries over to images defined by sampling functions like $\rho_{k,\mathscr{G}}$. To have a meaningful notion of smoothness, it is necessary to have a positive lower bound, $r_{\min}$, on r; given such a lower bound and a value for N, there is a finite range of values of k over which one can meaningfully discern the smoothness of the underlying function from the sampled values. For a fixed $r_{\min}$, this range increases along with the number of samples. Since this can only be done quantitatively within a well-specified collection of examples, we do not dwell further on this question.

We close this discussion with some representative examples. In Figure 1.5 we show examples of images defined by sampling functions $\rho_{k,\mathscr{G}}(\boldsymbol{x})$ that are supported in $[-1,1]^2$, with smoothness levels $0, 2$, and 4. These 128×128 images are zero-padded to be defined on a 256×256 grid. We use the lower bound $r > 0.01$, and $l = 100$ radial functions.

In our experiments, we also consider various constraints on the "known" support that are specified as "p-pixel neighborhoods" of the actual support, for $p = 0, 1, 2, \ldots$ This concept is defined in Definition 2.11. Examples of support constraints of this type are shown in Figure 1.6; these correspond to different sized neighborhoods of the exact supports of the objects shown in Figure 1.5. Comparing the support neighborhoods to the images in Figure 1.5

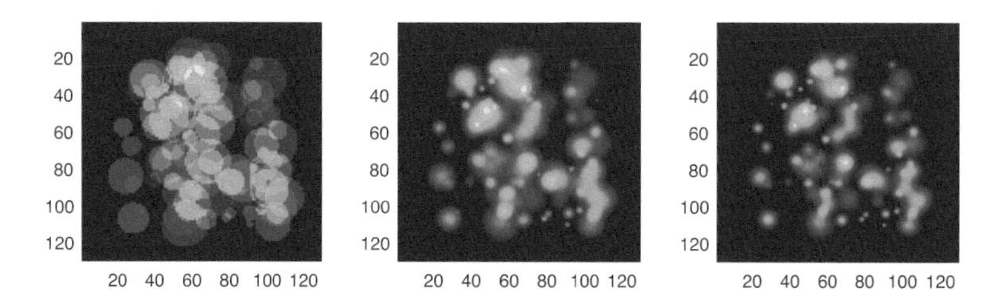

Figure 1.5 Images similar to those used in numerical experiments below. These are defined by sampling functions of the type defined in (1.46) with the smoothness levels: $k = 0, 2, 4$. The plots show the central 128×128 portion of the original 256×256 grid.

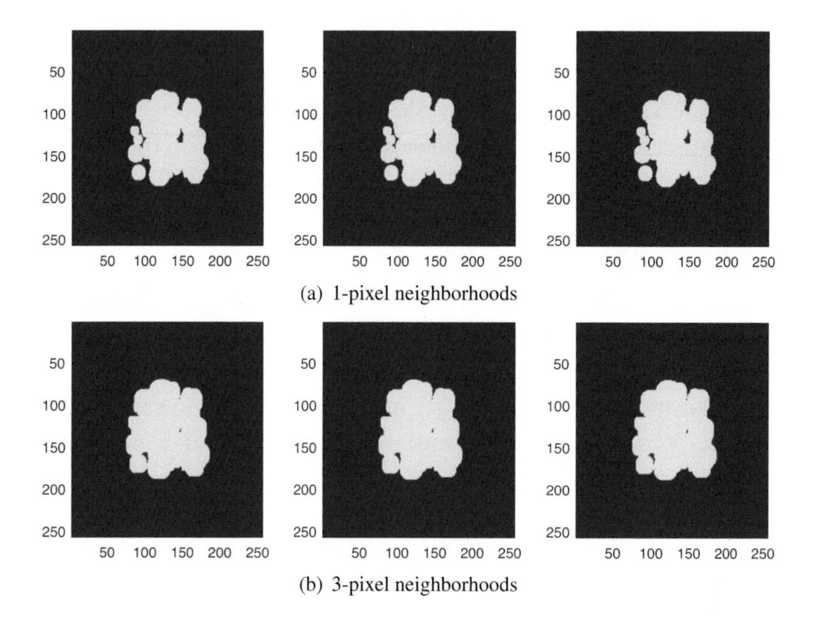

(a) 1-pixel neighborhoods

(b) 3-pixel neighborhoods

Figure 1.6 Different support constraints for the examples shown in Figure 1.5. The yellow regions are the p-pixel neighborhoods of the exact supports of these images for $p = 1, 3$. Note that these neighborhoods do not depend on k.

suggests that, for smoother images, the values of $|f_j|$ can be quite small in parts of the image's exact support. There are other approaches to estimating the support that are explored later in the book. These include finding the smallest rectangle that contains the object and using a sharp cutoff mask. As noted above, knowledge of the support and the manner in which the object tends to zero are important determinants in the behavior of iterative algorithms. For these examples we have used the definition of support given (1.54).

In numerical analysis, a standard way to obtain "smooth" examples is to convolve piecewise constant examples with samples of Gaussians, where the width of Gaussian controls the extent of the smoothing. There are several reasons that this is problematic in the present context. First, it is difficult to relate this notion of smoothness to standard notions of differentiability. Second, convolution spreads out the support of the image. Finally, as explained at the end of Section 1.2, convolution with an inversion symmetric image has the effect of dramatically altering the local geometry near to intersections between the magnitude torus and the linear subspace defined by a support constraint. Nonetheless it is sometimes useful to do this, and it is often done in practice, so we give examples similar to those above. These images are obtained by

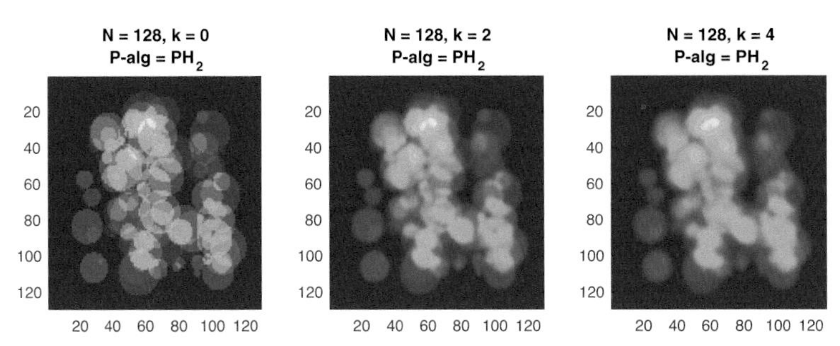

(a) Images defined by convolving a piecewise constant image with Gaussians of various widths. These are labeled $k = 0, 2, 4$.

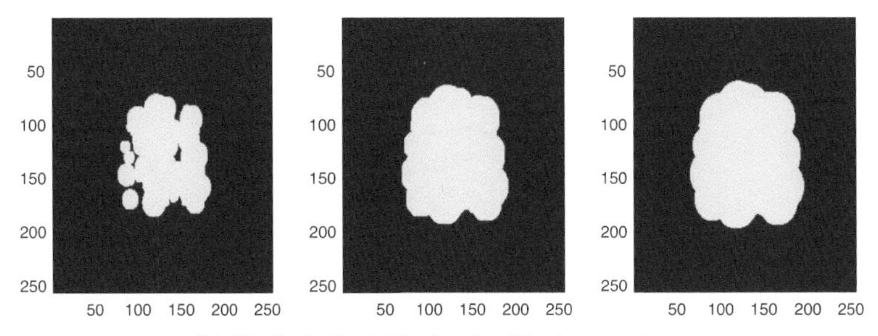

(b) The 1-pixel neighborhoods of the images above.

Figure 1.7 Examples of images defined by convolution with samples of Gaussians, and their 1-pixel neighborhoods.

convolution of the $k = 0$ example in Figure 1.5 with samples of Gaussians. Figure 1.7(a) shows these images for "$k = 0, 2, 4$," and Figure 1.7(b) show the corresponding 1-pixel neighborhoods. As noted, the k in these examples is not directly comparable to the k in the previous examples.

For a typical numerical experiment, we first choose an image, $\boldsymbol{f}_0$, with magnitude DFT data $\boldsymbol{a}_{f_0}$, and a support constraint, which is a subset $S \subset J$ that contains S_{f_0}. To initialize, we choose random phases, $\{e^{i\theta_k} : \boldsymbol{k} \in J\}$ and construct an image, $\boldsymbol{f}^{(0)}$, with DFT data $\widehat{f}_{\boldsymbol{k}}^{(0)} = a_{f_0;\boldsymbol{k}} e^{i\theta_k}$ for $\boldsymbol{k} \in J$. The sequence $\{\boldsymbol{f}^{(n)} : n = 0, 1, \ldots\}$ is then defined by iterating a map, starting at $\boldsymbol{f}^{(0)}$. If we are using a hybrid map, then we also need to construct a sequence of approximate reconstructions, $\{\boldsymbol{r}^{(n)}\}$, see (1.41).

In Figure 1.8 we show the results of 5,000 iterates of $D_{\mathbb{A}_{\boldsymbol{a}_{f_0}} B_S}$ for an image $\boldsymbol{f}_0$, which is a 256×256 sample of a function of the type defined in (1.46), with $k = 1$ and $l = 100$. To display the results of our experiments, we usually give a plot *in blue* of the *absolute errors* $\{\|\boldsymbol{r}^{(n)} - \boldsymbol{f}'_0\|_2\}$ where $\boldsymbol{f}'_0$ is the point in $\mathbb{A}_{\boldsymbol{a}_{f_0}} \cap B$ nearest to $\boldsymbol{r}^{(n)}$. We also show a plot *in red* of the *residuals*, which

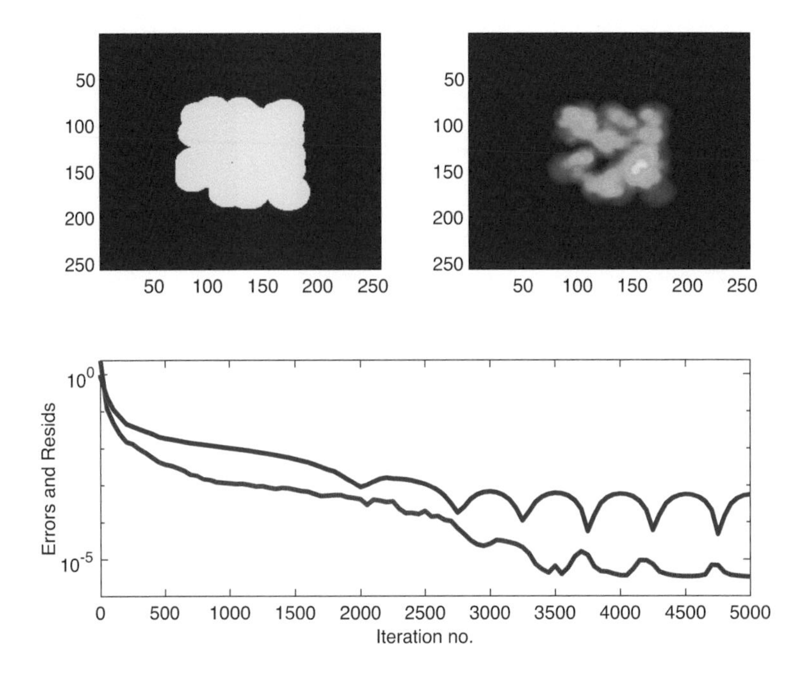

Figure 1.8 The data and results for a typical numerical experiment involving an iterative reconstruction algorithm. The first row shows the support mask, S, and the final reconstruction, $\boldsymbol{r}^{(5,000)}$. The second row contains semilog plots of the errors and residuals, with the blue plot showing the true errors, and the red plot showing the residuals.

are essentially the distance between the points in $\mathbb{A}_{\boldsymbol{a}_{f_0}}$ and B that are nearest to $\boldsymbol{r}^{(n)}$. In a real experiment, only the residuals are available. For hybrid maps, this is bounded above by $\|\boldsymbol{f}^{(n)} - \boldsymbol{f}^{(n+1)}\|_2$, which is what we display.

1.5 Comparison to the Continuum Phase Retrieval Problem

As noted above, this book focuses on the analysis of the discrete, classical, phase retrieval problem. There are many differences between this idealized problem and the real physical phase retrieval problem, using actual measured data. In the physical problem the measured data comes from a far field measurement of a solution to Maxwell's equations defined by a plane wave scattering off of an object of interest. In the Born approximation, the leading order term of the asymptotic expansion of such a solution, in powers of the wavelength, is proportional to the Fourier transform of the square of the refractive index of the scattering object. At X-ray energies, only the intensity can be measured, which, *to leading order*, is the squared modulus of the Fourier transform of this squared refractive index. The geometry of the sampling is determined by the Ewald sphere construction, with the frequency in the Fourier domain determined by the incoming X-ray energy and the direction of scatter (see (3) and Jacobsen 2019). The frequencies of the measured samples, therefore, lie on a sphere in $\mathbb{R}^3$ and not on a plane. Uniformly spaced samples collected on a measurement plane do not correspond to uniformly spaced sampled on a planar Cartesian grid in $\boldsymbol{k}$-space. Due to the intensity of the forward scattered field, a neighborhood of $\boldsymbol{k} = \boldsymbol{0}$ often cannot be measured at all.

From this brief discussion of the underlying physics, without any consideration of noise etc., it is already evident that the model we consider deviates from the actual physical problem in many ways. We do not consider this further, but instead consider the simpler question of the difference between collecting uniformly spaced samples of the continuum Fourier transform, $(\widehat{\rho}(\boldsymbol{\xi}_{\boldsymbol{k}}))$, of a function $\rho(\boldsymbol{x})$, versus the DFT of the samples, $(f_{\boldsymbol{j}})_{\boldsymbol{j}\in J}$, of ρ, see (1.6)–(1.7) The moduli of samples of the continuum Fourier transform are clearly a better model for the physical measurement, but this small change already produces significant obstacles to a careful analysis.

Suppose that $\operatorname{supp}\rho \subset \mathcal{K} \subset [-\frac{1}{2}, \frac{1}{2}]^d \subset [-1, 1]^d$. For $\boldsymbol{x} \in [-1, 1]^d$ the function $\rho(\boldsymbol{x})$ is given by

$$\rho(\boldsymbol{x}) = \frac{1}{2^d} \sum_{\boldsymbol{k}\in\mathbb{Z}^d} \widehat{\rho}\left(\frac{\boldsymbol{k}}{2}\right) e^{\pi i \langle \boldsymbol{x}, \boldsymbol{k}\rangle}. \tag{1.59}$$

The function defined by this sum vanishes for $\boldsymbol{x} \in [-1, 1]^d \setminus \mathcal{K}$. Unlike the function ρ, which is compactly supported in $\mathcal{K}$, the function defined by this sum is periodic of period 2 in each coordinate direction. For a fixed N, we let $S^{(N)} \subset J_N = [1 - N : N]^d$ such that

$$\frac{\boldsymbol{j}}{N} \in \mathcal{K}. \tag{1.60}$$

For $\boldsymbol{j} \in J_N$, let $f_{\boldsymbol{j}}^{(N)} = \rho\left(\frac{\boldsymbol{j}}{N}\right)$, and $(\widehat{f_{\boldsymbol{k}}}^{(N)})$ denote the DFT of these samples,

$$\widehat{f_{\boldsymbol{k}}}^{(N)} = \sum_{\boldsymbol{j}\in J_N} \rho\left(\frac{\boldsymbol{j}}{N}\right) e^{-\frac{2\pi i \langle \boldsymbol{j}, \boldsymbol{k}\rangle}{2N}}. \tag{1.61}$$

Up to a factor of $(\pi/N)^d$, this sum is a Riemann sum for the continuum Fourier data

$$\widehat{\rho}\left(\frac{\boldsymbol{k}}{2}\right) \qquad \text{for } \boldsymbol{k} \in J_N. \tag{1.62}$$

For ρ a piecewise continuous function,

$$\lim_{N\to\infty} (\pi/N)^d \widehat{f}_{\boldsymbol{k}}^{(N)} = \widehat{\rho}\left(\frac{\boldsymbol{k}}{2}\right). \tag{1.63}$$

The important difference between the DFT data and the samples of the continuum Fourier transform arises from the fact that, to fully specify a phase retrieval problem, one needs auxiliary information, such as an estimate on the support of ρ, (or $(f_{\boldsymbol{j}}^{(N)})$). If we apply the inverse DFT to the data $(\widehat{\boldsymbol{f}}_{\boldsymbol{k}}^{(N)})$ it exactly returns the sample sequence, $(f_{\boldsymbol{j}}^{(N)})$. In particular, we get a point in the intersection of the magnitude torus defined by $(|\widehat{f}_{\boldsymbol{k}}^{(N)}|)$ and the linear subspace $B_{S^{(N)}}$. The inverse DFT applied to the samples of the continuum transform with indices in J_N is given by

$$\tilde{f}^{(N)}\left(\frac{\boldsymbol{j}}{N}\right) = \frac{1}{|J_N|}\sum_{\boldsymbol{k}\in J_N} \widehat{\rho}\left(\frac{\boldsymbol{k}}{2}\right) e^{\frac{2\pi i\langle \boldsymbol{j},\boldsymbol{k}\rangle}{2N}}. \tag{1.64}$$

There is a constant C_N so that $C_N \tilde{f}^{(N)}\left(\frac{\boldsymbol{j}}{N}\right)$ is a partial sum for the Fourier series of ρ at $\boldsymbol{x} = \frac{\boldsymbol{j}}{N}$. What is crucial to note is that the samples $C_N \tilde{f}^{(N)}\left(\frac{\boldsymbol{j}}{N}\right)$, $\boldsymbol{j} \in J^{(N)}$ do not exactly equal $\rho\left(\frac{\boldsymbol{j}}{N}\right)$, $\boldsymbol{j} \in J^{(N)}$ and are, in fact, never supported in $S^{(N)}$. Generally, there is no sequence supported in $S^{(N)}$ with the magnitude Fourier data defined by the samples

$$\boldsymbol{a}_c^{(N)} = \left\{\left|\widehat{\rho}\left(\frac{\boldsymbol{k}}{2}\right)\right| : \boldsymbol{k} \in J_N\right\}.$$

Moreover, if the function ρ is nonnegative, the reconstructed values in (1.64) generally take both signs.

Thus, finitely many samples of magnitude continuum Fourier data and a standard support constraint, or nonnegativity constraint, are *always inconsistent*. That is,

$$\mathbb{A}_{\boldsymbol{a}_c^{(N)}} \cap B_{S^{(N)}} = \emptyset, \quad \text{or} \quad \mathbb{A}_{\boldsymbol{a}_c^{(N)}} \cap B_+ = \emptyset \quad \text{(respectively)}. \tag{1.65}$$

Moreover, in Barnett et al. (2020) the following result is proved: The set of functions supported in a fixed compact set, D, with a finite set of specified Fourier coefficients: $\widehat{\rho}(\boldsymbol{k}_j) = a_j$, for $j = 1, \ldots, M$, is infinite dimensional. The discrete model we have adopted for this text is perhaps the most realistic, simple model for the measurements in CDI that admits *exact* solutions and can therefore be carefully analyzed. As indicated above, the attainable quality of reconstructions, for the discrete problem, is related to the smoothness of the object one seeks to reconstruct. The analysis in this book shows that, for this discrete model, the highest quality reconstructions are obtainable only for sharply delineated objects with a precise support constraint.

1.6 Outline of the Book

The book is divided into three parts. Part I covers the theoretical foundations needed to understand the phase retrieval problem. It is largely concerned with the well-posedness of the discrete, classical, phase retrieval problem. Our focus is on the geometry near points of intersection between a magnitude torus, $\mathbb{A}_{\boldsymbol{a}}$, and a set, B, defined by auxiliary information. In this part, no reference is made to any particular algorithm. Its conclusions about the inherent difficulty of this problem apply to any algorithm for locating these intersection points.

In Part II, we apply these results to analyze the behavior of the iterative maps most commonly used for finding points in $\mathbb{A}_{\boldsymbol{a}} \cap B$. The basic algorithms are presented in Chapter 7 : alternating projection and hybrid iterative maps. We give a complete analysis of these algorithms in the case that A and B are affine subspaces of $\mathbb{R}^N$. This analysis makes clear how the "angles" between A and B at the point of intersection determine the local convergence properties of these algorithms, and is also the basic information needed to linearize these algorithms in case one or both of A, B is not linear. For the linear case the angle information is encoded in a matrix, H; the singular values of H are the cosines of these angles, and govern the convergence properties near to intersection points for algorithms based on either class of maps.

Part III discusses the statistical properties of hybrid iterative maps and a range of suggestions for improvements in the image reconstruction process. A principal cause for the failure of phase retrieval algorithms appears to be the nontransversality of the intersections between the magnitude torus $\mathbb{A}_{\boldsymbol{a}}$ and the set, B, defined by the auxiliary information. To improve the outcome of this process, therefore, requires modifications to the way in which the data is obtained. Several different approaches are outlined in Part III. We also consider an alternative optimization-based approach, which we call a *geometric Newton* method. This approach makes use of the very efficient description of the tangent bundle to a magnitude torus introduced in Chapter 2 and is quite different from the hybrid iterative maps commonly used.

1.A Appendix: Factoring Polynomials in Several Variables

In this appendix we present the basic facts about factoring polynomials in several variables, and show that the set of factorable, or reducible polynomials, is a lower dimensional subset. A polynomial in one variable, of degree n, is an expression of the form

$$p(z) = a_n z^n + a_{n-1} z^{n-1} + \cdots a_0, \tag{1.66}$$

where $\{a_0, \dots, a_n\}$ are complex numbers with $a_n \neq 0$. If $\{z_1, \dots, z_m\}$ are the roots of $p(z)$, with multiplicities $\{n_1, \dots, n_m\}$, then $p(z)$ can be factored as

$$p(z) = c(z - z_1)^{n_1} \dots (z - z_m)^{n_m}. \tag{1.67}$$

Where $c \in \mathbb{C}$ and $n_1 + \dots + n_m = n$. That is, any polynomial in a single variable can be factored, over the complex numbers, as a product of linear factors to positive powers.

Let $\mathscr{P}_{d,n}$ denote the set of polynomials in $d > 1$ variables, $z = (z_1, \dots, z_d)$, of degree n. Such a polynomial is an expression of the form

$$p(z) = \sum_{\boldsymbol{\alpha}: |\boldsymbol{\alpha}| \leq n} a_{\boldsymbol{\alpha}} z^{\boldsymbol{\alpha}}, \text{ where } z^{\boldsymbol{\alpha}} = z_1^{\alpha_1} \dots z_d^{\alpha_d}. \tag{1.68}$$

The sum is over nonnegative d-multi-indices $\boldsymbol{\alpha} = (\alpha_1, \dots, \alpha_d)$ with length

$$|\boldsymbol{\alpha}| = \alpha_1 + \dots + \alpha_d \leq n. \tag{1.69}$$

For polynomials, $p(z)$ in $d \geq 2$ variables, it is very unusual for there to exist polynomials, q_1, q_2, with positive degrees, so that

$$p(z) = q_1(z) q_2(z). \tag{1.70}$$

A polynomial that can be factored like this is called *reducible*, and a polynomial with no such factorization is called *irreducible*.

It is a classical fact that

$$\dim \mathscr{P}_{d,n} = \frac{(d+n)!}{d!n!} = \frac{(n+1)\dots(n+d)}{d!} = \sum_{j=1}^{d} b_j n^j + 1, \tag{1.71}$$

where the $\{b_j\}$ are positive (see Griffiths and Harris 1978). We let M denote the map from a pair of polynomials to their product

$$M(p,q) = p \cdot q. \tag{1.72}$$

If n_1, n_2 are positive integers, then

$$M : \mathscr{P}_{d,n_1} \times \mathscr{P}_{d,n_2} \to \mathscr{P}_{d,n_1+n_2}.$$

We let $\mathscr{P}_{d,n}^{\text{red}}$ denote the reducible polynomials in d-variables of degree n. Every reducible polynomial of degree n, i.e., with a nontrivial factorization as in (1.70), belongs to $M(\mathscr{P}_{d,n-l}, \mathscr{P}_{d,l})$, for some $1 \leq l < \lfloor \frac{n}{2} \rfloor$, and therefore,

$$\mathscr{P}_{d,n}^{\text{red}} \subset \bigcup_{l=1}^{\lfloor \frac{n}{2} \rfloor} M(\mathscr{P}_{d,n-l}, \mathscr{P}_{d,l}). \tag{1.73}$$

To show that the set of reducible polynomials is small, we show that if $d > 1$, and $1 \leq l \leq \lfloor \frac{n}{2} \rfloor$, then

$$\dim M(\mathscr{P}_{d,n-l}, \mathscr{P}_{d,l}) \leq \dim \mathscr{P}_{d,n-l} + \dim \mathscr{P}_{d,l} - 1 < \dim \mathscr{P}_{d,n}. \tag{1.74}$$

We subtract 1 in the middle because $M(\lambda p, \lambda^{-1} q) = M(p,q)$ for any nonzero complex number λ. For this purpose we use the formula for the dimension of $\mathscr{P}_{d,l}$ as a polynomial with positive coefficients, (1.71). From this formula it follows that (1.74) is equivalent to

$$\sum_{j=1}^{d} b_j [(n-l)^j + l^j] < \sum_{j=1}^{d} b_j n^j. \tag{1.75}$$

This is obvious as $(n-l)^j + l^j \leq n^j$ for $0 < j$ and all $0 \leq l \leq n$, with strict inequality if $0 < l < n$ and $j > 1$.

For $d = 2,\ 3$ we see that

$$\begin{aligned} \dim \mathscr{P}_{2,n} - (\dim \mathscr{P}_{2,n-l} + \dim \mathscr{P}_{2,l} - 1) &= l(n-l) \\ \dim \mathscr{P}_{3,n} - (\dim \mathscr{P}_{3,n-l} + \dim \mathscr{P}_{3,l} - 1) &= l(n-l)\left(\frac{n+4}{2}\right). \end{aligned} \tag{1.76}$$

From these formulæ we see that the largest dimensional component of reducible, degree n polynomials, in two variables, has codimension $(n-1)$, and codimension $(n-1)(n+4)/2$ in three variables. These are very low dimensional subsets, as n grows.

This leaves one subtle point to consider: the images $M(\mathscr{P}_{d,n-l}, \mathscr{P}_{d,l})$ may fail to be closed sets, and therefore their closures $\overline{M(\mathscr{P}_{d,n-l}, \mathscr{P}_{d,l})}$ could have larger dimensions. Using less elementary methods from algebraic geometry, one can show that, for each l, these sets are, in fact, closed. This shows that the set of irreducible polynomials, $\mathscr{P}_{d,n} \setminus \mathscr{P}_{d,n}^{\text{red}}$, is an open dense set whose complement is a very low dimensional subset of $\mathscr{P}_{d,n}$. The set $\mathscr{P}_{d,n}^{\text{red}}$ has $\lfloor n/2 \rfloor$ components; it would be quite interesting to know how this set is distributed as $n \to \infty$.

1.B Appendix: The Condition Number of a Problem

In numerical analysis, the condition number κ of a process (i.e., a function) is a measure of the maximum sensitivity of the computed quantity to changes in the input data. For a scalar function of a real variable, denoted by $f(x)$, it estimates the size of the change $f(x + \delta x) - f(x)$ for small δx. See

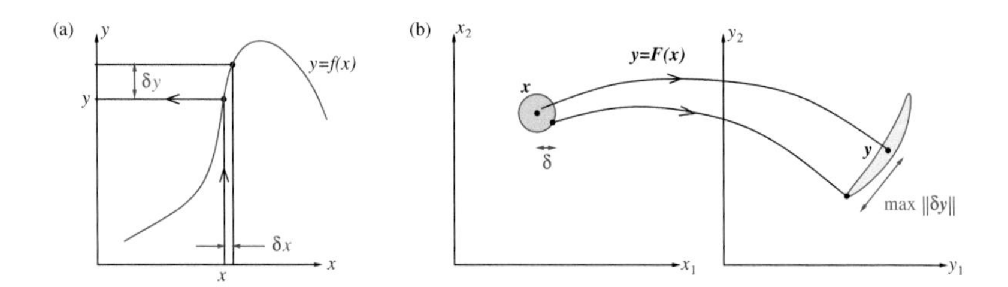

Figure 1.9 Illustration of the local sensitivity to small changes in input for (a) a scalar function $f : \mathbb{R} \to \mathbb{R}$, and (b) a vector function $f : \mathbb{R}^2 \to \mathbb{R}^2$. In each case the condition number $\kappa(x)$ measures the maximum amplification factor, or ratio of (relative) norms of output change to input change, in the limit $\delta \to 0^+$.

Figure 1.9(a). Assuming $x, f(x) \neq 0$ and that we seek the *relative* error, the (relative) condition number κ provides an estimate of the form

$$\left| \frac{f(x+\delta x) - f(x)}{f(x)} \right| \approx \kappa \left| \frac{\delta x}{x} \right|.$$

Letting $y = f(x)$, and $\delta y = f(x+\delta x) - f(x)$, it is defined by

$$\kappa(x) := \lim_{\delta \to 0^+} \sup_{|\delta x| \leq \delta} \frac{|\delta y|/|y|}{|\delta x|/|x|}. \tag{1.77}$$

Assuming $f(x)$ is differentiable, it is easy to check that

$$\kappa(x) = \left| \frac{x f'(x)}{f(x)} \right|.$$

Even in this simple context, one difficulty is already apparent – that the condition number is a local quantity, varying with the argument x.

The definition generalizes to vector functions $\mathbf{F} : \mathbb{R}^m \to \mathbb{R}^n$, once norms $\|\cdot\|$ are defined for the input and output spaces. Letting $\boldsymbol{y} = \mathbf{F}(\boldsymbol{x})$, and $\delta \boldsymbol{y} = \mathbf{F}(\boldsymbol{x} + \delta \boldsymbol{x}) - \mathbf{F}(\boldsymbol{x})$, then

$$\kappa(\boldsymbol{x}) := \lim_{\delta \to 0^+} \sup_{\|\delta \boldsymbol{x}\| \leq \delta} \frac{\|\delta \boldsymbol{y}\|/\|\boldsymbol{y}\|}{\|\delta \boldsymbol{x}\|/\|\boldsymbol{x}\|}. \tag{1.78}$$

It thus measures the maximum growth factor (relative to the input and output sizes), over all directions of infinitesimal input change. See Figure 1.9(b).

For example, if A is an invertible $n \times n$ matrix, and $\boldsymbol{y} = \mathbf{F}(\boldsymbol{x}) := A\boldsymbol{x}$, is the matrix multiplication function, then its condition number at $\boldsymbol{x}$ is $\kappa(\boldsymbol{x}) = \|A\| \|\boldsymbol{x}\|/\|\boldsymbol{y}\|$, where $\|A\|$ is the operator norm of A. Thus, κ does depend on the input data, but it is always bounded by $\|A\| \cdot \|A^{-1}\|$. This same bound clearly holds for the inverse map $\boldsymbol{y} = \mathbf{F}(\boldsymbol{x}) := A^{-1}\boldsymbol{x}$, which corresponds to solving the linear system $A\boldsymbol{y} = \boldsymbol{x}$ for the unknown vector $\boldsymbol{y}$, given known $\boldsymbol{x}$.

In the setting of phase retrieval, the maps are nonlinear, and the mapping of central interest is the solution operator, or local inverse of the measurement map, $\boldsymbol{f} = \mathscr{M}^{-1}_{f_0,S}(\boldsymbol{a})$, in the neighborhood of the data $\boldsymbol{a}_0 = \mathscr{M}(\boldsymbol{f}_0)$, defined by an image $\boldsymbol{f}_0 \in B_S$. This has the complication, beyond the previous Euclidean space example, that the map itself is only defined on a difficult to define, high codimensional subset of a Euclidean space. However, the definition (1.78) extends simply by restricting the input $\boldsymbol{a} + \delta\boldsymbol{a}$ to lie on this subset, while inheriting the norms from the embeddings into Euclidean spaces.

Furthermore, the condition number of $\mathscr{M}^{-1}_{f_0,S}$ is difficult to estimate a priori, both because it is a local property, and because we are solving a high-dimensional nonlinear inverse problem. The solution map is not available, even numerically, in any robust way. In this book we use mathematical analysis, plus, later in the book, suggestive numerical experiments, to study conditioning. Conditioning is closely related to the mathematical notion of well-posedness. These properties are discussed in greater detail in Chapter 3, in terms of a local linearization of this problem near a true intersection of $\mathbb{A}_{\boldsymbol{a}}$ with B_S.

It should be noted that the 1D function $f(x) = 1 + \sqrt{x^2 - 1}$ has an infinite condition number at the point $x = 1$. While this function fails to be Lipschitz continuous at 1 it is Hölder continuous of order $1/2$. In Chapter 3, we give rigorous criteria for when the condition number of $\mathscr{M}^{-1}_{f_0,S}$ is infinite. At such points $\mathscr{M}^{-1}_{f_0,S}$ fails to be Lipschitz continuous; nonetheless it does appear to be Hölder continuous, though its continuity properties are highly anisotropic.

PART I

Theoretical Foundations

2
The Geometry Near an Intersection

Let A and B be two subsets of $\mathbb{R}^N$; as noted in the Introduction, the local convergence behavior of iterative algorithms for finding points in $A \cap B$ is largely governed by the geometry near the intersection of A and B, with a crucial role being played by the angles between the subsets at intersection points. In the phase retrieval problem, the set A is a magnitude torus, $\mathbb{A}_{\boldsymbol{a}}$, and one can take for B the linear subspace, B_S, defined by a support condition. If $\boldsymbol{f}_0 \in \mathbb{A}_{\boldsymbol{a}} \cap B_S$, then the geometric question of interest is to understand the relationship between B_S and the fiber of the tangent bundle to $\mathbb{A}_{\boldsymbol{a}}$ through $\boldsymbol{f}_0$, $T_{\boldsymbol{f}_0}\mathbb{A}_{\boldsymbol{a}}$. In this chapter we study this geometric problem. In this book we often refer to $T_{\boldsymbol{f}_0}\mathbb{A}_{\boldsymbol{a}}$ as the tangent space to $\mathbb{A}_{\boldsymbol{a}}$ at $\boldsymbol{f}_0$. A rapid introduction to the concepts of the tangent bundle and tangent space are given in Appendix 2.A.

The unknown in our model problem is an image $\boldsymbol{f}_0 \in \mathbb{R}^J$, for a real image, or $\boldsymbol{f}_0 \in \mathbb{C}^J$, for a complex image. Here, $J \subset \mathbb{N}^d$ is a rectangular subset. In this chapter we consider general rectangular subsets

$$J = [1:N_1] \times \cdots \times [1:N_d], \tag{2.1}$$

and use the notation $\boldsymbol{N} = (N_1, \ldots, N_d)$ to denote the lengths of the sides of the rectangle defining the index set J. For images indexed by J the discrete Fourier transform (DFT), $\widehat{\boldsymbol{f}}$ is defined by (1.7); we also use the notation $\widehat{\boldsymbol{f}} = \mathscr{F}(\boldsymbol{f})$.

The set A is the magnitude torus specified by the magnitudes of the DFT coefficients, $\{a_{\boldsymbol{k}}\}$, of an image

$$\begin{aligned} \mathbb{A}_{\boldsymbol{a}} &= \{\boldsymbol{f} \in \mathbb{R}^J : |\widehat{f}_{\boldsymbol{k}}| = a_{\boldsymbol{k}} \quad \text{for } \boldsymbol{k} \in J\} \text{ for a real image, and} \\ \mathbb{A}_{\boldsymbol{a}} &= \{\boldsymbol{f} \in \mathbb{C}^J : |\widehat{f}_{\boldsymbol{k}}| = a_{\boldsymbol{k}} \quad \text{for } \boldsymbol{k} \in J\} \text{ for a complex image.} \end{aligned} \tag{2.2}$$

In the real case, the torus is represented in the DFT domain by

$$\widehat{\mathbb{A}}_{\boldsymbol{a}} = \{\widehat{\boldsymbol{f}} \in \mathbb{C}^J : |\widehat{f}_{\boldsymbol{k}}| = a_{\boldsymbol{k}} \quad \text{for } \boldsymbol{k} \in J\} \cap W_{\mathbb{R}}, \tag{2.3}$$

where $W_{\mathbb{R}}$ is the subspace of $\mathbb{C}^J$ containing the DFTs of real images, see (2.17) and (2.18); clearly $\mathbb{A}_{\boldsymbol{a}} = \mathscr{F}^{-1}(\widehat{\mathbb{A}}_{\boldsymbol{a}})$. In the complex case, the torus is represented in the DFT domain by

$$\widehat{\mathbb{A}}_{\boldsymbol{a}} = \{\widehat{\boldsymbol{f}} \in \mathbb{C}^J : |\widehat{f}_{\boldsymbol{k}}| = a_{\boldsymbol{k}} \quad \text{for } \boldsymbol{k} \in J\}, \tag{2.4}$$

and as before, $\mathbb{A}_{\boldsymbol{a}} = \mathscr{F}^{-1}(\widehat{\mathbb{A}}_{\boldsymbol{a}})$. In both cases, the set $\widehat{\mathbb{A}}_{\boldsymbol{a}}$ is a product of round circles, centered at 0, lying in complex coordinate lines. Since $\mathscr{F}$ is a linear isometry (up to a scale factor), the set $\mathbb{A}_{\boldsymbol{a}}$ is again a torus, though the action of $U(1) \times \cdots \times U(1)$ is no longer diagonal in the standard coordinates.

The set $\mathbb{A}_{\boldsymbol{a}}$ is obviously not convex. If $\boldsymbol{g} \in \mathbb{C}^J$, then it has a unique nearest point, in the Euclidean sense, to $\widehat{\mathbb{A}}_{\boldsymbol{a}}$ if and only if $g_{\boldsymbol{k}} \neq 0$ whenever $a_{\boldsymbol{k}} \neq 0$. Since $\mathscr{F}$ is a unitary map, up to a constant, the projection map $P_{\mathbb{A}_{\boldsymbol{a}}}$ is defined on the complement of a finite collection of linear subspaces of codimension 2 (corresponding to $\boldsymbol{g}$ with any coordinate $g_{\boldsymbol{k}} = 0$, but $a_{\boldsymbol{k}} \neq 0$). There is no way to extend $P_{\mathbb{A}_{\boldsymbol{a}}}$ to be continuous at these exceptional points.

If $S \subset J$, then we set

$$B_S = \{\boldsymbol{f} \in \mathbb{R}^J : f_{\boldsymbol{j}} = 0 \quad \text{for } \boldsymbol{j} \notin S\} \text{ for real images, and} \tag{2.5}$$

$$B_S = \{\boldsymbol{f} \in \mathbb{C}^J : f_{\boldsymbol{j}} = 0 \quad \text{for } \boldsymbol{j} \notin S\} \text{ for complex images.} \tag{2.6}$$

Note that B_S is a linear subspace, and the map, P_{B_S}, is the orthogonal projection onto B_S. Assuming that the dimension $d \geq 2$, and the subset S, containing the support of the unknown image, satisfies $S \subset R \subset J$, where R is a rectangular subset with side lengths at most half those of J, then generically the magnitude DFT data determines a single point on $\mathbb{A}_{\boldsymbol{a}}$ with small support, up to trivial associates (Hayes 1982). The intersection $\mathbb{A}_{\boldsymbol{a}} \cap B_S \neq \emptyset$ consists of those trivial associates with support in S.

In Section 7.2.2 we analyze algorithms for finding points in $A \cap B$, based on iterating maps, for the case of a pair of linear subspaces. For A and B linear, the convergence rate is determined by the "angles between A and B." These angles can be defined in various ways. The simplest way is to let R be a matrix whose columns are an orthonormal basis for A, Q a matrix whose columns are an orthonormal basis for B, and define $H = R^t Q$. The singular values of H are the cosines of the angles between A and B. More precisely, there are triples $\{(\boldsymbol{u}_j, \boldsymbol{v}_j, \sigma_j) : j = 1, \ldots, \min\{\dim A, \dim B\}\}$, where $\{\widetilde{\boldsymbol{u}}_j = Q\boldsymbol{u}_j\} \subset A, \{\widetilde{\boldsymbol{v}}_j = R\boldsymbol{v}_j\} \subset B$, $\{\sigma_j\} \subset [0,1]$, and

$$H\boldsymbol{u}_j = \sigma_j \boldsymbol{v}_j \text{ and } H^t \boldsymbol{v}_j = \sigma_j \boldsymbol{u}_j. \tag{2.7}$$

Since $\{\sigma_j\} \subset [0,1]$, there is a set $\{\theta_j\} \subset [0, \frac{\pi}{2}]$, so that

$$\sigma_j = \cos\theta_j. \tag{2.8}$$

For each j, θ_j is the angle between $\widetilde{\boldsymbol{u}}_j$ and $\widetilde{\boldsymbol{v}}_j$.

This can be described more invariantly, in terms of π_A and π_B, the orthogonal projections onto A and B, by

$$\pi_A \widetilde{\boldsymbol{v}}_j = \sigma_j \widetilde{\boldsymbol{u}}_j \text{ and } \pi_B \widetilde{\boldsymbol{u}}_j = \sigma_j \widetilde{\boldsymbol{v}}_j. \tag{2.9}$$

These vectors and the angles between them have simple variational characterizations:

$$\cos\theta_j = \max_{U \subset A,\ \dim U = j} \quad \min_{\widetilde{\boldsymbol{u}} \in U \setminus \{\boldsymbol{0}\}} \quad \frac{\langle \widetilde{\boldsymbol{u}}, \pi_B \widetilde{\boldsymbol{u}} \rangle}{\|\widetilde{\boldsymbol{u}}\|^2} \tag{2.10}$$

$$= \min_{U \subset A,\ \dim U = N+1-j} \quad \max_{\widetilde{\boldsymbol{u}} \in U \setminus \{\boldsymbol{0}\}} \quad \frac{\langle \widetilde{\boldsymbol{u}}, \pi_B \widetilde{\boldsymbol{u}} \rangle}{\|\widetilde{\boldsymbol{u}}\|^2}. \tag{2.11}$$

Here, we assume that $N = \dim A$. There is an analogous formula with π_B replaced by π_A and $U \subset A$ replaced by $V \subset B$. Evidently

$$\dim A \cap B = |\{j : \sigma_j = 1\}|.$$

If, for example, $\dim A > \dim B$, then $\dim A \cap B^{\perp} \geq \dim A - \dim B$.

If one (or both) of the subsets is nonlinear, then it is possible for the intersections to be isolated but not transversal. That is, for an isolated point $\boldsymbol{f} \in A \cap B$, the tangent spaces, $T_{\boldsymbol{f}} A$ and $T_{\boldsymbol{f}} B$, may intersect in a positive dimensional subspace. In Section 7.3 we show that the linearized analysis near such an intersection point corresponds to having a positive dimensional subspace on which the linearized map is the identity. In the context of phase retrieval, a nontransversal intersection usually leads to dramatically slower convergence and, very often, stagnation for algorithms that search for points in $\mathbb{A} \cap B$. There are, however, other geometric settings where algorithms of the type used in phase retrieval "defeat" the nontransversality of the intersection between A and B and find points in $A \cap B$ very rapidly. For a very simple example of this phenomenon, see Section 8.1.1.

As before, we assume that images $\boldsymbol{f}$, indexed by J, are extended to be periodic in all dimensions with the periods given by the vector $\boldsymbol{N} = (N_1, \ldots, N_d)$. Recall that for $\boldsymbol{v} = (v_1, \ldots, v_d) \in \mathbb{Z}^d$, we define the translate of $\boldsymbol{f}$ by $\boldsymbol{v}$ to be the image

$$f_{\boldsymbol{j}}^{(\boldsymbol{v})} \stackrel{d}{=} f_{\boldsymbol{j}-\boldsymbol{v}}. \tag{2.12}$$

For definiteness, the notation $S_{\boldsymbol{f}^{(\boldsymbol{v})}}$ still refers to the subset $\boldsymbol{j} \in J$ where (the J-periodically extended) $f_{\boldsymbol{j}}^{(\boldsymbol{v})} \neq 0$. If $\boldsymbol{f} \in \mathbb{A}_{\boldsymbol{a}}$, then each translate, $\boldsymbol{f}^{(\boldsymbol{v})}$, also belongs to $\mathbb{A}_{\boldsymbol{a}}$; whether or not $\boldsymbol{f}^{(\boldsymbol{v})} \in B_S$ depends on S and the choice of $\boldsymbol{v}$. For an image $\boldsymbol{f}$, the inverted image, $\check{\boldsymbol{f}}$, is defined by

$$\check{f}_{\boldsymbol{j}} \stackrel{d}{=} f_{-\boldsymbol{j}} \text{ for a real image and } \check{f}_{\boldsymbol{j}} \stackrel{d}{=} \overline{f_{-\boldsymbol{j}}} \text{ for a complex image.} \tag{2.13}$$

If $f \in \mathbb{A}_a$, then $\check{f} \in \mathbb{A}_a$; if $f \in \mathbb{A}_a \cap B_S$, then, depending upon f and the choice of S, $\check{f}$ may or may not belong to B_S.

2.1 The Tangent Space to the Magnitude Torus

A special feature of the classical phase retrieval problem is that the tangent bundle to the magnitude torus has two very simple, explicit, and rapidly computable descriptions. Assuming f_0 is a point of intersection, this allows for a thorough discussion of the intersections of the tangent spaces, $T_{f_0}\mathbb{A}_a \cap T_{f_0}B$. Various nontrivial properties of these intersections follow from these descriptions.

The fact that the projections onto the fibers of the tangent and normal bundles can be computed very quickly allows us to do realistic numerical experiments and may lead to new algorithms for recovering the phase information. For example, suppose that B is a linear subspace, and $f \in \mathbb{A}_a$ is a real image. The best linear approximation to $\mathbb{A}_a$ in a neighborhood of f is, by definition, the fiber of the tangent bundle, $T_f\mathbb{A}_a \subset \mathbb{R}^J$. Starting from f, the linear problem that best approximates the problem of finding a point in $\mathbb{A}_a \cap B$ is that of finding a point in $T_f\mathbb{A}_a \cap B$ or, if this intersection is empty, finding a pair of points that minimizes the distance between these affine subspaces. Used iteratively, this approach then constitutes a geometric "Newton's method" for the phase retrieval problem. This is analyzed in greater detail in Section 13.3.

Remark 2.1 We use the somewhat nonstandard notation, $T_f\mathbb{A}_a$ and $N_f\mathbb{A}_a$, to denote the fibers, at f, of the tangent and normal bundles to $\mathbb{A}_a$ thought of as *affine subspaces* of the ambient vector space $\mathbb{R}^J$ (or $\mathbb{C}^J$) in which $\mathbb{A}_a$ lies. These concepts are briefly introduced in Appendix 2.A. The notation $T_f^0\mathbb{A}_a$ and $N_f^0\mathbb{A}_a$ denote the underlying vector spaces, so that

$$T_f\mathbb{A}_a = f + T_f^0\mathbb{A}_a \text{ and } N_f\mathbb{A}_a = f + N_f^0\mathbb{A}_a. \tag{2.14}$$

We use the phrases "tangent space at f" and "normal space at f" to refer to either the affine model, or the underlying vector space. The meaning should be clear from the context.

For complex images, the $\dim \mathbb{A}_a = |J'|$ where $J' = \{\boldsymbol{k} \in J : a_{\boldsymbol{k}} \neq 0\}$. In the DFT domain we can describe $\mathbb{A}_a$ parametrically as

$$\widehat{\mathbb{A}}_a = \{(e^{i\theta_{\boldsymbol{k}}} a_{\boldsymbol{k}}) : \theta_{\boldsymbol{k}} \in [0, 2\pi) \quad \text{for } \boldsymbol{k} \in J\}. \tag{2.15}$$

The tangent space at $\widehat{\boldsymbol{f}} \in \widehat{\mathbb{A}}_{\boldsymbol{a}}$ is then spanned, over $\mathbb{R}$, by the vectors

$$i\widehat{f}_{\boldsymbol{k}}\boldsymbol{e}_{\boldsymbol{k}} \quad \text{for } \boldsymbol{k} \in J'. \tag{2.16}$$

Here, $\boldsymbol{e}_{\boldsymbol{k}}$ is the basis vector for $\mathbb{C}^J$ that is 1 for the index $\boldsymbol{k}$ and otherwise 0. Each of these vectors lies in a complex line. We identify the complex number $z = a + ib$ with the tangent vector $a\partial_x + b\partial_y$ in the real plane underlying the complex line. For later reference $i(a+ib) \leftrightarrow a\partial_y - b\partial_x$, which is the 90° counterclockwise rotation of the vector $a\partial_x + b\partial_y$.

When the images *are* real, then the dimension is cut roughly in half because of well-known symmetries in the DFT. These can be described in terms of relationships among the indices. To each index $\boldsymbol{k} \in J$, there is a unique conjugate index $\boldsymbol{k}' \in J$, given by

$$k'_l = 2 - k_l \mod N_l, \text{ for } l = 1, \ldots, d, \tag{2.17}$$

so that

$$\widehat{f}_{\boldsymbol{k}'} = \overline{\widehat{f}_{\boldsymbol{k}}}. \tag{2.18}$$

For a few indices it may happen that $\boldsymbol{k} \equiv \boldsymbol{k}'$; for these indices the DFT components are necessarily real. This implies $\widehat{f}_{\boldsymbol{k}} = \pm a_{\boldsymbol{k}}$, and these coefficients contribute nothing to $\dim \mathbb{A}_{\boldsymbol{a}}$. When such $a_{\boldsymbol{k}} \neq 0$, the magnitude torus has several connected components corresponding to the different choices of signs for these real DFT coefficients.

If $\mathbb{A}_{\boldsymbol{a}}$ is a torus composed of real images, then its image under the DFT can be parameterized by

$$\widehat{\mathbb{A}}_{\boldsymbol{a}} = \{(e^{i\theta_{\boldsymbol{k}}}a_{\boldsymbol{k}}) : \text{ where } a_{\boldsymbol{k}} = a_{\boldsymbol{k}'} \text{ and } \theta_{\boldsymbol{k}} = -\theta_{\boldsymbol{k}'} \in [0, 2\pi), \text{ for } \boldsymbol{k} \in J\}. \tag{2.19}$$

Since $\mathscr{F}$ is a unitary map, up to a scalar factor, the fiber of the tangent bundle at $\boldsymbol{f} \in \mathbb{A}_{\boldsymbol{a}}$, $T^0_{\boldsymbol{f}}\mathbb{A}$, is spanned by the vectors

$$\boldsymbol{t}_j = \mathscr{F}^{-1}(\widehat{\boldsymbol{t}}_j) \text{ where } \widehat{\boldsymbol{t}}_j = \left[i\frac{\widehat{f}_j\boldsymbol{e}_j - \overline{\widehat{f}_j}\boldsymbol{e}_{j'}}{\sqrt{2|\widehat{f}_j|}} \right] \text{ for } \boldsymbol{j} \in J, \text{ with } \widehat{f}_j \neq 0, \tag{2.20}$$

and the fiber of the normal bundle, $N^0_{\boldsymbol{f}}\mathbb{A}$, by

$$\boldsymbol{n}_j = \mathscr{F}^{-1}(\widehat{\boldsymbol{n}}_j) \text{ where } \widehat{\boldsymbol{n}}_j = \left[\frac{\widehat{f}_j\boldsymbol{e}_j + \overline{\widehat{f}_j}\boldsymbol{e}_{j'}}{\sqrt{2|\widehat{f}_j|}} \right] \text{ for } \boldsymbol{j} \in J, \text{ with } \widehat{f}_j \neq 0. \tag{2.21}$$

Theorem 1.6 follows from this fact; a detailed proof is given below. If $\widehat{f}_j = \widehat{f}_{j'} = 0$, then $\mathscr{F}^{-1}(\boldsymbol{e}_j + \boldsymbol{e}_{j'})$, and $\mathscr{F}^{-1}(i(\boldsymbol{e}_j - \boldsymbol{e}_{j'}))$ are also basis vectors for $N_f^0 \mathbb{A}_{\boldsymbol{a}}$.

Remark 2.2 Recall that, for real images, $\widehat{\mathbb{A}}_{\boldsymbol{a}}$ is defined to lie in the subspace $W_{\mathbb{R}} \subset \mathbb{C}^J$. The fiber of the normal bundle of $\widehat{\mathbb{A}}_{\boldsymbol{a}}$ at every point also contains the subspace $W_{\mathbb{R}}^{\perp}$, but it plays no role in our analysis, so no further mention of this subspace will be made.

If we identify $\mathbb{R}^{2n}$ with $\mathbb{C}^n$,

$$(x_1, y_1, \ldots, x_n, y_n) \leftrightarrow (x_1 + iy_1, \ldots, x_n + iy_n), \tag{2.22}$$

then the real inner product on $\mathbb{R}^{2n}$ is given by

$$\boldsymbol{u} \cdot \boldsymbol{v} = \langle \boldsymbol{u}, \boldsymbol{v} \rangle = \operatorname{Re}(\boldsymbol{w}, \boldsymbol{z}), \tag{2.23}$$

where $\boldsymbol{w}, \boldsymbol{z}$ are the complex vectors representing $\boldsymbol{u}, \boldsymbol{v}$ and

$$(\boldsymbol{w}, \boldsymbol{z}) = \sum_{j=1}^{n} w_j \bar{z}_j. \tag{2.24}$$

With this is mind, it follows easily from these formulæ and the fact that $\mathscr{F}$ is unitary (up to a constant factor), that the $\{\boldsymbol{t}_j\}$ are orthogonal to the $\{\boldsymbol{n}_j\}$.

Each nontrivial tangent direction appears twice in (2.20), and if $\boldsymbol{j} = \boldsymbol{j}'$, then $\boldsymbol{t}_j = 0$. In this case $\widehat{f}_j$ is real and $\boldsymbol{n}_j = \mathscr{F}^{-1}(\boldsymbol{e}_j) \in N_f^0 \mathbb{A}$. We let $J_t \subset J$ denote a choice of indices that lists each nonzero tangent direction once; if $\boldsymbol{j}_1 \neq \boldsymbol{j}_2$ are elements of J_t, then $\boldsymbol{t}_{j_1}$ is orthogonal to $\boldsymbol{t}_{j_2}$. The set

$$\{\boldsymbol{t}_j : \boldsymbol{j} \in J_t\} \tag{2.25}$$

is then an orthonormal basis for $T_f^0 \mathbb{A}_{\boldsymbol{a}}$. Generically, the cardinality of J_t depends on the parities of the dimensions of J in each of the underlying d directions. If each dimension is even and equal to $2N$, then $|J_t| = 2^{d-1}(N^d - 1)$. Of course, if $\widehat{f}_{\boldsymbol{k}} = 0$, for some $\boldsymbol{k}$, then this lowers the $\dim \mathbb{A}_{\boldsymbol{a}}$, but generically all DFT coefficients are nonzero. Note, however, that if $\boldsymbol{g}$ is a one-dimensional sequence of length $2N$, which satisfies

$$g_j = g_{2N-j+1}, \tag{2.26}$$

then the DFT coefficient

$$\widehat{g}_{N+1} = 0. \tag{2.27}$$

As the index $N + 1$ is self-conjugate, in and of itself, this fact would be unexceptional, but if a d-dimensional sequence can be factored as $f_j = g_{j_1} h_{\tilde{j}}$, with $\widetilde{\boldsymbol{j}} = (j_2, \ldots, j_d)$ and $\boldsymbol{g}$ satisfying (2.26), then, elementary properties of the d-dimensional DFT imply that

$$\widehat{f}_{(N+1)\widetilde{\boldsymbol{k}}} = 0 \text{ for all } \widetilde{\boldsymbol{k}}. \tag{2.28}$$

Hence, symmetries of the image can force many DFT coefficients to vanish.

In all cases, a basis for the tangent space can be described (in the DFT domain) by the list of the indices in J_t paired with $|J_t|$ complex numbers of modulus 1. Notice that an orthonormal basis for a subspace of this dimension generically requires a matrix with dimensions $|J_t| \times |J|$. Hence, these subspaces admit a *highly* compressed representation, which requires only the DFT to access and use. In Appendix 2.B we show how to use the basis for $N_{\boldsymbol{f}}^0 \mathbb{A}_{\boldsymbol{a}}$ in (2.21) to give a formula for the orthogonal projection onto this subspace that is essentially a diagonal matrix conjugated by the DFT.

There are also rather different, explicit descriptions of $T_{\boldsymbol{f}}^0 \mathbb{A}_{\boldsymbol{a}}$ and $N_{\boldsymbol{f}}^0 \mathbb{A}_{\boldsymbol{a}}$ in the image domain.

Lemma 2.3 *Let $\boldsymbol{f} \in \mathbb{A}_{\boldsymbol{a}}$ and $\boldsymbol{v} \in \mathbb{N}^d$, then, for real images,*

$$\boldsymbol{\tau}^{(\boldsymbol{v})} = \boldsymbol{f}^{(\boldsymbol{v})} - \boldsymbol{f}^{(-\boldsymbol{v})} \in T_{\boldsymbol{f}}^0 \mathbb{A}_{\boldsymbol{a}} \text{ and } \boldsymbol{\nu}^{(\boldsymbol{v})} = \boldsymbol{f}^{(\boldsymbol{v})} + \boldsymbol{f}^{(-\boldsymbol{v})} \in N_{\boldsymbol{f}}^0 \mathbb{A}_{\boldsymbol{a}}. \tag{2.29}$$

For complex images,

$$\begin{gathered}\boldsymbol{\tau}^{(\boldsymbol{v})} = \boldsymbol{f}^{(\boldsymbol{v})} - \boldsymbol{f}^{(-\boldsymbol{v})} \in T_{\boldsymbol{f}}^0 \mathbb{A}_{\boldsymbol{a}} \text{ and } \boldsymbol{\nu}^{(\boldsymbol{v})} = \boldsymbol{f}^{(\boldsymbol{v})} + \boldsymbol{f}^{(-\boldsymbol{v})} \in N_{\boldsymbol{f}}^0 \mathbb{A}_{\boldsymbol{a}}, \\ i\boldsymbol{\tau}^{(\boldsymbol{v})} \in N_{\boldsymbol{f}}^0 \mathbb{A}_{\boldsymbol{a}} \text{ and } i\boldsymbol{\nu}^{(\boldsymbol{v})} \in T_{\boldsymbol{f}}^0 \mathbb{A}_{\boldsymbol{a}}.\end{gathered} \tag{2.30}$$

Remark 2.4 This is a rather remarkable result. While it is not possible to say whether an image $\boldsymbol{f}$ belongs to $\mathbb{A}_{\boldsymbol{a}}$ without computing its DFT, given that $\boldsymbol{f} \in \mathbb{A}_{\boldsymbol{a}}$, the lemma gives a completely explicit description, entirely in the image domain, of the tangent space $T_{\boldsymbol{f}}^0 \mathbb{A}_{\boldsymbol{a}}$.

Proof To prove the lemma we just need to understand how $\widehat{\boldsymbol{f}}^{(\boldsymbol{v})}$ is related to $\widehat{\boldsymbol{f}}$. Recall that the DFT is given by the formula (1.7), which implies that

$$\widehat{f}_{\boldsymbol{k}}^{(\boldsymbol{v})} = \exp\left(-\frac{2\pi i(\boldsymbol{k}-\boldsymbol{1}) \cdot \boldsymbol{v}}{\boldsymbol{N}}\right) \widehat{f}_{\boldsymbol{k}}, \tag{2.31}$$

where $\boldsymbol{N} = (N_1, \ldots, N_d)$, $\boldsymbol{1} = (1, \ldots, 1)$, and

$$\frac{\boldsymbol{k} \cdot \boldsymbol{v}}{\boldsymbol{N}} \stackrel{d}{=} \sum_{l=1}^{d} \frac{\boldsymbol{k}_l \boldsymbol{v}_l}{N_l}. \tag{2.32}$$

Using this formula for real images, and equation (2.17) we see that

$$
\begin{aligned}
\widehat{f}_{\boldsymbol{k}}^{(\boldsymbol{v})} - \widehat{f}_{\boldsymbol{k}}^{(-\boldsymbol{v})} &= -2i \sin\left(\frac{2\pi(\boldsymbol{k}-\mathbf{1})\cdot\boldsymbol{v}}{\boldsymbol{N}}\right)\widehat{f}_{\boldsymbol{k}},\\
\widehat{f}_{\boldsymbol{k}'}^{(\boldsymbol{v})} - \widehat{f}_{\boldsymbol{k}'}^{(-\boldsymbol{v})} &= -2i \sin\left(\frac{2\pi(\boldsymbol{k}'-\mathbf{1})\cdot\boldsymbol{v}}{\boldsymbol{N}}\right)\widehat{f}_{\boldsymbol{k}'} = 2i \sin\left(\frac{2\pi(\boldsymbol{k}-\mathbf{1})\cdot\boldsymbol{v}}{\boldsymbol{N}}\right)\overline{\widehat{f}_{\boldsymbol{k}}},
\end{aligned}
\tag{2.33}
$$

and

$$
\begin{aligned}
\widehat{f}_{\boldsymbol{k}}^{(\boldsymbol{v})} + \widehat{f}_{\boldsymbol{k}}^{(-\boldsymbol{v})} &= 2\cos\left(\frac{2\pi(\boldsymbol{k}-\mathbf{1})\cdot\boldsymbol{v}}{\boldsymbol{N}}\right)\widehat{f}_{\boldsymbol{k}},\\
\widehat{f}_{\boldsymbol{k}'}^{(\boldsymbol{v})} + \widehat{f}_{\boldsymbol{k}'}^{(-\boldsymbol{v})} &= 2\cos\left(\frac{2\pi(\boldsymbol{k}'-\mathbf{1})\cdot\boldsymbol{v}}{\boldsymbol{N}}\right)\widehat{f}_{\boldsymbol{k}'} = 2\cos\left(\frac{2\pi(\boldsymbol{k}-\mathbf{1})\cdot\boldsymbol{v}}{\boldsymbol{N}}\right)\overline{\widehat{f}_{\boldsymbol{k}}}.
\end{aligned}
\tag{2.34}
$$

These relations show that the vectors $\widehat{\boldsymbol{f}}^{(\boldsymbol{v})} - \widehat{\boldsymbol{f}}^{(-\boldsymbol{v})}$ are sums, with real coefficients, of the vectors $\{\widehat{\boldsymbol{t}}_{\boldsymbol{k}} : \boldsymbol{k} \in J_t\}$, and therefore, $\boldsymbol{f}^{(\boldsymbol{v})} - \boldsymbol{f}^{(-\boldsymbol{v})} \in T_{\boldsymbol{f}}^0 \mathbb{A}_{\boldsymbol{a}}$. In the DFT representation it is also clear that the vectors $\boldsymbol{f}^{(\boldsymbol{v})} + \boldsymbol{f}^{(-\boldsymbol{v})}$ are orthogonal to the span of the $\{\boldsymbol{t}_{\boldsymbol{k}} : \boldsymbol{k} \in J_t\}$, hence, in $N_{\boldsymbol{f}}^0 \mathbb{A}_{\boldsymbol{a}}$.

The proof in the complex case is similar, but easier, and is left to the reader. □

Each point $\boldsymbol{j} \in J$ defines a translation vector $(j_1, \ldots, j_d)$, and the set of such vectors evidently has the same cardinality as J itself. Under a reasonable genericity condition this lemma gives a second description for the tangent space to a magnitude torus.

Lemma 2.5 *Suppose that $\boldsymbol{f}$ is a vector such that $\widehat{f}_{\boldsymbol{k}} \neq 0$ for any choice of $\boldsymbol{k} \in J$; let $\mathbb{A}_{\boldsymbol{a}}$ be the magnitude torus defined in the DFT domain by the magnitude data $\{|\widehat{f}_{\boldsymbol{k}}| : \boldsymbol{k} \in J\}$. For real images, the vectors $\{\boldsymbol{f}^{(\boldsymbol{j})} : \boldsymbol{j} \in J\}$ are a basis for $\mathbb{R}^J$; moreover, the vectors $\{\boldsymbol{\tau}^{(\boldsymbol{j})} : \boldsymbol{j} \in J\}$ span $T_{\boldsymbol{f}}^0 \mathbb{A}_{\boldsymbol{a}}$ and the vectors $\{\boldsymbol{\nu}^{(\boldsymbol{j})} : \boldsymbol{j} \in J\}$ span $N_{\boldsymbol{f}}^0 \mathbb{A}_{\boldsymbol{a}}$.*

For complex images, the vectors $\{\boldsymbol{f}^{(\boldsymbol{j})} : \boldsymbol{j} \in J\}$ are a basis for $\mathbb{C}^J$; moreover, the vectors $\{\boldsymbol{\tau}^{(\boldsymbol{j})}, i\boldsymbol{\nu}^{(\boldsymbol{j})} : \boldsymbol{j} \in J\}$ span $T_{\boldsymbol{f}}^0 \mathbb{A}_{\boldsymbol{a}}$ and the vectors $\{\boldsymbol{\nu}^{(\boldsymbol{j})}, i\boldsymbol{\tau}^{(\boldsymbol{j})} : \boldsymbol{j} \in J\}$ span $N_{\boldsymbol{f}}^0 \mathbb{A}_{\boldsymbol{a}}$.

Remark 2.6 The first statement is a discrete version of Wiener's L^1-Tauberian theorem.

Proof The first part of the argument works equally well for real or complex images. It clearly suffices to show that the DFTs $\{\widehat{\boldsymbol{f}^{(\boldsymbol{k})}} : \boldsymbol{k} \in J\}$ are a basis. To see this, we observe that

$$\widehat{\boldsymbol{f}^{(\boldsymbol{k})}}_{\boldsymbol{j}} = \exp\left(-\frac{2\pi i(\boldsymbol{j}-\mathbf{1})\cdot\boldsymbol{k}}{\boldsymbol{N}}\right)\widehat{f}_{\boldsymbol{j}}. \tag{2.35}$$

If $\widehat{\boldsymbol{g}}$ is orthogonal to all of these vectors, that is

$$\sum_{\boldsymbol{j}\in J}\exp\left(-\frac{2\pi i(\boldsymbol{j}-\mathbf{1})\cdot\boldsymbol{k}}{\boldsymbol{N}}\right)\widehat{f}_{\boldsymbol{j}}\overline{\widehat{g}_{\boldsymbol{j}}} = 0 \quad \text{for all } \boldsymbol{k}\in J, \tag{2.36}$$

then, as the DFT is invertible, the vector with entries $\widehat{f}_{\boldsymbol{j}}\overline{\widehat{g}_{\boldsymbol{j}}}$ must vanish. If $\widehat{f}_{\boldsymbol{j}} \neq 0$ for every $\boldsymbol{j} \in J$, then clearly this implies that $\boldsymbol{g} = 0$. Note that if $\boldsymbol{f}$ and $\boldsymbol{g}$ are real, then the sums in (2.36) are real as well.

To prove the final assertions for real images, we see that $\{\boldsymbol{\tau}^{(\boldsymbol{j})} : \boldsymbol{j} \in J\}$ together with $\{\boldsymbol{\nu}^{(\boldsymbol{j})} : \boldsymbol{j} \in J\}$ is again a spanning set for $\mathbb{R}^J$. The vectors in the first set are tangent to $\mathbb{A}_{\boldsymbol{a}}$ at $\boldsymbol{f}$ and those in the second set are normal to it. Hence, the dimension of the span of the first set is at most $\dim \mathbb{A}_{\boldsymbol{a}}$, and that of the second set is at most $|J| - \dim \mathbb{A}_{\boldsymbol{a}}$. Since the union spans a vector space of dimension exactly $|J|$, it follows that, in both cases, the upper bound must be attained.

As before, we leave the proof in the complex case to the interested reader. □

Remark 2.7 The vectors $\{\boldsymbol{\tau}^{(\boldsymbol{j})} : \boldsymbol{j} \in J\}$ still span $T^0_{\boldsymbol{f}}\mathbb{A}_{\boldsymbol{a}}$ even if, for a subset of indices J_0, the DFT coefficients vanish. If we let $W \subset \mathbb{R}^J$ denote the subspace

$$W = \{\boldsymbol{f} : \widehat{f}_{\boldsymbol{k}} = 0 \quad \text{for } \boldsymbol{k} \in J_0\},$$

then $\mathbb{A}_{\boldsymbol{a}} \subset W$, and this subspace is invariant under translations and inversions. The proof of Lemma 2.5 shows that $\{\boldsymbol{\tau}^{(\boldsymbol{j})} : \boldsymbol{j} \in J\}$ spans $T^0_{\boldsymbol{f}}\mathbb{A}_{\boldsymbol{a}} \subset W$, and $\{\boldsymbol{\nu}^{(\boldsymbol{j})} : \boldsymbol{j} \in J\}$ spans $N^0_{\boldsymbol{f}}\mathbb{A}_{\boldsymbol{a}} \cap W$. The subspace $W^{\perp} \subset N^0_{\boldsymbol{f}}\mathbb{A}_{\boldsymbol{a}}$, and $N^0_{\boldsymbol{f}}\mathbb{A}_{\boldsymbol{a}} = N^0_{\boldsymbol{f}}\mathbb{A}_{\boldsymbol{a}} \cap W \oplus W^{\perp}$.

Remark 2.8 It is also interesting to note that if $\boldsymbol{f}_0$ is a nonnegative image, then Lemma 2.3 implies that $N^0_{\boldsymbol{f}_0}\mathbb{A}_{\boldsymbol{a}}$ has a basis consisting of nonnegative vectors.

2.2 The Intersection of the Tangent Bundle and the Support Constraint

Suppose now that f_0 is a real image indexed by J. The support constraint is usually defined by a set $S \supset S_{f_0}$ that is somewhat larger than S_{f_0}. For a generic choice of magnitude vector, $\boldsymbol{a}$, the totality of the intersection set $\mathbb{A}_{\boldsymbol{a}} \cap B_S$ is obtained by translating a fixed choice of $\boldsymbol{f}_0 \in \mathbb{A}_{\boldsymbol{a}} \cap B_S$, and possibly its reflection, $\check{\boldsymbol{f}}_0$, by a finite set of integer vectors. We let

$$\mathscr{T}_{f_0,S} = \{\boldsymbol{v} \in \mathbb{Z}^d : \boldsymbol{f}_0^{(\boldsymbol{v})} \in B_S\} \text{ and } \check{\mathscr{T}}_{f_0,S} = \{\boldsymbol{v} \in \mathbb{Z}^d : \check{\boldsymbol{f}}_0^{(\boldsymbol{v})} \in B_S\}. \tag{2.37}$$

Corollary 2.9 *Let* $f_0 \in \mathbb{A}_{\boldsymbol{a}} \cap B_S$*, and suppose that the directions* $\boldsymbol{v}$ *and* $-\boldsymbol{v}$ *both belong to* $\mathscr{T}_{f_0,S}$*, then*

$$\boldsymbol{f}_0^{(\boldsymbol{v})} - \boldsymbol{f}_0^{(-\boldsymbol{v})} \in T^0_{f_0}\mathbb{A}_{\boldsymbol{a}} \cap B_S. \tag{2.38}$$

A similar result holds for $\check{\mathscr{T}}_{f_0,S}$.

Proof If both $\boldsymbol{v}$ and $-\boldsymbol{v}$ belong to $\mathscr{T}_{f_0,S}$, then $S_{f_0^{(\boldsymbol{v})}}$ and $S_{f_0^{(-\boldsymbol{v})}}$ are contained in S and

$$\boldsymbol{f}_0^{(\boldsymbol{v})} - \boldsymbol{f}_0^{(-\boldsymbol{v})} \in T^0_{f_0}\mathbb{A}_{\boldsymbol{a}},$$

and therefore, it belongs to $T^0_{f_0}\mathbb{A}_{\boldsymbol{a}} \cap B_S$. □

If $f_0 \in \mathbb{A}_{\boldsymbol{a}} \cap B_S$, then the points $\{\boldsymbol{f}_0^{(\boldsymbol{v})} : \boldsymbol{v} \in \mathscr{T}_{f_0,S}\} \cup \{\check{\boldsymbol{f}}_0^{(\boldsymbol{v})} : \boldsymbol{v} \in \check{\mathscr{T}}_{f_0,S}\}$ are all allowable solutions to the phase retrieval problem; compared to machine precision, these points are typically separated from one another by a substantial distance, see Figure 8.9. The Corollary shows that if there is "wiggle room" in the support constraint S, as reflected by a positive cardinality of the set

$$\mathscr{T}^{\text{sym}}_{f_0,S} = \{\boldsymbol{v} \in \mathscr{T}_{f_0,S} : -\boldsymbol{v} \in \mathscr{T}_{f_0,S}\},$$

then the intersection between $\mathbb{A}_{\boldsymbol{a}}$ and B_S cannot be transversal at f_0. The more wiggle room there is, the greater the dimension of $T_{f_0}\mathbb{A}_{\boldsymbol{a}} \cap B_S$ becomes. It should also be noted that for vectors $\boldsymbol{v} \in \mathscr{T}^{\text{sym}}_{f_0,S}$ the dimensions of $T_{f_0^{(\boldsymbol{v})}}\mathbb{A}_{\boldsymbol{a}} \cap B_S$ depend on $\boldsymbol{v}$.

Remark 2.10 If f_0 is complex, then similar considerations apply. With the set $\mathscr{T}^{\text{sym}}_{f_0,S}$ as defined above, for $\boldsymbol{v} \in \mathscr{T}^{\text{sym}}_{f_0,S}$, the vectors

$$\boldsymbol{f}_0^{(\boldsymbol{v})} - \boldsymbol{f}_0^{(-\boldsymbol{v})} \text{ and } i(\boldsymbol{f}_0^{(\boldsymbol{v})} + \boldsymbol{f}_0^{(-\boldsymbol{v})}) \in T^0_{f_0}\mathbb{A}_{\boldsymbol{a}} \cap B_S, \tag{2.39}$$

once again leading to a nontransversal intersection between $\mathbb{A}_{\boldsymbol{a}}$ and B_S at f_0.

Definition 2.11 For a subset $U \subset J$, extended J-periodically to $\mathbb{Z}^d$, the p-pixel neighborhood of U is defined to be

$$U_p = \{\boldsymbol{j} + \boldsymbol{v} : \boldsymbol{j} \in U, \boldsymbol{v} \in C_p\}, \tag{2.40}$$

where $C_p = \{(i, j) : |i| \le p, |j| \le p\}$, which equals $\boldsymbol{v} \in \mathbb{Z}^d$ with $\|\boldsymbol{v}\|_\infty \le p$.

Alternately, one can define the p-pixel neighborhood of a set, U by

$$U_p = \{\boldsymbol{j} : \min_{\boldsymbol{l} \in U} \|\boldsymbol{j} - \boldsymbol{l}\|_\infty \le p\}. \tag{2.41}$$

If $d = 2$, then the cardinality of set C_1 is 9, and, arguing inductively, we easily show that $|C_p \setminus \{\boldsymbol{0}\}| = 4p(p+1)$. Let $S_{f,p}$ denote the p-pixel neighborhood of S_f. For each $\boldsymbol{v} \in C_p$, the support of the translates $S_{f^{(\pm \boldsymbol{v})}} \subset S_{f,p}$. Hence, from Lemma 2.3 it follows that

$$\dim T_f \mathbb{A}_{\boldsymbol{a}} \cap B_{S_{f,p}} \ge \frac{|C_p \setminus \{\boldsymbol{0}\}|}{2} = 2p(p+1). \tag{2.42}$$

This formula gives the exact dimensions of these intersections for generic images.

For a vector $\boldsymbol{v} \in \mathcal{T}^{\text{sym}}_{f_0, S}$ the normal vector $\boldsymbol{\nu}^{(\boldsymbol{v})}$ is also supported in S. The span of these vectors is, therefore, a subspace of $N^0_{f_0} \mathbb{A}_{\boldsymbol{a}} \cap B_S$, which is again positive dimensional. When $T^0_{f_0} \mathbb{A}_{\boldsymbol{a}} \cap B_S$, and $N^0_{f_0} \mathbb{A}_{\boldsymbol{a}} \cap B_S$, are nontrivial, they play a significant role in the asymptotic behavior of hybrid iterative map-based algorithms.

Remark 2.12 In a recent paper, Barnett and Doppelt used these explicit descriptions of $T^0_{f_0} \mathbb{A}_{\boldsymbol{a}}$ and $N^0_{f_0} \mathbb{A}_{\boldsymbol{a}}$ to show that, for certain $\boldsymbol{v} \in J$, and $p \ge 0$, there are curves, $\boldsymbol{f}_{\boldsymbol{v},p}(t)$ lying in $B_{S_{2p}}$ with the following properties:

$$\begin{aligned} \boldsymbol{f}_{\boldsymbol{v},p}(0) &= \boldsymbol{f}_0, \\ \partial_t \boldsymbol{f}_{\boldsymbol{v},p}(0) &= \boldsymbol{\tau}^{(\boldsymbol{v})}, \\ \|\boldsymbol{f}_{\boldsymbol{v},p}(t) - \boldsymbol{f}_0\| &\ge C|t| \text{ for a } C > 0, \\ \operatorname{dist}(\boldsymbol{f}_{\boldsymbol{v},p}(t), \mathbb{A}_{\boldsymbol{a}}) &\le C|t|^{2p+2}. \end{aligned} \tag{2.43}$$

That is to say: these special curves, which lie in $B_{S_{2p}}$ make order $(2p+1)$-contact with the magnitude torus. For more details (see Doppelt and Barnett 2020, pers. comm.). A similar constructions appears in Fienup (2020, pers. comm.).

Using our description of the basis for $T^0_{f_0} \mathbb{A}_{\boldsymbol{a}}$ it is possible to prove some further results about the intersections $T_{f_0} \mathbb{A}_{\boldsymbol{a}} \cap B_S$. In this discussion we restrict ourselves to real images, though the inequality (2.62) remains true if f is complex and φ real.

To state these results we require a little further notation: for an image f, indexed by J, we define

$$\boldsymbol{a}_f = (|\widehat{f}_k|)_{k \in J}, \tag{2.44}$$

and

$$\boldsymbol{\tau}_f^{(\boldsymbol{v})} = \boldsymbol{f}^{(\boldsymbol{v})} - \boldsymbol{f}^{(-\boldsymbol{v})}, \quad \text{for } \boldsymbol{v} \in J. \tag{2.45}$$

Finally, let $J_t^f \subset J$ denote a choice of indices so that $\{\boldsymbol{\tau}_f^{(\boldsymbol{v})} : \boldsymbol{v} \in J_t^f\}$ is a basis for $T_f^0 \mathbb{A}_{\boldsymbol{a}_f}$. For generic real images, with $a_k \neq 0$ for all $\boldsymbol{k} \in J$, there is a universal subset J_t^u that can be used for this purpose; generally, J_t^f can be taken as a subset of J_t^u. For $\boldsymbol{\alpha} \in \mathbb{R}^{J_t^f}$ define the vector $\boldsymbol{\tau}_f^{\boldsymbol{\alpha}} \in T_f^0 \mathbb{A}_{\boldsymbol{a}_f}$ by

$$\boldsymbol{\tau}_f^{\boldsymbol{\alpha}} = \sum_{\boldsymbol{v} \in J_t^f} \alpha_{\boldsymbol{v}} \boldsymbol{\tau}_f^{(\boldsymbol{v})}. \tag{2.46}$$

Recall that if f and φ are two images indexed by J, then their (periodic) convolution is defined by

$$(\boldsymbol{f} * \boldsymbol{\varphi})_j = \sum_{l \in J} f_{j-l} \varphi_l. \tag{2.47}$$

From formula (2.47) it follows easily that, for any $\boldsymbol{v} \in J$,

$$(\boldsymbol{f} * \boldsymbol{\varphi})^{(\boldsymbol{v})} = \boldsymbol{f}^{(\boldsymbol{v})} * \boldsymbol{\varphi} = \boldsymbol{f} * \boldsymbol{\varphi}^{(\boldsymbol{v})}, \tag{2.48}$$

which implies that

$$\boldsymbol{\tau}_{f*\varphi}^{(\boldsymbol{v})} = \left[\boldsymbol{\tau}_f^{(\boldsymbol{v})}\right] * \boldsymbol{\varphi} = \boldsymbol{f} * \left[\boldsymbol{\tau}_\varphi^{(\boldsymbol{v})}\right], \tag{2.49}$$

is a vector in $T_{f*\varphi}^0 \mathbb{A}_{\boldsymbol{a}_{f*\varphi}}$. These formulæ give obvious relations between the tangent bundles to the magnitude tori defined by $\boldsymbol{f} * \boldsymbol{\varphi}$ and those defined by $\boldsymbol{f}$ and $\boldsymbol{\varphi}$. Properly understood

$$T_{f*\varphi}^0 \mathbb{A}_{\boldsymbol{a}_{f*\varphi}} = T_f^0 \mathbb{A}_{\boldsymbol{a}_f} * \boldsymbol{\varphi} = \boldsymbol{f} * T_\varphi^0 \mathbb{A}_{\boldsymbol{a}_\varphi}. \tag{2.50}$$

They can also be used to investigate their intersections with various subspaces defined by support conditions.

Let $S \subset J$, and let $\boldsymbol{\alpha} \in \mathbb{R}^{J_t^f}$ be chosen so that the tangent vector $\boldsymbol{\tau}_f^{\boldsymbol{\alpha}}$, defined in (2.46), satisfies

$$\tau_{f,\boldsymbol{j}}^{\boldsymbol{\alpha}} = 0 \quad \text{for } \boldsymbol{j} \notin S. \tag{2.51}$$

If $\boldsymbol{i} \in J$, then the shifted vector

$$\tau^{\boldsymbol{\alpha}}_{f,\boldsymbol{j}-\boldsymbol{i}} = 0 \quad \text{for } \boldsymbol{j} \notin S + \boldsymbol{i}. \tag{2.52}$$

It follows from (2.48) that convolving the vector $\boldsymbol{\tau}^{\boldsymbol{\alpha}}_f$ with $\boldsymbol{\varphi}$, produces a vector in the fiber of $T_{f*\varphi}\mathbb{A}_{\boldsymbol{a}_{f*\varphi}}$,

$$\tau^{\boldsymbol{\alpha}}_{f*\varphi,\boldsymbol{j}} = \sum_{\boldsymbol{i}\in S_\varphi} \tau^{\boldsymbol{\alpha}}_{f,\boldsymbol{j}-\boldsymbol{i}}\varphi(\boldsymbol{i}). \tag{2.53}$$

From (2.52) it follows that

$$\tau^{\boldsymbol{\alpha}}_{f*\varphi,\boldsymbol{j}} = 0 \quad \text{for } \boldsymbol{j} \in \bigcap_{\boldsymbol{i}\in S_\varphi} [S+\boldsymbol{i}]^c = \left[\bigcup_{\boldsymbol{i}\in S_\varphi} S+\boldsymbol{i}\right]^c = [S+S_\varphi]^c, \tag{2.54}$$

where $S + S_\varphi = \{\boldsymbol{j} + \boldsymbol{i} : \boldsymbol{j} \in S \text{ and } \boldsymbol{i} \in S_\varphi\}$. This proves the following proposition:

Proposition 2.13 *Let f and $\boldsymbol{\varphi}$ be images indexed by J, and let $\boldsymbol{\tau}^{\boldsymbol{\alpha}}_f \in T^0_f\mathbb{A}_{\boldsymbol{a}_f}$, as defined in* (2.46)*, have its support in S, then $\boldsymbol{\tau}^{\boldsymbol{\alpha}}_f * \boldsymbol{\varphi} \in T^0_{f*\varphi}\mathbb{A}_{\boldsymbol{a}_{f*\varphi}}$, is supported in $S + S_\varphi$.*

Remark 2.14 It is possible that $S + S_\varphi = J$, though in typical applications to phase retrieval, both S and S_φ are small subsets of J and their sum is as well.

The representation in (2.46) allows us to identify the fiber of the tangent space $T^0_f\mathbb{A}_{\boldsymbol{a}_f}$ with a set of vectors in $\mathbb{R}^J$ in a manner that makes no explicit reference to the image, or the torus. As noted above, the index set J^f_t may always be taken to be a subset of the universal index set J^u_t. The construction leading up to Proposition 2.13 is entirely symmetric in the two images: if $\boldsymbol{\beta} \in J^\varphi_t$ is such that the tangent vector, $\boldsymbol{\tau}^{\boldsymbol{\beta}}_\varphi$ to $\mathbb{A}_{\boldsymbol{a}_\varphi}$, has support in S, then $f * \boldsymbol{\tau}^{\boldsymbol{\beta}}_\varphi \in T^0_{f*\varphi}\mathbb{A}_{\boldsymbol{a}_{f*\varphi}}$ with support in $S + S_f$. Note that if f and $\boldsymbol{\varphi}$ are nonnegative, then

$$S_{f*\varphi} = S_f + S_\varphi, \tag{2.55}$$

see Section 4.1; in all cases $S_{f*\varphi} \subset S_f + S_\varphi$.

Suppose that $\boldsymbol{\alpha} \in J^f_t$ is such that $\boldsymbol{\tau}^{\boldsymbol{\alpha}}_f \in T^0_f\mathbb{A}_{\boldsymbol{a}_f}$ is supported in $S_{f,p}$, then $\boldsymbol{\tau}^{\boldsymbol{\alpha}}_f * \boldsymbol{\varphi} \in T_{f*\varphi}\mathbb{A}_{\boldsymbol{a}_{f*\varphi}}$ is supported in $S_{f,p} + S_\varphi \subset S_{f*\varphi,p}$. By identifying the intersections

$$T_f\mathbb{A}_{\boldsymbol{a}_f} \cap B_{S_{f,p}} \text{ and } T_\varphi\mathbb{A}_{\boldsymbol{a}_\varphi} \cap B_{S_{\varphi,p}} \tag{2.56}$$

with subsets of J_t^u, we can then apply Proposition 2.13 to establish that

$$\dim T_{f*\varphi}\mathbb{A}_{a_{f*\varphi}} \cap B_{S_{f*\varphi},p} \geq \dim T_f\mathbb{A}_{a_f} \cap B_{S_f,p} + \dim T_\varphi\mathbb{A}_{a_\varphi} \cap B_{S_\varphi,p} - \dim[T_f\mathbb{A}_{a_f} \cap B_{S_f,p}] \cap [T_\varphi\mathbb{A}_{a_\varphi} \cap B_{S_\varphi,p}]. \quad (2.57)$$

The intersection on the far right of (2.57) is to be understood in terms of the vectors $\boldsymbol{\alpha} \in J_t^u$ used to represent elements of these two vector spaces. For generic images these two subspaces are identical, and therefore,

$$\dim T_{f*\varphi}\mathbb{A}_{a_{f*\varphi}} \cap B_{S_{f*\varphi},p} = \dim T_f\mathbb{A}_{a_f} \cap B_{S_f,p} = \dim T_\varphi\mathbb{A}_{a_\varphi} \cap B_{S_\varphi,p}. \quad (2.58)$$

There is a class of nongeneric images, which arise frequently in applications, for which $\dim T_f\mathbb{A}_{a_f} \cap B_{S_f,p} > 2p(p+1)$. Our final result in this section addresses this issue.

Proposition 2.15 *Let $\boldsymbol{\varphi}$ be an image indexed by J, extended periodically to $\mathbb{Z}^d$, which satisfies $\varphi_{\boldsymbol{j}} = \varphi_{-\boldsymbol{j}}$. Let $J_t^\varphi \subset J_t^u$, denote a set of indices so that $\{\boldsymbol{\tau}_\varphi^{(\boldsymbol{v})} : \boldsymbol{v} \in J_t^\varphi\}$ is a basis for $T_\varphi^0\mathbb{A}_{a_\varphi}$, and let $S \subset J$, be an inversion symmetric subset (if $\boldsymbol{j} \in S$, then $-\boldsymbol{j} \in S$). The set of vectors $\boldsymbol{\alpha} \in J_t^\varphi$ such that*

$$\tau_{\varphi,\boldsymbol{j}}^{\boldsymbol{\alpha}} = 0 \quad \text{for } \boldsymbol{j} \in S^c \quad (2.59)$$

is a linear subspace of dimension at least $|J_t^\varphi| - \frac{|S^c|}{2}$.

Proof Let $\boldsymbol{\alpha} \in J_t^\varphi$; the inversion symmetry of $\boldsymbol{\varphi}$ implies that $\tau_{\varphi,\boldsymbol{j}}^{(\boldsymbol{v})} = -\tau_{\varphi,-\boldsymbol{j}}^{(\boldsymbol{v})}$, for all $\boldsymbol{v}, \boldsymbol{j}$, and therefore,

$$\tau_{\varphi,\boldsymbol{j}}^{\boldsymbol{\alpha}} = -\tau_{\varphi,-\boldsymbol{j}}^{\boldsymbol{\alpha}} \quad \text{for any } \boldsymbol{\alpha}. \quad (2.60)$$

From this relation it is clear that $\tau_{\varphi,\boldsymbol{j}}^{\boldsymbol{\alpha}} = 0$, if and only if $\tau_{\varphi,-\boldsymbol{j}}^{\boldsymbol{\alpha}} = 0$. The map $\boldsymbol{\alpha} \mapsto \boldsymbol{\tau}_\varphi^{\boldsymbol{\alpha}}$ is linear and the equations in (2.59) therefore constitute, at most, $\frac{|S^c|}{2}$ independent linear conditions. Elementary linear algebra implies that the solution space to this homogeneous equation has dimension at least $|J_t^\varphi| - \frac{|S^c|}{2}$. □

Remark 2.16 This result is also true for images that are inversion symmetric with respect to any point in J.

Suppose, for example, that $J = [-N : N]^2$ and $\boldsymbol{\varphi}$ is an inversion symmetric, 2D image with support in a set $S_\varphi \subset [-\frac{N}{2} : \frac{N}{2}]^2$. If $\widehat{\varphi}(\boldsymbol{k}) \neq 0$ for any $\boldsymbol{k}$, then $|J_t^\varphi| \approx (2N+1)^2/2 - 1$, and therefore, the set of $\boldsymbol{\alpha} \in J_t^\varphi$ for which $\boldsymbol{\tau}_{\boldsymbol{\alpha}}^\varphi$ has support in S_φ has dimension about $|S_\varphi|/2$. That is,

$$\dim T_\varphi\mathbb{A}_{a_\varphi} \cap B_{S_\varphi} \approx \frac{|S_\varphi|}{2}. \quad (2.61)$$

For a generic image this dimension is 0. For an inversion symmetric image, with at least a few points in its support, even the intersection of $\mathbb{A}_{a_\varphi}$ and B_{S_φ} fails to be transversal! If $J = [1:2N]^2$, then an inversion symmetric image has approximately $4N$ DFT coefficients that vanish for symmetry reasons, which reduces $|J_t^\varphi|$ by about $2N$. For this case, the set of $\boldsymbol{\alpha} \in J_t^\varphi$ for which $\boldsymbol{\tau}_{\boldsymbol{\alpha}}^\varphi$ has support in S_φ has dimension about $|S_\varphi|/2 - 2N$.

In applications to phase retrieval this result coupled with (2.57) shows that convolution with an inversion symmetric image exacerbates the problem of nontransversality in that

$$\dim T_{f*\varphi}\mathbb{A}_{a_{f*\varphi}} \cap B_{S_{f*\varphi},p} \gg \dim T_f \mathbb{A}_{a_f} \cap B_{S_f,p}. \tag{2.62}$$

If $|S_\varphi|$ is large enough, then this will hold even for $p = 0$. In this setting, the standard practice of apodizing measured data with a Gaussian, or indeed the DFT of any inversion symmetric image, becomes inadvisable (see Example 2.18). In the context of astronomical observations, Fienup and Kowalczyk study the effects on the phase retrieval problem of cutting off the magnitude DFT data with a succession of functions with wider and wider supports (see Fienup and Kowalczyk 1990) .

Remark 2.17 Suppose that the image f_0 is obtained as samples of a continuous, real valued, function ρ with compact support in $(0,1)^d$:

$$f_{0\boldsymbol{j}} = \rho\left(\frac{\boldsymbol{j}}{N}\right). \tag{2.63}$$

We now suppose that ρ is extended to be periodic in all of $\mathbb{R}^d$, so that

$$f_{0\boldsymbol{j}}^{(\boldsymbol{v})} = \rho\left(\frac{\boldsymbol{j}-\boldsymbol{v}}{N}\right) \tag{2.64}$$

has the same meaning as above. Denote the measured magnitude DFT data by $\boldsymbol{a}_N$. If $N \to \infty$ so that $\boldsymbol{j}/N \to \boldsymbol{x}$, and $\boldsymbol{v}/N \to \boldsymbol{y}$, then the right-hand side in (2.64) tends to $\rho(\boldsymbol{x}-\boldsymbol{y})$. There is also a well-defined limit, $\mathbb{A}_c$, of the tori $\{\mathbb{A}_{\boldsymbol{a}_N}\}$, as $N \to \infty$, as the subset of $L^1([0,1]^d)$ consisting of functions g such that the Fourier coefficients satisfy

$$|\widehat{g}(\boldsymbol{k})| = \left| \int\limits_{[0,1]^d} g(\boldsymbol{x}) e^{-2\pi i \boldsymbol{k}\cdot\boldsymbol{x}} d\boldsymbol{x} \right| = |\widehat{\rho}(\boldsymbol{k})| \quad \text{for all } \boldsymbol{k} \in \mathbb{Z}^d. \tag{2.65}$$

In the limit, Lemma 2.3 states that the "vectors" $\{\rho(\cdot+\boldsymbol{y}) - \rho(\cdot-\boldsymbol{y}) : \boldsymbol{y} \in [0,1]^d\}$ belong to the tangent space to the torus at ρ, i.e., $T_\rho^0\mathbb{A}_c$, and the vectors $\{\rho(\cdot+\boldsymbol{y}) + \rho(\cdot-\boldsymbol{y}) : \boldsymbol{y} \in [0,1]^d\}$ are normal to the torus at this point.

These facts are easily verified by direct computation in the Fourier domain. There is also clearly a sense in which the spans of these vectors are dense in the corresponding vector spaces. As before, if f is nonnegative, then the fiber of the normal bundle $N^0_\rho \mathbb{A}_c$ has a spanning set that consists entirely of nonnegative vectors.

2.3 Numerical Examples

As follows from Corollary 3.6 below, the nontransversality of the intersection $\mathbb{A}_{\boldsymbol{a}} \cap B_S$ makes the problem of finding these intersection points ill conditioned. Its effects on practical algorithms are hard to gauge as these are entangled with related phenomena whose effects are sometimes more evident: very small, but nonzero angles between $T_{f_0}\mathbb{A}_{\boldsymbol{a}}$ and B_S lead to slow convergence and, often times, outright stagnation for essentially all known algorithms. Moreover, nonzero critical points of the distance function $d_{\mathbb{A}_{\boldsymbol{a}} B_S}$ can produce attracting basins that also prevent algorithms from finding true intersection points. In this section we examine how $\dim T_{f_0}\mathbb{A}_{\boldsymbol{a}} \cap B_S$ and the small angles depend on the properties of the image and the choice of support constraint.

The results of Section 2.2 suggest that, for a generic image, f_0, the dimension of $T_{f_0}\mathbb{A}_{\boldsymbol{a}} \cap B_S$ is largely determined by the accuracy of S as an estimate for S_{f_0}, and this is indeed the case. For $S = S_{f_0,p}$

$$\dim T_{f_0}\mathbb{A}_{\boldsymbol{a}} \cap B_{S_{f_0,p}} = 2p(p+1), \tag{2.66}$$

for generic images. If $\boldsymbol{v} \in C_p \setminus \{\boldsymbol{0}\}$, then $f_0^{(\boldsymbol{v})} \in B_{S_{f_0,p}}$ and the value of

$$\dim T_{f_0^{(\boldsymbol{v})}}\mathbb{A}_{\boldsymbol{a}} \cap B_{S_{f_0,p}} < 2p(p+1) \tag{2.67}$$

can be deduced from a purely combinatorial analysis. For example, in the extreme cases $\boldsymbol{v} = (\pm p, \pm p)$, these dimensions are 0. In Example 8.7 we see that even an intersection of dimension 1 can produce stagnation.

As noted earlier, if R and Q are matrices whose columns are orthonormal bases for $T^0_{f_0}\mathbb{A}_{\boldsymbol{a}}$ and B_S, respectively, then the angles between these two subspaces are obtained by computing the reduced singular value decomposition (SVD)

$$R^t Q = U\Sigma V^t. \tag{2.68}$$

The columns of the matrices RU and QV are again orthonormal bases for the respective subspaces. The diagonal entries of Σ are the inner products between columns of RU and those of QV, hence, they give the cosines of the “extremal” angles between these collections of vectors, see (2.10). The dimension of the intersection of these subspaces equals the number of singular values that are equal to 1. The matrix $H = R^t Q$ appears below in our linearized analyses of both the alternating projection (AP) algorithm, and the hybrid iterative map algorithms when using support as the auxiliary information. As noted above, singular values very close to 1 lead to very slow convergence, or even stagnation, for all of these algorithms. In our experiments we see that the number of singular values very close to 1 increases monotonically with smoothness of the image and the looseness of the support constraint.

For the experiments in this section, we generate 2D images of the form

$$f_{0j} = \sum_{i=1}^{M} m_i \psi_k \left(\frac{\| \boldsymbol{j} - \boldsymbol{c}_i \|}{r_i} \right), \tag{2.69}$$

where the magnitudes, centers, and radii, $\{(m_i, \boldsymbol{c}_i, r_i)\}$, are selected randomly and

$$\psi_k(r) = (1 - x^2)^k \chi_{[-1,1]}(x). \tag{2.70}$$

In these examples the magnitudes $\{m_i\}$ are real. As noted above, for integral k this is a $\mathscr{C}^{k-1}$-function whose $(k-1)$st derivative is Lipschitz continuous. For noninteger k, the function has $[|k|]$ continuous derivatives, and the highest-order derivative is $(k - [|k|])$-Hölder continuous. Examples are shown in Figure 1.5. The centers and radii are selected so that the terms of the sum have support in a specified subrectangle of a given rectangle. The larger rectangle has edge lengths equal to twice those of the subregion.

In the first set of examples the larger square is 64×64 pixels. The relatively small image size is used so that we are able to directly compute the SVD of $R^t Q$. The tangent space to the magnitude torus has dimension 2046. If S_{f_0} is the support of the image f_0, then $S_{f_0, p}$ denotes its p-pixel neighborhood. We also consider the behavior of the singular values of H when we use support constraints defined by the p-pixel neighborhoods of R_{f_0}, the smallest rectangle containing S_{f_0}. For nonnegative images this rectangle can be determined from the magnitude DFT data via the autocorrelation function (see Fienup et al. 1982, 1990). Finally, we give examples that demonstrate the increase in dimension of $T_{f_0 * \varphi} \mathbb{A}_{\boldsymbol{a}_{f_0 * \varphi}} \cap B_{S_{f_0 * \varphi}, p}$, predicted in Section 2.2, that results from convolving with an inversion symmetric image, $\boldsymbol{\varphi}$.

As noted in the Introduction (Section 1.4), in the context of finite precision arithmetic, the definition of the support of a smooth image requires the choice of a cutoff threshold for the image magnitude, namely the parameter

ϵ in (1.53). In Example 2.18, we assume $\epsilon = 10^{-14}$ and define the support constraints in terms of the set

$$S_{f_0}^{10^{-14}} = \{j : |f_{0j}| > 10^{-14}\}. \tag{2.71}$$

To simplify the notation we sometimes write S_f instead of the more accurate $S_f^{10^{-14}}$. In the final example we consider the effect on $\dim T_{f_0}\mathbb{A}_a \cap B_S$ of choosing different thresholds to define the support of f_0.

With $\epsilon_{\text{mach}} = 10^{-16}$ two directions are equal if the angle between them is less than 10^{-8} radians, i.e., $|1 - \cos\theta| \leq 10^{-16}$.

Example 2.18 We now examine relationships, beyond the exact intersections of $T_{f_0}\mathbb{A}_a$ and B_{S_p}, in five randomly selected examples, with $k = 0, 2, 4, 6$ and $p = 0, 1, 2, 3$. In Figure 2.1 we show plots of $\log_{10}(1 - \sigma_n)$, where $\{\sigma_n\}$ are the

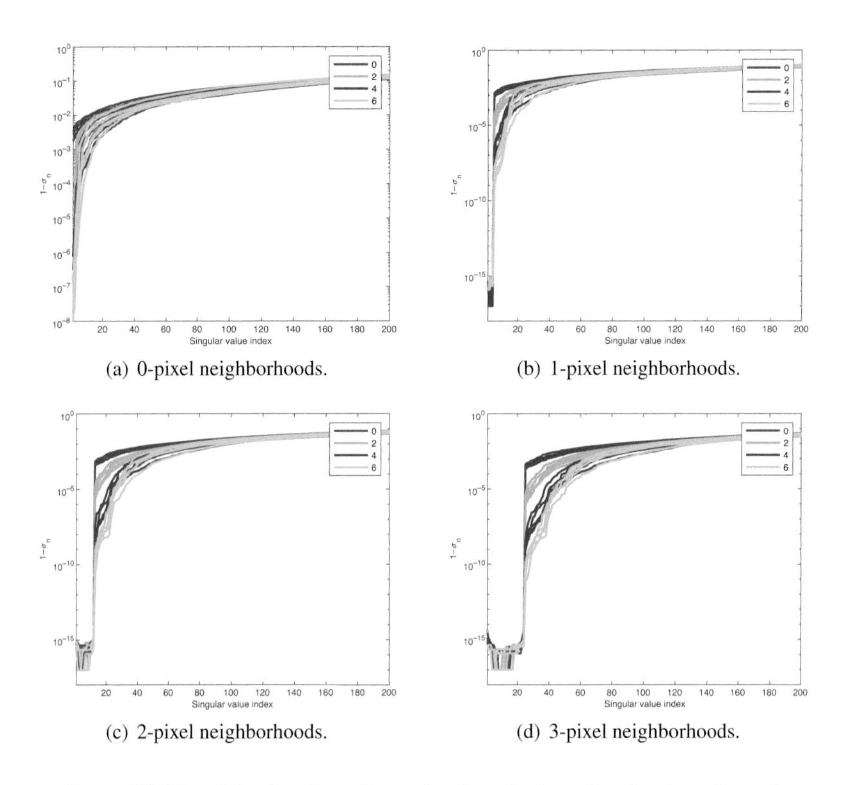

(a) 0-pixel neighborhoods.

(b) 1-pixel neighborhoods.

(c) 2-pixel neighborhoods.

(d) 3-pixel neighborhoods.

Figure 2.1 The plots show $\log_{10}(1 - \sigma_n)$, where $\{\sigma_n\}$ are the singular values of $H = R^t Q$, for varying degrees of smoothness $k = 0, 2, 4, 6$ (corresponding to the colors: red, green, blue, and cyan). For each k, plots are produced by creating five random 64×64 examples with support constraint defined by $S_{f_0,p}$, for $p = 0, 1, 2, 3$, corresponding to (a), (b), (c), and (d), respectively.

singular values of H, for different sized support neighborhoods ($p = 0, 1, 2, 3$) and different degrees of smoothness. Notice that the dimensions of exact intersections, i.e., $|\{n : \sigma_n = 1\}| = 2p(p+1)$, but the number of $\{\sigma_n\}$ close to 1 increases with both k and p.

It may be unrealistic to expect to find approximate support regions that are p-pixel neighborhoods of the exact support S_{f_0}. On the other hand, for nonnegative images, from the magnitude DFT data one can determine the smallest rectangle $R_{f_0} \supset S_{f_0}$, see Proposition 4.8 in Section 4.1. In this set of examples, we use the same procedure employed to obtain Figure 2.1, with the rectangular sets $S = R_{f_0,0}, R_{f_0,1}, R_{f_0,2}$, and $R_{f_0,3}$ used to define support constraints. As before, we use five 64×64 images at four levels of smoothness, $k = 0, 2, 4, 6$ and compute the singular values of H. The dimensions of the exact intersections are again given by $2p(p+1)$. For fixed values of (k, p) there appear to be many more singular values close to 1 than in the previous examples, and greater dependence on the choice of image. This is not too surprising as S_{f_0} contains much more information about the image than R_{f_0}. Figure 2.2 shows the results of these experiments.

The foregoing two experiments indicate that a smoother image tends to have more singular values very close to 1. Intuitively, this follows from the fact that the smoother the function, the less sharply defined is its boundary. The p-pixel support condition, therefore, contains large numbers of vectors which lie approximately in both $T_{f_0}\mathbb{A}_{\boldsymbol{a}}$ and B_{f_0,S_p}.

The last experiment in this example explores the effects on the dimension of $T_{f_0}\mathbb{A}_{\boldsymbol{a}} \cap B_{f_0,S_p}$ of smoothing the image by convolving with an inversion symmetric image. For these experiments we use images defined by (2.69) with $k = 0$, which are then convolved with $\psi_m(\|\boldsymbol{j}\|/R)$, with $m = 2, 4, 6$. These kernels all have the same support. In Figure 2.3 we show the $\{\log_{10}(1 - \sigma_n)\}$ for these values of m and $p = 0, 1, 2, 3$. For reference we also show the results without convolution. As predicted, when the image is convolved the $\dim T_{f_0}\mathbb{A}_{\boldsymbol{a}} \cap B_{S_{f_0,p}}$ is increased by a fixed amount that depends only on p, not m. This increase is $44, 64, 84$, and 104 for $p = 0, 1, 2$, and 3, respectively.

Example 2.19 We now consider the effect of the choice of threshold, ϵ, in the definition of $S^{\epsilon}_{f_0}$, see (1.53), on the $\dim T_{f_0}\mathbb{A}_{\boldsymbol{a}} \cap B_S$. We let $S^{\epsilon}_{f_0,p}$ denote the p-pixel neighborhood of

$$S^{\epsilon}_{f_0} = \{\boldsymbol{j} : |f_{0\boldsymbol{j}}| \geq \epsilon\}.$$

Even in finite precision arithmetic, the intersections $\mathbb{A}_{\boldsymbol{a}} \cap B_{S^{\epsilon}}$ are, strictly speaking, often empty. The minimum distance between these sets is comparable to ϵ. Nonetheless, we can still compute the dimensions of the intersections

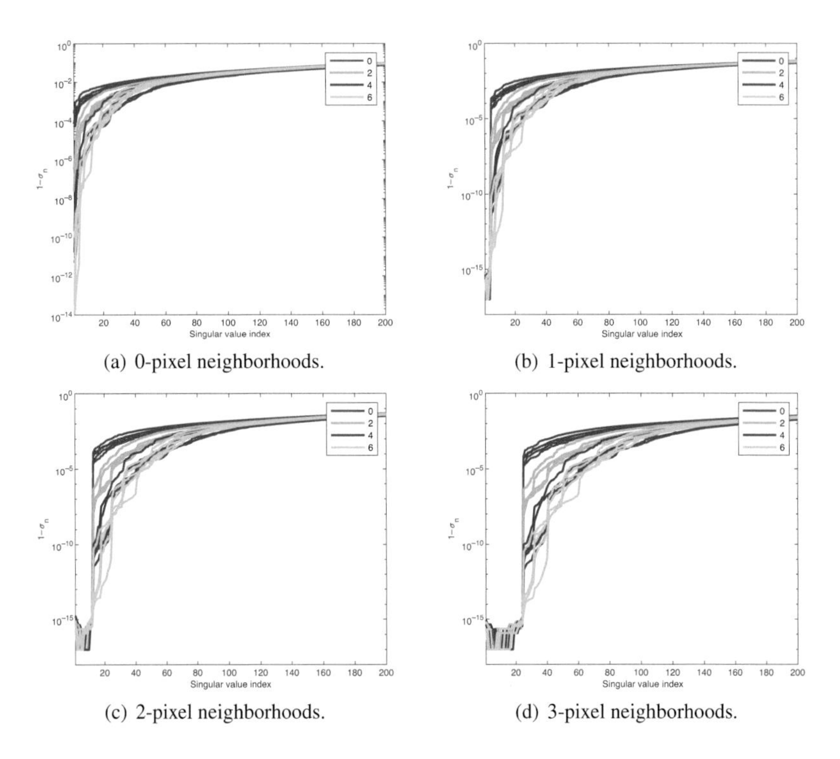

(a) 0-pixel neighborhoods.

(b) 1-pixel neighborhoods.

(c) 2-pixel neighborhoods.

(d) 3-pixel neighborhoods.

Figure 2.2 The plots show $\log_{10}(1-\sigma_n)$, where $\{\sigma_n\}$ are the singular values of $H = R^t Q$, for $k = 0, 2, 4, 6$ (corresponding to the colors: red, green, blue, and cyan) and rectangular support constraints, $R_{f_0,p}$. For each k, plots are produced by creating five random 64×64 examples with support constraint defined by $R_{f_0,p}$, for $p = 0, 1, 2, 3$, corresponding to (a), (b), (c), and (d), respectively.

of the linear subspaces $T_{f_0}\mathbb{A}_{\boldsymbol{a}}$ and $B_{S^{\epsilon}_{f_0,p}}$, and the behavior of the small angles between them. These quantities behave largely as expected: for given (k, p) the number of $\{\sigma_n\}$ very close to 1 decreases as ϵ increases. For a given smoothness, k, the sets $S^{\epsilon}_{f_0,p}$ stabilize at some value of ϵ. The value at which this occurs depends on how finely the image is sampled. For coarsely sampled images ($N = 32, 64$) and small values of k ($k = 1, 2, 3$) these neighborhoods are fixed after $\epsilon \leq 10^{-6}$, and therefore, the angles between $\dim T_{f_0}\mathbb{A}_{\boldsymbol{a}}$ and $B_{S^{\epsilon}_p}$ are as well.

To see an effect from the choice of threshold we need to work with more finely sampled images. We use a single 256×256 image, at three levels of smoothness $k = 1, 3, 5$, and seven thresholds for the support,

$$\epsilon \in \{10^{-4},\ 10^{-6},\ 10^{-8},\ 10^{-10},\ 10^{-12},\ 10^{-14},\ 10^{-16}\}.$$

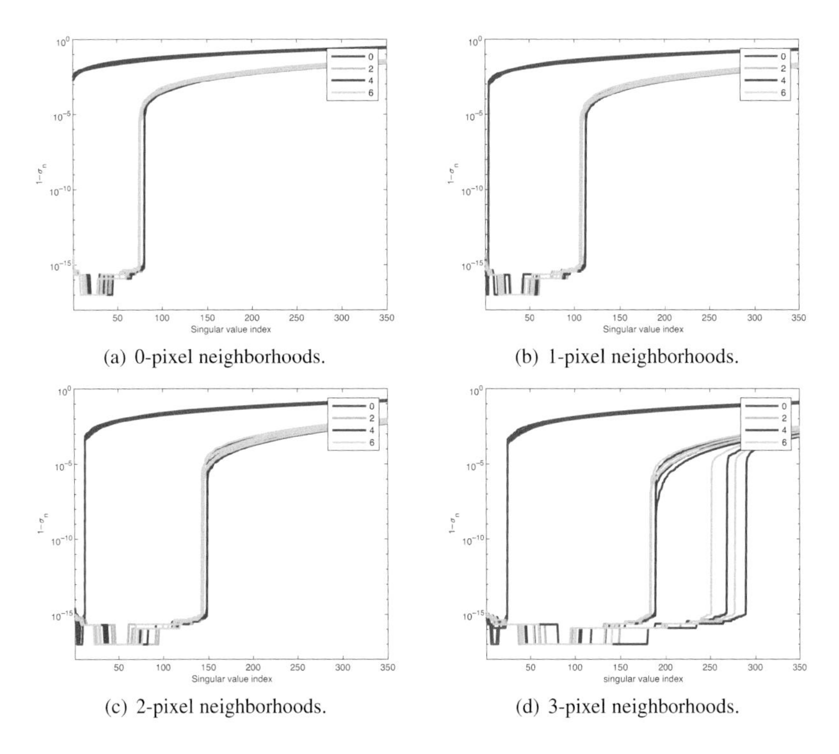

(a) 0-pixel neighborhoods.
(b) 1-pixel neighborhoods.
(c) 2-pixel neighborhoods.
(d) 3-pixel neighborhoods.

Figure 2.3 The plots show $\log_{10}(1-\sigma_n)$, where $\{\sigma_n\}$ are the singular values of $H = R^t Q$, after convolving with ψ_m with $m = 2, 4, 6$ (corresponding to the colors: green, blue, and cyan) and support constraints $S_{f_0,p}$. The red plots (labeled $m = 0$) are the singular values without convolution. For each m, p, plots are produced by creating five random 64×64 examples.

These images are too large to directly compute the SVD of H. Instead, we compute the angles between the subspaces

$$\mathcal{V}_{f_0,12} = \text{span}\{\boldsymbol{\tau}_{f_0}^{(\boldsymbol{v})} : \boldsymbol{v} \in J \text{ with } \|\boldsymbol{v}\|_\infty \leq 12\} \subset T_{f_0}\mathbb{A}_{\boldsymbol{a}} \tag{2.72}$$

and $B_{S^\epsilon_{f_0,p}}$, for $p = 2$. From the variational characterization of the $\sigma_j = \cos\theta_j$ given in (2.10), replacing $T_{f_0}\mathbb{A}_{\boldsymbol{a}}$ with the subspace $\mathcal{V}_{f_0,12}$ leads to singular values that are possibly smaller, and therefore, angles that are possibly larger than the exact values. Empirically, we have found that using a neighborhood of size 12 gives a very good approximation to the true angles.

The results of these experiments are summarized in Figure 2.4. When the smoothness parameter $k = 1$, the support neighborhood is more or less unchanged after $\epsilon = 10^{-3}$. For $k = 3$, the support neighborhood is more or less unchanged after $\epsilon = 10^{-6}$. For $k = 5$, the support neighborhood is more or less unchanged after $\epsilon = 10^{-8}$. This experiment shows that the choice of

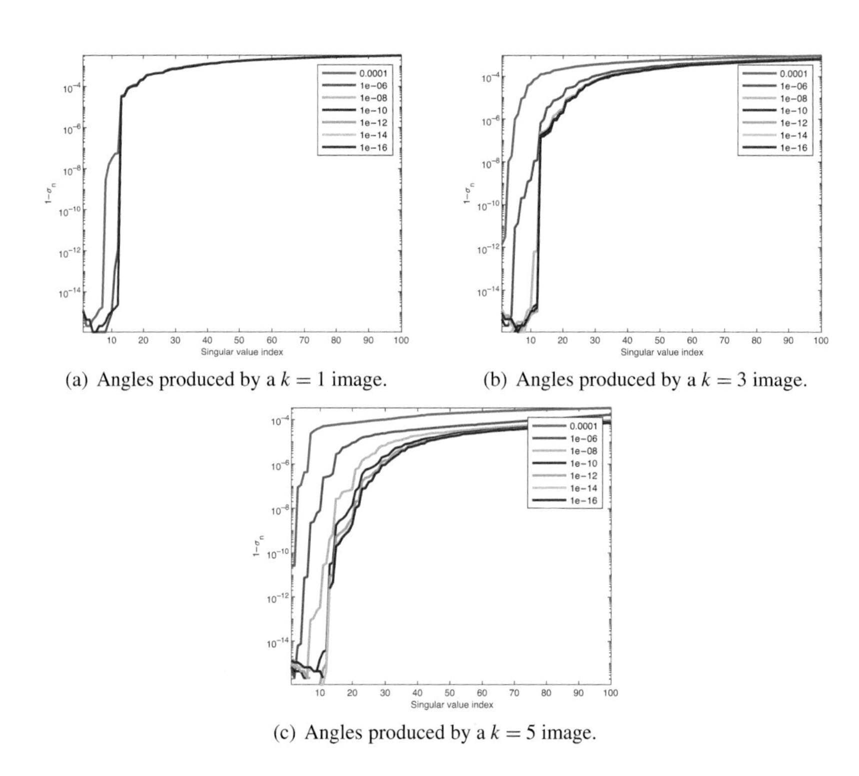

(a) Angles produced by a $k = 1$ image.

(b) Angles produced by a $k = 3$ image.

(c) Angles produced by a $k = 5$ image.

Figure 2.4 The plots show $\log_{10}(1 - \sigma_n)$, where $\{\sigma_n\}$ are the singular values of $H = R^t Q$, with R an orthonormal basis for $\mathcal{V}_{f_0,12}$, and support thresholds $\epsilon \in \{10^{-4},\ 10^{-6},\ 10^{-8},\ 10^{-10},\ 10^{-12},\ 10^{-14},\ 10^{-16}\}$ (corresponding to the colors: blue, orange, yellow, purple, green, cyan, and magenta), as indicated in the figure legends. We perform this experiment for three smoothness levels, $k = 1, 3, 5$.

support threshold ϵ has an effect on the geometry near to the intersections (or almost intersections) of $\mathbb{A}_{\boldsymbol{a}}$ and $B_{S_{f_0,p}}$, but this effect is not dramatic unless one uses a very large value, e.g., $\epsilon = 10^{-4}$. The smoother the image, the larger the effect. Increasing ϵ is tempting as it makes the intersection between $\mathbb{A}_{\boldsymbol{a}}$ and $B_{S^{\epsilon}_{f_0,p}}$ "more transversal." On the other hand the data that defines the problem, $\{|\widehat{f_k}|\}$ and $S^{\epsilon}_{f_0,p}$, become more and more inconsistent with the increase in ϵ.

2.A Appendix: The Tangent and Normal Bundles for Submanifolds of $\mathbb{R}^N$

In this appendix we review some basic concepts from the differential geometry of smooth submanifolds of Euclidean spaces. More complete treatments can be found in Spivak (1965, 1979) and Lee (2018).

Definition 2.20 A subset M of $\mathbb{R}^n$ is a smooth submanifold of dimension m if for each $\boldsymbol{x} \in M$, there is

(i) An $r > 0$ such that the open neighborhood $U_{\boldsymbol{x}} = B_r(\boldsymbol{x}) \cap M$, is homeomorphic to a ball in $\mathbb{R}^m$. Here, $B_r(\boldsymbol{x})$ is the open ball of radius r in $\mathbb{R}^n$.
(ii) A $\mathscr{C}^\infty$-homeomorphism $\phi_{\boldsymbol{x}} : B_1 \to U_{\boldsymbol{x}}$, where B_1 is the open unit ball in $\mathbb{R}^m$.
(iii) For each $\boldsymbol{y} \in B_1$, the first derivative, $d\phi_{\boldsymbol{x}}(\boldsymbol{y}) : \mathbb{R}^m \to \mathbb{R}^n$, is an injective linear map

The pair $(U_{\boldsymbol{x}}, \phi_{\boldsymbol{x}})$ is a *local coordinate chart* for M near to $\boldsymbol{x}$.

While the coordinates charts are not unique, for each $\boldsymbol{y} \in B_1$ the image of the linear map $d\phi_{\boldsymbol{x}}(\boldsymbol{y})$ is an m-dimensional subspace of $\mathbb{R}^n$, which does not depend on the choice of $\phi_{\boldsymbol{x}}$, satisfying the conditions above. If $\widetilde{\boldsymbol{x}} = \phi_{\boldsymbol{x}}(\boldsymbol{y})$, then we denote this image by

$$d\phi_{\boldsymbol{x}}(\boldsymbol{y})\mathbb{R}^m \stackrel{d}{=} T^0_{\widetilde{\boldsymbol{x}}}M. \tag{2.73}$$

The vector space $T^0_{\widetilde{\boldsymbol{x}}}M$ is called the *fiber of the tangent bundle* to M through $\widetilde{\boldsymbol{x}}$; the affine subspace $\widetilde{\boldsymbol{x}} + T^0_{\widetilde{\boldsymbol{x}}}M$ is clearly the best linear approximation to M through $\widetilde{\boldsymbol{x}}$. We often refer to this affine subspace as the tangent space to M at $\widetilde{\boldsymbol{x}}$. If $\gamma : (-1,1) \to M$ is a smooth curve with $\gamma(0) = \widetilde{\boldsymbol{x}}$, then the chain rule implies that the tangent vector to this curve satisfies

$$\gamma'(0) \in T^0_{\widetilde{\boldsymbol{x}}}M. \tag{2.74}$$

This allows us to define the tangent bundle itself as a submanifold of $\mathbb{R}^n \times \mathbb{R}^n$.

Definition 2.21 If $M \subset \mathbb{R}^n$ is a smooth submanifold, then the set

$$TM = \bigcup_{\boldsymbol{x} \in M} \{\boldsymbol{x}\} \times T^0_{\boldsymbol{x}}M \tag{2.75}$$

is a submanifold of $\mathbb{R}^n \times \mathbb{R}^n$, which is called the *tangent bundle* of M. The projection onto the first factor defines a smooth map $\pi_1 : TM \to M$, with $\pi_1^{-1}(\boldsymbol{x}) = \{\boldsymbol{x}\} \times T^0_{\boldsymbol{x}}M$.

To show that TM is a smooth submanifold we need to give coordinate charts. If $(U_{\boldsymbol{x}}, \phi_{\boldsymbol{x}})$ is a coordinate chart for M, then $(U_{\boldsymbol{x}} \times \mathbb{R}^m, (\phi_{\boldsymbol{x}}, d\phi_{\boldsymbol{x}}))$ gives a coordinate chart for TM over the set $\pi_1^{-1}(U_{\boldsymbol{x}})$. As noted above, for $\boldsymbol{x} \in M$ the affine subspace $T_{\boldsymbol{x}}M = \boldsymbol{x} + T^0_{\boldsymbol{x}}M$ is the best "linear" approximation to M at the point $\boldsymbol{x}$, which means that if we choose a vector $\boldsymbol{v} \in T^0_{\boldsymbol{x}}M$ then

$$\operatorname{dist}(\boldsymbol{x} + t\boldsymbol{v}, M) = O(t^2) \text{ as } t \to 0. \tag{2.76}$$

Definition 2.22 The *fiber of the normal bundle* to M at $\boldsymbol{x}$ is the orthogonal complement in $\mathbb{R}^n$ to the fiber of tangent bundle at $\boldsymbol{x}$:

$$N_{\boldsymbol{x}}^0 M = \left[T_{\boldsymbol{x}}^0 M\right]^{\perp}. \tag{2.77}$$

The *normal bundle* itself is the smooth submanifold of $\mathbb{R}^n \times \mathbb{R}^n$ defined by

$$NM = \bigcup_{x \in M} \{x\} \times N_{\boldsymbol{x}}^0 M. \tag{2.78}$$

The projection onto the first factor $\pi_1 : NM \to M$ is a smooth map with $\pi_1^{-1}(\boldsymbol{x}) = \{\boldsymbol{x}\} \times N_{\boldsymbol{x}}^0 M$, the fiber of the normal bundle to M at $\boldsymbol{x}$.

We leave to the interested reader the problem of constructing local coordinate charts for NM. The property of the normal bundle that is analogous to (2.76) is: let $\boldsymbol{v} \in N_{\boldsymbol{x}}^0 M$ be a unit vector; for t sufficiently small the point in M closest to $\boldsymbol{x} + t\boldsymbol{v}$ is $\boldsymbol{x}$, and

$$\operatorname{dist}(\boldsymbol{x} + t\boldsymbol{v}, M) = |t|. \tag{2.79}$$

We now turn to the concept of transversality, beginning with the case of a pair of linear subspaces.

Definition 2.23 Two linear subspaces V and W of $\mathbb{R}^n$, with $\dim V + \dim W \leq n$, are *transversal* if $V \cap W = \{\mathbf{0}\}$.

This leads easily to a notion of transversality for a pair of submanifolds that intersect at a point.

Definition 2.24 Let M_1 and M_2 be two smooth submanifolds of $\mathbb{R}^n$ that have an **isolated** intersection at $\boldsymbol{x}$. This intersection is said to be *transversal* if the intersection of the vector spaces $T_{\boldsymbol{x}}^0 M_1$ and $T_{\boldsymbol{x}}^0 M_2$ is transversal.

If the intersection is transversal, then near to $\boldsymbol{x}$, the subset $M_1 \cup M_2$ is well-approximated by the "linear" model $T_{\boldsymbol{x}} M_1 \cup T_{\boldsymbol{x}} M_2$, near to $\boldsymbol{x} = T_{\boldsymbol{x}} M_1 \cap T_{\boldsymbol{x}} M_2$. We let B_1^m denote the unit ball in $\mathbb{R}^m$. Assume that $\dim M_1 = m_1$ and $\dim M_2 = m_2$; one can show that, if the intersection at $\boldsymbol{x}$ between M_1 and M_2 is transversal, then there are $r_1 > 0, r_2 > 0$ and a smooth, injective map,

$$\Phi : B_1^{m_1} \times B_1^{m_2} \longrightarrow \mathbb{R}^n,$$

with $d\Phi(\mathbf{0}, \mathbf{0})$ injective, such that

$$\Phi(B_1^{m_1} \times \{\mathbf{0}\}) = M_1 \cap B_{r_1}^n(\boldsymbol{x}) \text{ and } \Phi(\{\mathbf{0}\} \times B_1^{m_2}) = M_2 \cap B_{r_2}^n(\boldsymbol{x}). \tag{2.80}$$

If the intersection is nontransversal, then $\dim T_{\boldsymbol{x}} M_1 \cap T_{\boldsymbol{x}} M_2 > 0$, which precludes $T_{\boldsymbol{x}} M_1 \cup T_{\boldsymbol{x}} M_2$ from providing a good model for $M_1 \cup M_2$ near to $\boldsymbol{x}$.

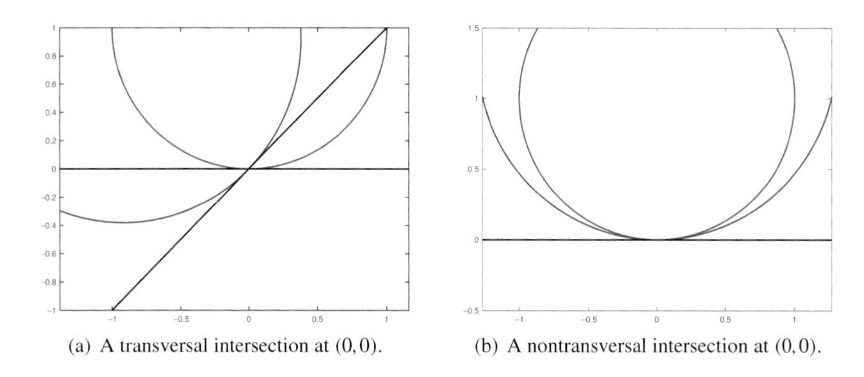

(a) A transversal intersection at $(0,0)$.

(b) A nontransversal intersection at $(0,0)$.

Figure 2.5 Plots showing a transversal and a non-transversal intersection of submanifolds of $\mathbb{R}^2$, along with the tangent lines to the two submanifolds through $(0,0)$, which are shown in black.

Figure 2.5 shows examples of transversal and nontransversal intersections along with their tangent spaces at the point of intersection.

2.B Appendix: Fast Projections onto the Tangent and Normal Bundles

The vector spaces $T_f^0\mathbb{A}_a$ and $N_f^0\mathbb{A}_a$ are orthogonal complements, so it is only necessary to find a fast projection onto one of them. We write the projection onto the fiber of the normal bundle as

$$P_{N_f^0\mathbb{A}_a}(\boldsymbol{g}) = \mathscr{F}^{-1} \circ \widehat{P}_N \circ \mathscr{F}(\boldsymbol{g}), \text{ for } \boldsymbol{g} \in \mathbb{R}^J. \tag{2.81}$$

Using the basis for $N_f^0\mathbb{A}_a$ given in (2.21) we give a fast algorithm to implement $\widehat{P}_N$.

If $\boldsymbol{f}$ is a real image, then $\widehat{\boldsymbol{f}} = \mathscr{F}(\boldsymbol{f})$ denotes its DFT. Recall that if $\boldsymbol{k}'$ is the index conjugate to $\boldsymbol{k}$, then $\widehat{f}_{\boldsymbol{k}'} = \overline{\widehat{f}_{\boldsymbol{k}}}$. For $\boldsymbol{k} \in J$ with $\boldsymbol{k} \neq \boldsymbol{k}'$, we let

$$\widehat{\boldsymbol{n}}_{\boldsymbol{k}} = \frac{1}{\sqrt{2}|\widehat{f}_{\boldsymbol{k}}|}\left[\widehat{f}_{\boldsymbol{k}}\boldsymbol{e}_{\boldsymbol{k}} + \overline{\widehat{f}_{\boldsymbol{k}}}\boldsymbol{e}_{\boldsymbol{k}'}\right], \text{ assuming that } \widehat{f}_{\boldsymbol{k}} \neq 0. \tag{2.82}$$

If $\boldsymbol{k} \neq \boldsymbol{k}'$, but $\widehat{f}_{\boldsymbol{k}} = 0$, then $\widehat{\boldsymbol{n}}_{\boldsymbol{k}} = \boldsymbol{e}_{\boldsymbol{k}}$, $\widehat{\boldsymbol{n}}_{\boldsymbol{k}'} = \boldsymbol{e}_{\boldsymbol{k}'}$; if $\boldsymbol{k} = \boldsymbol{k}'$, then $\widehat{\boldsymbol{n}}_{\boldsymbol{k}} = \boldsymbol{e}_{\boldsymbol{k}}$.

Let $J_n \subset J$ be a list of indices that contains one of $\boldsymbol{k} \neq \boldsymbol{k}'$, but not both, and $J_e \subset J$ a list of the indices for which $\boldsymbol{k} = \boldsymbol{k}'$. The vectors $\{\widehat{\boldsymbol{n}}_{\boldsymbol{k}} : \boldsymbol{k} \in J_n \cup J_e\}$ are an orthonormal basis, in the DFT domain, for $N_f^0\mathbb{A}_{\boldsymbol{a}}$. Assuming that $\widehat{f}_{\boldsymbol{k}} \neq 0$, for any $\boldsymbol{k}$, the projection $\widehat{P}_N$ takes the form

$$\widehat{P}_N(\widehat{\boldsymbol{g}}) = \sum_{\boldsymbol{k} \in J_N \cup J_e} \langle \widehat{\boldsymbol{g}}, \widehat{\boldsymbol{n}}_{\boldsymbol{k}} \rangle \widehat{\boldsymbol{n}}_{\boldsymbol{k}} \boldsymbol{e}_{\boldsymbol{k}}. \tag{2.83}$$

If $\boldsymbol{k} = \boldsymbol{k}'$, then $\widehat{g}_{\boldsymbol{k}}$ is real and $\langle \widehat{\boldsymbol{g}}, \widehat{\boldsymbol{n}}_{\boldsymbol{k}} \rangle = \widehat{g}_{\boldsymbol{k}}$. Otherwise

$$\langle \widehat{\boldsymbol{g}}, \widehat{\boldsymbol{n}}_{\boldsymbol{k}} \rangle = \frac{\widehat{g}_{\boldsymbol{k}} \overline{\widehat{f}_{\boldsymbol{k}}} + \overline{\widehat{g}_{\boldsymbol{k}}} \widehat{f}_{\boldsymbol{k}}}{\sqrt{2}|\widehat{f}_{\boldsymbol{k}}|} = \frac{\sqrt{2}\operatorname{Re}(\widehat{g}_{\boldsymbol{k}} \overline{\widehat{f}_{\boldsymbol{k}}})}{|\widehat{f}_{\boldsymbol{k}}|} \text{ if } \widehat{f}_{\boldsymbol{k}} \neq 0. \tag{2.84}$$

Since $\operatorname{Re}(\widehat{g}_{\boldsymbol{k}} \overline{\widehat{f}_{\boldsymbol{k}}}) = \operatorname{Re}(\overline{\widehat{g}_{\boldsymbol{k}}} \widehat{f}_{\boldsymbol{k}})$ we can rewrite $\widehat{P}_N(\widehat{\boldsymbol{g}})$ as

$$\widehat{P}_N(\widehat{\boldsymbol{g}}) = \sum_{\boldsymbol{k} \in J} \frac{\operatorname{Re}(\widehat{g}_{\boldsymbol{k}} \overline{\widehat{f}_{\boldsymbol{k}}})}{|\widehat{f}_{\boldsymbol{k}}|} \frac{\widehat{f}_{\boldsymbol{k}}}{|\widehat{f}_{\boldsymbol{k}}|} \boldsymbol{e}_{\boldsymbol{k}}. \tag{2.85}$$

Let $\Lambda^{\widehat{f}}$ denote the $J \times J$ diagonal matrix with

$$\Lambda^{\widehat{f}}_{\boldsymbol{kk}} = \frac{\widehat{f}_{\boldsymbol{k}}}{|\widehat{f}_{\boldsymbol{k}}|}, \text{ for } \boldsymbol{k} \in J, \tag{2.86}$$

we can rewrite $\widehat{P}_N(\widehat{\boldsymbol{g}})$ as

$$\widehat{P}_N(\widehat{\boldsymbol{g}}) = \Lambda^{\widehat{f}} \operatorname{Re}[\overline{\Lambda^{\widehat{f}}} \widehat{\boldsymbol{g}}]. \tag{2.87}$$

If $\boldsymbol{f}$ has some DFT coefficients that vanish, then this method has to be slightly modified. Let J_0 be a list of the indices for which $\widehat{f}_{\boldsymbol{k}} = 0$. We define $\Lambda^{\widehat{f}}_{\boldsymbol{kk}}$ to be zero for $\boldsymbol{k} \in J_0$, and then

$$\widehat{P}_N(\widehat{\boldsymbol{g}}) = \Lambda^{\widehat{f}} \operatorname{Re}[\overline{\Lambda^{\widehat{f}}} \widehat{\boldsymbol{g}}] + \sum_{\boldsymbol{k} \in J_0} \widehat{g}_{\boldsymbol{k}} \boldsymbol{e}_{\boldsymbol{k}}. \tag{2.88}$$

In all cases

$$P_{N_f^0\mathbb{A}_a}(\boldsymbol{g}) = \mathscr{F}^{-1}\left[\widehat{P}_N \widehat{\boldsymbol{g}}\right]. \tag{2.89}$$

Evidently the computational work in evaluating $P_{N_f^0\mathbb{A}_a}(\boldsymbol{g})$ is dominated by two applications of the DFT. The orthogonal projection onto $T_f^0\mathbb{A}_{\boldsymbol{a}}$ is given by

$$P_{T_f^0\mathbb{A}_a}(\boldsymbol{g}) = \boldsymbol{g} - P_{N_f^0\mathbb{A}_a}(\boldsymbol{g}). \tag{2.90}$$

3
Well-Posedness

Phase retrieval is an example of an inverse problem, in which one seeks to recover a function f in some class $\mathscr{X}$, given measurements $\mathscr{M}(f)$ in some class $\mathscr{Y}$. The measurement map $\mathscr{N}: \mathscr{X} \to \mathscr{Y}$ is often not surjective. Nonetheless one can seek to invert $\mathscr{N}$ on its range, $\mathscr{M}(\mathscr{X})$. An inverse problem is said to be well posed if:

(i) Each point in $\mathscr{M}(\mathscr{X})$ uniquely specifies a point in $\mathscr{X}$.
(ii) The inverse map $\mathscr{N}^{-1} : \mathscr{M}(\mathscr{X}) \to \mathscr{X}$ is "continuous."

The spaces $\mathscr{X}$ and $\mathscr{Y}$ are usually Hilbert spaces with norms $\|\cdot\|_{\mathscr{X}}$ and $\|\cdot\|_{\mathscr{Y}}$, respectively. From a computational perspective, the inverse map is "usefully" continuous if it satisfies a Lipschitz estimate, which is usually expressed by saying that there is a constant $C > 0$ so that

$$C\|\boldsymbol{f}_1 - \boldsymbol{f}_2\|_{\mathscr{X}} \leq \|\mathscr{M}(\boldsymbol{f}_1) - \mathscr{M}(\boldsymbol{f}_2)\|_{\mathscr{Y}} \text{ for all } \boldsymbol{f}_1, \boldsymbol{f}_2 \in \mathscr{X}. \tag{3.1}$$

If the measurements are nonlinear, then it is appropriate to replace this global discussion with a local discussion near a pair $(\boldsymbol{f}, \mathscr{M}(\boldsymbol{f}))$. This is the case for the phase retrieval problem which, even when solvable, typically has multiple well-separated but equivalent solutions.

Other weaker notions of continuity (i.e., Hölder continuity) imply that the inverse mapping is ill conditioned, and will lose many digits of accuracy in any finite precision approximation (see John 1960; Epstein and Schotland 2008). Suppose that there is a $p > 1$ so that $\mathscr{N}$ satisfies an estimate of the form

$$C\|\boldsymbol{f}_1 - \boldsymbol{f}_2\|_{\mathscr{X}}^p \leq \|\mathscr{M}(\boldsymbol{f}_1) - \mathscr{M}(\boldsymbol{f}_2)\|_{\mathscr{Y}} \text{ for all } \boldsymbol{f}_1, \boldsymbol{f}_2 \in \mathscr{X}; \tag{3.2}$$

which implies that the inverse is $\frac{1}{p}$-Hölder continuous

$$\|\mathscr{N}^{-1}(\boldsymbol{y}_1) - \mathscr{N}^{-1}(\boldsymbol{y}_2)\|_{\mathscr{X}} \leq C\|\boldsymbol{y}_1 - \boldsymbol{y}_2\|_{\mathscr{Y}}^{\frac{1}{p}}, \tag{3.3}$$

for $\boldsymbol{y}_1, \boldsymbol{y}_2$ in the range of $\mathscr{N}$. If $\boldsymbol{y}$ is the exact data and $\boldsymbol{y}+\delta\boldsymbol{y}$ is the measured data, then, putting aside the question of whether $\boldsymbol{y}+\delta\boldsymbol{y}$ is in the range of $\mathscr{N}$, we see that

$$\|\mathscr{N}^{-1}(\boldsymbol{y}) - \mathscr{N}^{-1}(\boldsymbol{y}+\delta\boldsymbol{y})\|_{\mathscr{X}} \leq C\|\delta\boldsymbol{y}\|_{\mathscr{Y}}^{\frac{1}{p}}. \tag{3.4}$$

That is, if the measurements are accurate to order ϵ, then the reconstruction can be expected to be accurate to order $\epsilon^{\frac{1}{p}}$, which is to say that the reconstruction process loses the fraction $\frac{p-1}{p}$ of the digits available in the measurements. Moreover, if the constant is $C \gg 1$, then the extent of this ill conditioning is further magnified.

For the discrete, classical, phase retrieval problem we see below (Section 3.1) that one expects an estimate of this sort with $p = 2$. In this context, the objects are discrete images indexed by a finite lattice, J, i.e., $\mathscr{X} = \mathbb{R}^J$ or $\mathbb{C}^J$. The basic measurement is the magnitude discrete Fourier transform (DFT) data, $\mathscr{M}(\boldsymbol{f}) = |\widehat{\boldsymbol{f}}| \in \mathbb{R}_+^J$. As noted earlier in the Introduction, without some auxiliary information, these measurements do not specify a discrete subset of $\mathbb{R}^J$ (or $\mathbb{C}^J$) but rather an entire magnitude torus $\mathbb{A}_{\boldsymbol{a}}$. Suppose that the auxiliary information is an a priori estimate, S, for the support, $S_{\boldsymbol{f}}$ of $\boldsymbol{f}$. Provided that the side length of S in each dimension is less than half that of J, the magnitude DFT data, along with the information that $S_{\boldsymbol{f}} \subset S$, define a finite subset of $\mathbb{R}^J$. That is, the intersection $\mathbb{A}_{\boldsymbol{a}} \cap B_S$ is finite. In this setting, let us define the restricted measurement map as $\widetilde{\mathscr{M}} \overset{d}{=} \mathscr{M} \restriction_{B_S}$. We work below with this map as it generically has local inverses defined on its range.

If the image, $\boldsymbol{f}$, is complex, then the intersections $\mathbb{A}_{\boldsymbol{a}} \cap B_S$ are never isolated points. This is because the full circle of images $\{e^{i\theta}\boldsymbol{f} : \theta \in [0, 2\pi)\} \subset \mathbb{A}_{\boldsymbol{a}} \cap B_S$. As this would considerably complicate the discussion, in this chapter we restrict our attention to the case of real images. Similar results hold for the complex case but are technically more involved to state and prove.

For most choices of magnitude data $\boldsymbol{a} \in \mathbb{R}_+^J$, the intersection of $\mathscr{M}^{-1}(\boldsymbol{a})$ with B_S is the empty set. With the assumptions above, when this intersection is non-empty, it is finite, and generically consists of those trivial associates of a single point in the inverse image that satisfy the support condition. When the Z-transform of $\boldsymbol{f} \in \mathbb{A}_{\boldsymbol{a}} \cap B_S$ is reducible, the inverse image of $\widetilde{\mathscr{M}}$ remains a finite set. Since the set of irreducible polynomials is open and dense, see Appendix 1.A, when the Z-transform of $\boldsymbol{f} \in \mathbb{A}_{\boldsymbol{a}} \cap B_S$ is irreducible, the map $\widetilde{\mathscr{M}}$ is invertible on its range near to $\widetilde{\mathscr{M}}(\boldsymbol{f})$. In this chapter we consider the well-posedness of the discrete, classical, phase retrieval problem in this circumstance.

We first consider the continuity properties of the local inverse, showing that if $f_0 \in \mathbb{A}_a \cap B_S$, then the local inverse satisfies a Lipschitz estimate if and only if this intersection is transversal at f_0. This shows that, irrespective of the algorithm used, the phase retrieval problem is computationally hard whenever the intersection is nontransversal.

We next turn to *numerical* nonuniqueness phenomena, which may well be more consequential than the true nonuniqueness described above. To describe these phenomena, we need to choose an $\epsilon > 0$, which, in a purely mathematical context, can be thought of as a number comparable to machine precision. In an applied setting, ϵ can be thought of as a measure of the accuracy of the measurements. For an image f, recall that its ϵ-support is defined as

$$S_f^{\epsilon} = \{ j : |f_j| \geq \epsilon \| f \| \}. \tag{3.5}$$

We say that f has support contained in S, to precision ϵ, if $S_f^{\epsilon} \subset S$.

Definition 3.1 If (a, S) is data for the phase retrieval problem for which there are two images f_1, f_2 such that

(i) $\|f_1\|_2 = \|f_2\|_2$ and the distance to the trivial associate of f_2 closest to f_1 is much greater than $\epsilon \| f_1 \|_2$;
(ii) $S_{f_i}^{\epsilon} \subset S$ for $i = 1, 2$; and
(iii) $\|\mathscr{M}(f_i) - a\|_2 < \epsilon \|a\|_2$, for $i = 1, 2$,

then we say that the solution to the phase retrieval problem defined by (a, S) is ϵ-nonunique.

In Section 3.2 we explain two different mechanisms leading to data defining ϵ-nonunique phase retrieval problems. One of these phenomena is a universal consequence of the existence of nongeneric data for which the solution is not unique, up to trivial associates, whereas the second is more closely connected to properties of the DFT itself.

Remark 3.2 The material in Section 3.1 is clearly related to, and to some extent, inspired by the results in Cahill et al. (2016) and Alaifari et al. (2019). A different analysis of the uniqueness and conditioning of the phase retrieval problem appears in Sanz (1985). This work focuses on the algebro-geometric aspects of the problem.

3.1 Conditioning and Transversality

In Chapter 2, we saw that $\dim T_{f_0}\mathbb{A}_{\boldsymbol{a}} \cap B_S$ is often positive, and that a less restrictive support condition increases this dimension. In this section we consider the consequences of nontransversality of the intersection $\mathbb{A}_{\boldsymbol{a}} \cap B_S$ on the conditioning of the phase retrieval problem.

In Cahill et al. (2016), the authors examine the stability properties of the *generalized* phase retrieval problem defined by a frame, $\Phi = \{\boldsymbol{\varphi}_n : n \in \mathscr{I}\}$ for a Hilbert space $\mathscr{H}$. They define the map $\mathscr{N}_\Phi : \mathscr{H} \to \mathbb{R}_+^{|\mathscr{I}|}$, by setting

$$\mathscr{N}_\Phi(\boldsymbol{x}) = (|\langle \boldsymbol{x}, \boldsymbol{\varphi}_n \rangle| : n \in \mathscr{I}). \tag{3.6}$$

They say that the frame Φ "does phase retrieval" if this map is invertible, up to an "overall" phase. If $\mathscr{H}$ is a complex Hilbert space, then this means that $\mathscr{N}_\Phi(\boldsymbol{x}) = \mathscr{N}_\Phi(\boldsymbol{y})$ if and only if $\boldsymbol{x} = e^{i\theta}\boldsymbol{y}$ for a $\theta \in \mathbb{R}$; if $\mathscr{H}$ is real, then only the overall sign is ambiguous. We denote the equivalence classes defined by these relations as $\mathscr{H}/\sim$. Among other things, they show that if $\dim \mathscr{H} < \infty$, and Φ is a frame that does phase retrieval, then, in the complex case, there is a constant $0 < C$ so that

$$C \inf_{\theta \in \mathbb{R}} \|\boldsymbol{x} - e^{i\theta}\boldsymbol{y}\|_{\mathscr{H}} \leq \|\mathscr{N}_\Phi(\boldsymbol{x}) - \mathscr{N}_\Phi(\boldsymbol{y})\|_{\mathscr{H}}. \tag{3.7}$$

That is, the inverse of $\mathscr{N}_\Phi$ is Lipschitz continuous from it range to $\mathscr{H}/\sim$.

Cahill et al. (2016) distinguish between real and complex phase retrieval problems. In the complex case, $\mathscr{H}$ is a Hilbert space over $\mathbb{C}$ and the unobserved measurements $\{\langle \boldsymbol{x}, \boldsymbol{\varphi}_n \rangle\}$ are complex valued. In the real case everything is real. For the complex case a phase, i.e., a point on the unit circle in $\mathbb{C}$, must be recovered for each measurement, and the overall image is only determined up to a global phase. In the real case, only the signs need to be determined, and the image is specified only up to a global sign. The problem we are considering does not quite fit into this framework. While it is true that we only consider real images, so there is only a global sign ambiguity in the reconstructed image (up to trivial associates), the unobserved measurements, $\{\widehat{f}_{\boldsymbol{k}} : \boldsymbol{k} \in J\}$, are complex numbers. Thus, we must actually retrieve a point on the unit circle for each of the measured values $\{|\widehat{f}_{\boldsymbol{k}}| : \boldsymbol{k} \in J\}$.

In this chapter we show that the transversality properties at an intersection of B_S with $\mathbb{A}_{\boldsymbol{a}}$ determine whether or not the map $\mathscr{M} : B_S \to \mathbb{R}_+^J$,

$$\mathscr{M} : \boldsymbol{f} \longrightarrow |\widehat{\boldsymbol{f}}| = \{|\widehat{f}_{\boldsymbol{k}}| : \boldsymbol{k} \in J\}, \tag{3.8}$$

has a Lipschitz local inverse near to a point $\mathscr{M}(\boldsymbol{f})$ for $\boldsymbol{f} \in B_S \cap \mathbb{A}_{\boldsymbol{a}}$. This, in turn, determines whether or not the problem of reconstructing the phase is infinitesimally well-conditioned. The forward map, which is actually defined on all of $\mathbb{R}^J$, is clearly continuous in the ℓ_2-topology: the Parseval relation, and triangle inequality imply that

$$\|\mathscr{M}(\boldsymbol{f}_1) - \mathscr{M}(\boldsymbol{f}_2)\|_2 \leq \|\boldsymbol{f}_1 - \boldsymbol{f}_2\|_2. \tag{3.9}$$

The map $\boldsymbol{f} \mapsto \mathscr{M}(\boldsymbol{f})$ is not invertible on $\mathbb{R}^J$, as the inverse image of a point in $\boldsymbol{a} \in \mathbb{R}_+^J$ is the magnitude torus, $\mathbb{A}_{\boldsymbol{a}}$, which has dimension about half of $|J|$. If we fix a sufficiently small set, $S \subset J$, then for $\boldsymbol{f} \in B_S$, the inverse image $\mathscr{M}^{-1}(\mathscr{M}(\boldsymbol{f})) \cap B_S$ is a finite set. For a generic choice of $\boldsymbol{f}$ this set consists of plus or minus those trivial associates, $\{\pm \boldsymbol{f}^{(\boldsymbol{v})}\} \cup \{\pm \check{\boldsymbol{f}}^{(\boldsymbol{v}')}\}$, with support in S. If S is a good enough estimate for $S_{\boldsymbol{f}}$, then $\pm\boldsymbol{f}$ may be the unique points in this set; in any case, the set of preimages is finite. Note, however that the cardinality of $\mathscr{M}^{-1}(\mathscr{M}(\boldsymbol{f})) \cap B_S$ generally depends on $\boldsymbol{f} \in B_S$. Nonetheless, for $\boldsymbol{f} \in B_S$ and $0 < \delta$ small enough, we can consider the local inverse $\widetilde{\mathscr{M}}^{-1}$ of $\widetilde{\mathscr{M}}$, defined on the image of a neighborhood of $\boldsymbol{f}$,

$$\widetilde{\mathscr{M}}(B_\delta(\boldsymbol{f})) = \{\mathscr{M}(\boldsymbol{g}) : \boldsymbol{g} \in B_S \text{ and } \|\boldsymbol{f} - \boldsymbol{g}\|_2 < \delta\},$$

for which $\widetilde{\mathscr{M}}^{-1}(\widetilde{\mathscr{M}}(\boldsymbol{f})) = \boldsymbol{f}$. In what follows we assume that $\delta < \|\boldsymbol{f}\|$ so that we can ignore the sign ambiguity.

In contrast to the result in Cahill et al. (2016) we have the following:

Theorem 3.3 *For an image $\boldsymbol{f}$, let $\boldsymbol{a}_{\boldsymbol{f}} = \mathscr{M}(\boldsymbol{f})$ define magnitude torus $\mathbb{A}_{\boldsymbol{a}_{\boldsymbol{f}}}$. If $\boldsymbol{f} \in B_S$ and $T_{\boldsymbol{f}}\mathbb{A}_{\boldsymbol{a}_{\boldsymbol{f}}} \cap B_s \neq \{\boldsymbol{f}\}$, then there exists a constant C, a number $0 < \eta$, and a unit vector $\boldsymbol{v} \in B_S$ such that*

$$\|\mathscr{M}(\boldsymbol{f}) - \mathscr{M}(\boldsymbol{f} + t\boldsymbol{v})\|_2 \leq C|t|^2 \textit{ for } |t| < \eta. \tag{3.10}$$

Remark 3.4 Note that $\|\boldsymbol{f} - (\boldsymbol{f} + t\boldsymbol{v})\|_2 = |t|$, and therefore at a point where the intersection $B_S \cap \mathbb{A}_{\boldsymbol{a}_{\boldsymbol{f}}}$ is nontransversal, the local inverse *cannot* be uniformly Lipschitz continuous. Such an estimate actually holds for small perturbations, $\boldsymbol{v}$, belonging to a subspace of dimension equal to $\dim T_{\boldsymbol{f}}\mathbb{A}_{\boldsymbol{a}_{\boldsymbol{f}}} \cap B_s$. For directions satisfying the estimate in (3.10), the best uniform estimate that the local inverse of $\widetilde{\mathscr{M}}$ can possibly satisfy is a $\frac{1}{2}$-Hölder estimate

$$\|\widetilde{\mathscr{M}}^{-1}(|\widehat{\boldsymbol{f}}_1|) - \widetilde{\mathscr{M}}^{-1}(|\widehat{\boldsymbol{f}}_2|)\| \leq c\||\widehat{\boldsymbol{f}}_1| - |\widehat{\boldsymbol{f}}_2|\|^{\frac{1}{2}}. \tag{3.11}$$

In Fritz John's terminology this is called a well-behaved, improperly posed problem (see John 1960). In the field of inverse problems, it is called *weakly* ill posed. Just as it is possible to compute square roots of numbers near to zero with high relative accuracy, this fact alone does not rule out the possibility of an algorithm that, given highly accurate data, computes $\widetilde{\mathscr{M}}^{-1}(|\widehat{\boldsymbol{f}}|)$ with high relative accuracy. It does render the process of phase reconstruction highly sensitive to measurement errors. The linearization of this inverse problem is the problem of finding points in $B_S \cap T_f \mathbb{A}_{a_f}$, a positive dimensional subspace. Strictly speaking, the linearized problem, therefore, has an infinite condition number, which is a reflection of the fact that the linearization does not give an adequate description of the problem near to a nontransversal intersection.

Using the result of Barnett and Doppelt described in Remark 2.12, one can show that if $S_{f,2p} \subset S$, then there is a constant C and a sequence $\{\delta f_n\} \subset B_S$ so that $\delta f_n \to \mathbf{0}$, and

$$\|\mathcal{M}(f + \delta f_n) - \mathcal{M}(f)\|_2 \leq C\|\delta f_n\|^{2p+2}. \tag{3.12}$$

In these directions the local inverse $\widetilde{\mathcal{M}}^{-1}$ is, at best, $1/(2p+2)$-Hölder continuous.

Proof [Theorem 3.3] Since $T_f \mathbb{A}_{a_f} \cap B_s \neq \{f\}$, there is a unit vector $v \in B_S$ so that $f + tv$ belongs to this intersection; the line $\{f + tv : t \in \mathbb{R}\} \subset B_S$. From the definition of tangency, it follows that there is a constant C and a $0 < \eta$ so that

$$\operatorname{dist}(f + tv, \mathbb{A}_{a_f}) \leq C|t|^2 \quad \text{for } |t| \leq \eta. \tag{3.13}$$

Let f_t be a point on $\mathbb{A}_{a_f}$ which realizes this minimum distance. Of course $\mathcal{M}(f_t) = \mathcal{M}(f)$, and therefore, using (3.9), we obtain the estimate

$$\begin{aligned} \|\mathcal{M}(f + tv) - \mathcal{M}(f)\|_2 &= \|\mathcal{M}(f + tv) - \mathcal{M}(f_t)\|_2 \\ &\leq \|f + tv - f_t\|_2 \\ &\leq C|t|^2 \quad \text{for } |t| \leq \eta. \end{aligned} \tag{3.14}$$

□

This begs the question of what happens when $T_f \mathbb{A}_{a_f} \cap B_S = \{f\}$ alone. In this case we can prove a local version of a Lipschitz lower bound like that given in (3.7). As noted above, for a generic image f with sufficiently small support, the only other images, $\widetilde{f}$, for which $\mathcal{M}(f) = \mathcal{M}(\widetilde{f})$ are the trivial associates of $\pm f$. For $0 < \eta$, sufficiently small, the images

$$\{f + \delta f : \delta f \in B_S \text{ and } \|\delta f\|_2 < \eta\},$$

are again generic, and the distances to the closest trivial associates, among this collection of images, is bounded below by a positive constant. As noted above, for images in this collection, the map $\delta f \mapsto \mathcal{M}(f + \delta f)$ is locally invertible. We now show that, generically, if the intersection of $\mathbb{A}_{a_f}$ and B_S is transversal at f, then the local inverse, which takes $\mathcal{M}(f)$ to f, satisfies a Lipschitz lower bound.

Theorem 3.5 *For an image* $\boldsymbol{f}$ *we let* $\mathbb{A}_{\boldsymbol{a}_f}$ *denote the magnitude torus defined by* $\mathscr{M}(\boldsymbol{f})$*. Suppose that* $\widehat{f}_{\boldsymbol{k}} \neq 0$ *for any* $\boldsymbol{k} \in J$*. If* $\boldsymbol{f} \in B_S$ *and* $T_{\boldsymbol{f}} \mathbb{A}_{\boldsymbol{a}_f} \cap B_s = \{\boldsymbol{f}\}$*, then there are positive constants,* η, C*, so that, if* $\delta \boldsymbol{f} \in B_S$ *and* $\|\delta \boldsymbol{f}\|_2 < \eta$*, then*

$$C\|\delta \boldsymbol{f}\|_2 \leq \|\mathscr{M}(\boldsymbol{f}) - \mathscr{M}(\boldsymbol{f} + \delta \boldsymbol{f})\|_2. \tag{3.15}$$

These theorems have the following corollary:

Corollary 3.6 *Let* $S \subset J$ *be a set such that* $\mathscr{M}^{-1}(\boldsymbol{a}) \cap B_S$ *is finite. If* $\boldsymbol{f} \in B_S$ *satisfies* $\widehat{f}_{\boldsymbol{k}} \neq 0$ *for any* $\boldsymbol{k} \in J$*, then the local inverse* $\widetilde{\mathscr{M}}^{-1}$ *defined near to* $\mathscr{M}(\boldsymbol{f})$ *is Lipschitz if and only if* $T_{\boldsymbol{f}} \mathbb{A}_{\boldsymbol{a}_f} \cap B_S = \{\boldsymbol{f}\}$*.*

Remark 3.7 Since the conditioning of a nonlinear inverse problem is determined by the linearization of the map we are trying to invert, this result says that, independently of the algorithm used to find the inverse, the discrete, classical, phase retrieval problem may be locally well conditioned at a point $\mathscr{M}(\boldsymbol{f})$ if and only if $T_{\boldsymbol{f}} \mathbb{A}_{\boldsymbol{a}_f} \cap B_S = \{\boldsymbol{f}\}$.

Proof [Theorem 3.5] In the proof of this theorem we use a different normalization for the DFT than that used elsewhere in the book. Here we assume that the DFT defined in (1.7) is multiplied by a constant so that $\boldsymbol{f} \mapsto \widehat{\boldsymbol{f}}$ is an ℓ_2-isometry, that is $\|\boldsymbol{f}\|_2 = \|\widehat{\boldsymbol{f}}\|_2$.

Recall that if $\boldsymbol{g} \in \mathbb{A}_{\boldsymbol{a}_f}$, then we let $T_{\boldsymbol{g}}^0 \mathbb{A}_{\boldsymbol{a}_f}$ and $N_{\boldsymbol{g}}^0 \mathbb{A}_{\boldsymbol{a}_f}$ denote the linear subspaces underlying the fibers of the tangent and normal bundles to $\mathbb{A}_{\boldsymbol{a}_f}$ at $\boldsymbol{g}$. Let π_t and π_n denote the orthogonal projections onto $T_{\boldsymbol{f}}^0 \mathbb{A}_{\boldsymbol{a}_f}$ and $N_{\boldsymbol{f}}^0 \mathbb{A}_{\boldsymbol{a}_f}$, respectively. The hypothesis of the theorem implies that the map $\pi_n : B_S \to N_{\boldsymbol{f}}^0 \mathbb{A}_{\boldsymbol{a}_f}$ is injective, and so has an inverse on its range. From this we see that there are constants C_1, C_2 so that, for $\delta \boldsymbol{f} \in B_S$, we have the estimates

$$\|\pi_t(\delta \boldsymbol{f})\|_2 \leq C_1 \|\pi_n(\delta \boldsymbol{f})\|_2 \text{ and } \|\delta \boldsymbol{f}\|_2 \leq C_2 \|\pi_n(\delta \boldsymbol{f})\|_2. \tag{3.16}$$

Suppose that $\boldsymbol{f}, \boldsymbol{g} \in B_S$, then the distance $\|\mathscr{M}(\boldsymbol{f}) - \mathscr{M}(\boldsymbol{g})\|_2$ can be expressed in terms of the distance between $\boldsymbol{g}$ and the torus $\mathbb{A}_{\boldsymbol{a}_f}$. If $\widehat{\boldsymbol{f}}$ and $\widehat{\boldsymbol{g}}$ are the associated DFTs, then

$$\begin{aligned} \|\mathscr{M}(\boldsymbol{f}) - \mathscr{M}(\boldsymbol{g})\|_2^2 &= \sum_{\boldsymbol{k} \in J} ||\widehat{f}_{\boldsymbol{k}}| - |\widehat{g}_{\boldsymbol{k}}||^2 \\ &= \sum_{\boldsymbol{k} \in J} ||\widehat{f}_{\boldsymbol{k}}| e^{i\theta_{\boldsymbol{k}}} - |\widehat{g}_{\boldsymbol{k}}| e^{i\theta_{\boldsymbol{k}}}|^2 \end{aligned} \tag{3.17}$$

for any choice of phases $\{e^{i\theta_{\boldsymbol{k}}} : \boldsymbol{k} \in J\}$. We can choose the phases that define the point $\boldsymbol{g}$ (in the second sum) and then it is clear that the point $(|\widehat{f}_{\boldsymbol{k}}| e^{i\theta_{\boldsymbol{k}}})_{\boldsymbol{k} \in J}$ is the point in $\mathbb{A}_{\boldsymbol{a}_f}$ nearest to $\boldsymbol{g}$, hence,

$$\|\mathcal{M}(\boldsymbol{f}) - \mathcal{M}(\boldsymbol{g})\|_2 = \text{dist}(\boldsymbol{g}, \mathbb{A}_{\boldsymbol{a}_f}). \tag{3.18}$$

In order to estimate the $\text{dist}(\boldsymbol{g}, \mathbb{A}_{\boldsymbol{a}_f})$ we define the function

$$F(\boldsymbol{v}_t, \boldsymbol{v}_n) = \text{dist}(\boldsymbol{f} + \boldsymbol{v}_t + \boldsymbol{v}_n, \mathbb{A}_{\boldsymbol{a}_f}) \text{ with } \boldsymbol{v}_t \in T_{\boldsymbol{f}}^0 \mathbb{A}_{\boldsymbol{a}_f} \text{ and } \boldsymbol{v}_n \in N_{\boldsymbol{f}}^0 \mathbb{A}_{\boldsymbol{a}_f}. \tag{3.19}$$

For sufficiently small $\boldsymbol{v}_n$ we have

$$F(0, \boldsymbol{v}_n) = \|\boldsymbol{v}_n\|_2. \tag{3.20}$$

F is a Lipschitz function of its arguments and we can show that, for small enough vectors $(\boldsymbol{v}_t, \boldsymbol{v}_n)$, we have the estimate that

$$|\partial_s F(s\boldsymbol{v}_t, \boldsymbol{v}_n)| \leq \frac{4s}{m}\|\boldsymbol{v}_t\|_2^2 \quad \text{for } s \in [0, 1], \tag{3.21}$$

where $m = \min\{|\widehat{f}_{\boldsymbol{k}}| : \boldsymbol{k} \in J\}$. We prove this in Lemma 3.8 below.

Using these observations we see that the fundamental theorem of calculus applies and we can express $F(\boldsymbol{v}_t, \boldsymbol{v}_n)$ as

$$F(\boldsymbol{v}_t, \boldsymbol{v}_n) = \|\boldsymbol{v}_n\|_2 + \int_0^1 \partial_s F(s\boldsymbol{v}_t, \boldsymbol{v}_n) ds. \tag{3.22}$$

The estimate in (3.21) implies that

$$F(\boldsymbol{v}_t, \boldsymbol{v}_n) \geq \|\boldsymbol{v}_n\|_2 - \frac{2}{m}\|\boldsymbol{v}_t\|_2^2. \tag{3.23}$$

If $\delta \boldsymbol{f} \in B_S$ is represented by $\boldsymbol{v}_t + \boldsymbol{v}_n$, then this estimate and those in (3.16) show that

$$F(\boldsymbol{v}_t, \boldsymbol{v}_n) \geq \|\boldsymbol{v}_n\|_2 (1 - \frac{2C_1^2}{m}\|\boldsymbol{v}_n\|) \geq \frac{\|\delta \boldsymbol{f}\|_2}{C_2}(1 - \frac{2C_1^2}{m}\|\boldsymbol{v}_n\|). \tag{3.24}$$

Let $\eta = \frac{m}{4C_1^2}$, then for $\|\boldsymbol{v}_n\|_2 < \eta$, we have the estimate that

$$\frac{\|\delta \boldsymbol{f}\|_2}{2C_2} \leq \|\mathcal{M}(\boldsymbol{f}) - \mathcal{M}(\boldsymbol{f} + \delta \boldsymbol{f})\|_2. \tag{3.25}$$

As $\delta \boldsymbol{f} = \boldsymbol{v}_t + \boldsymbol{v}_n$ is an orthogonal decomposition, if $\|\delta \boldsymbol{f}\|_2 < \eta$, then $\|\boldsymbol{v}_n\|_2 < \eta$, as well. But for the verification of (3.21), this completes the proof of the theorem. □

Here we prove (3.21).

Lemma 3.8 *Let $\boldsymbol{f}$ be a real image and $\mathbb{A}_{\boldsymbol{a}_f}$ the magnitude torus defined by $\boldsymbol{a}_f$. Suppose that $\widehat{f}_{\boldsymbol{k}} \neq 0$ for any $\boldsymbol{k} \in J$. For $\boldsymbol{g} \in \mathbb{A}_{\boldsymbol{a}_f}$, let $T_{\boldsymbol{g}}^0 \mathbb{A}_{\boldsymbol{a}_f}$ and $N_{\boldsymbol{g}}^0 \mathbb{A}_{\boldsymbol{a}_f}$ denote the vector spaces underlying the fibers of the tangent and normal bundle* *to* $\mathbb{A}_{\boldsymbol{a}_f}$ *at* $\boldsymbol{g}$. *There is a* $0 < \eta$ *so that, for* $\|\boldsymbol{v}_t + \boldsymbol{v}_n\|_2 < \eta$, *the function* $F : T^0_{\boldsymbol{g}}\mathbb{A}_{\boldsymbol{a}_f} \times N^0_{\boldsymbol{g}}\mathbb{A}_{\boldsymbol{a}_f} \to [0,\infty)$ *defined by*

$$F(\boldsymbol{v}_t, \boldsymbol{v}_n) = \operatorname{dist}(\boldsymbol{g} + \boldsymbol{v}_t + \boldsymbol{v}_n, \mathbb{A}_{\boldsymbol{a}_f}) \tag{3.26}$$

satisfies

$$\begin{aligned} F(\boldsymbol{0}, \boldsymbol{v}_n) &= \|\boldsymbol{v}_n\|_2, \\ |\partial_s F(s\boldsymbol{v}_t, \boldsymbol{v}_n)| &\leq \frac{4s}{m}\|\boldsymbol{v}_t\|_2^2 \qquad \textit{for } s \in [0,1], \end{aligned} \tag{3.27}$$

where $m = \min\{|\widehat{f}_{\boldsymbol{k}}| : \boldsymbol{k} \in J\}$.

Remark 3.9 As F is a Lipschitz function, the derivative $\partial_s F(s\boldsymbol{v}_t, \boldsymbol{v}_n)$ is defined for almost every $s \in [0,1]$, and this derivative satisfies the fundamental theorem of calculus. The inequality is satisfied whenever the derivative is defined. The constant m is the same for any point $\boldsymbol{g} \in \mathbb{A}_{\boldsymbol{a}_f}$.

Proof To prove this result we use an explicit description of the vector spaces $T^0_{\boldsymbol{g}}\mathbb{A}_{\boldsymbol{a}_f}, N^0_{\boldsymbol{g}}\mathbb{A}_{\boldsymbol{a}_f}$, in the DFT domain, given in (2.20). There is a set of indices J_t, so that, for $\boldsymbol{k} \in J_t$, the conjugate index $\boldsymbol{k}' \neq \boldsymbol{k}$. Moreover, the tangent bundle, in the DFT representation, is spanned by

$$\boldsymbol{t}_{\boldsymbol{k}} = i\left[\widehat{g}_{\boldsymbol{k}}\boldsymbol{e}_{\boldsymbol{k}} - \overline{\widehat{g}_{\boldsymbol{k}}}\boldsymbol{e}_{\boldsymbol{k}'}\right] \quad \text{for } \boldsymbol{k} \in J_t. \tag{3.28}$$

The vectors

$$\boldsymbol{n}_{\boldsymbol{k}} = \left[\widehat{g}_{\boldsymbol{k}}\boldsymbol{e}_{\boldsymbol{k}} + \overline{\widehat{g}_{\boldsymbol{k}}}\boldsymbol{e}_{\boldsymbol{k}'}\right] \quad \text{for } \boldsymbol{k} \in J_t \tag{3.29}$$

lie in the normal bundle. To get a spanning set we need to add in the basis vectors $\boldsymbol{e}_{\boldsymbol{k}}$ for indices that are self-conjugate; we denote these indices by J_n^0.

Since the DFT is an isometry it suffices to study the $\operatorname{dist}(\widehat{\boldsymbol{g}} + \widehat{\boldsymbol{v}_t} + \widehat{\boldsymbol{v}_n}, \widehat{\mathbb{A}_{\boldsymbol{a}_f}})$. We let $F(\hat{\boldsymbol{v}}_t, \hat{\boldsymbol{v}}_n)$ denote the function F in the DFT representation. In this representation, the torus is a product of circles lying in the coordinate planes defined by the pairs $(\boldsymbol{e}_{\boldsymbol{k}}, \boldsymbol{e}_{\boldsymbol{k}'})$. To find the nearest point it, therefore, suffices to work one coordinate plane at a time, which we now do, omitting the basis vectors $(\boldsymbol{e}_{\boldsymbol{k}}, \boldsymbol{e}_{\boldsymbol{k}'})$.

For a pair $(p_{\boldsymbol{k}}, q_{\boldsymbol{k}})$ of real numbers we need to find the closest point on the circle $\{(e^{i\theta}\widehat{g}_{\boldsymbol{k}}, e^{-i\theta}\overline{\widehat{g}_{\boldsymbol{k}}}) : \theta \in [0, 2\pi)\}$, to the point

$$w_{\boldsymbol{k}} = (\widehat{g}_{\boldsymbol{k}}, \overline{\widehat{g}_{\boldsymbol{k}}}) + ip_{\boldsymbol{k}}(\widehat{g}_{\boldsymbol{k}}, -\overline{\widehat{g}_{\boldsymbol{k}}}) + q_{\boldsymbol{k}}(\widehat{g}_{\boldsymbol{k}}, \overline{\widehat{g}_{\boldsymbol{k}}}). \tag{3.30}$$

Evidently, that point is

$$\widetilde{w}_{\boldsymbol{k}} = \frac{((1+q_{\boldsymbol{k}}+ip_{\boldsymbol{k}})\widehat{g}_{\boldsymbol{k}}, \overline{(1+q_{\boldsymbol{k}}+ip_{\boldsymbol{k}})\widehat{g}_{\boldsymbol{k}}})}{\sqrt{(1+q_{\boldsymbol{k}})^2 + p_{\boldsymbol{k}}^2}}. \tag{3.31}$$

The distance between these points is given by

$$\|w_k - \widetilde{w}_k\|_2 = \sqrt{2}|1 - \sqrt{(1+q_k)^2 + p_k^2}||\widehat{g}_k|. \tag{3.32}$$

An elementary calculation shows that

$$\partial_{p_k}\|w_k - \widetilde{w}_k\|_2 = \pm\sqrt{2}|\widehat{g}_k|\frac{p_k}{\sqrt{(1+q_k)^2 + p_k^2}}. \tag{3.33}$$

An arbitrary normal vector takes the form

$$\hat{\boldsymbol{v}}_n = \sum_{k \in J_t} q_k \boldsymbol{n}_k + \sum_{k \in J_n^0} r_k \widehat{g}_k \boldsymbol{e}_k, \tag{3.34}$$

and a tangent vector takes the form

$$\hat{\boldsymbol{v}}_t = \sum_{k \in J_t} p_k \boldsymbol{t}_k, \tag{3.35}$$

where $(\boldsymbol{p}, \boldsymbol{q}, \boldsymbol{r})$ is a real vector. Note that

$$\|\hat{\boldsymbol{v}}_t\|^2 = 2\sum_{k \in J_t} |p_k|^2 |\widehat{g}_k|^2. \tag{3.36}$$

Using the planar calculation from above, we see that

$$F(\hat{\boldsymbol{v}}_t, \hat{\boldsymbol{v}}_n) = \left[\sum_{k \in J_t} \|w_k - \widetilde{w}_k\|_2^2 + \sum_{k \in J_n^0} |\widehat{g}_k|^2 (1 - r_k)^2\right]^{\frac{1}{2}}. \tag{3.37}$$

Differentiating with respect to p_k gives

$$\partial_{p_k} F(\hat{\boldsymbol{v}}_t, \hat{\boldsymbol{v}}_n) = \frac{\partial_{p_k}\|w_k - \widetilde{w}_k\|^2}{2F(\hat{\boldsymbol{v}}_t, \hat{\boldsymbol{v}}_n)} = \pm\sqrt{2}\frac{|\widehat{g}_k|\|w_k - \widetilde{w}_k\| p_k}{\sqrt{(1+q_k)^2 + p_k^2} \cdot F(\hat{\boldsymbol{v}}_t, \hat{\boldsymbol{v}}_n)}. \tag{3.38}$$

If $\frac{1}{\sqrt{2}} - 1 < q_k$ for every $\boldsymbol{k}$, then we see that

$$|\partial_{p_k} F(\hat{\boldsymbol{v}}_t, \hat{\boldsymbol{v}}_n)| \leq 2|\widehat{g}_k||p_k|. \tag{3.39}$$

As

$$\partial_s F(s\hat{\boldsymbol{v}}_t, \hat{\boldsymbol{v}}_n) = \sum_{k \in J_t} \partial_{p_k} F(s\hat{\boldsymbol{v}}_t, \hat{\boldsymbol{v}}_n) p_k, \tag{3.40}$$

if $m = \min\{|\widehat{g}_k| : \boldsymbol{k} \in J_t\}$, then equations (3.36) and (3.39) imply that

$$|\partial_s F(s\hat{\boldsymbol{v}}_t, \hat{\boldsymbol{v}}_n)| \leq \frac{4s}{m}\|\hat{\boldsymbol{v}}_t\|^2. \tag{3.41}$$

Returning to the image-space representation of F and using the fact that the DFT is an isometry completes the proof of (3.27). □

Remark 3.10 The results in this chapter are largely unchanged if the image is allowed to be complex: the problem of locally inverting $\widetilde{\mathscr{M}}$ is well conditioned if and only if $T_f \mathbb{A}_{\boldsymbol{a}} \cap B_S = \{f + itf : t \in R\}$.

Remark 3.11 The ill-posedness of the phase retrieval problem, in the case that the intersections of $\mathbb{A}_{\boldsymbol{a}}$ and B_S are nontransversal, suggests that algorithms for approximating $\widetilde{\mathscr{M}}^{-1}$ should display considerable sensitivity to initial conditions as well as to noise, or other inaccuracies in the measured data.

Example 3.12 In this experiment we examine the ratios

$$\frac{\|\mathscr{M}(f) - \mathscr{M}(f + \delta f)\|}{\|\delta f\|^q} \quad \text{for } q = 1, 2.$$

For $\delta f \in T_f \mathbb{A}_{\boldsymbol{a}} \cap B_S$ we use $q = 2$, and for randomly selected δf we use $q = 1$. The reference image, f, is a 256×256 piecewise constant image, and $S = S_{f,4}$ is a 4-pixel neighborhood. Figure 3.1(a) shows a histogram of these ratios with $q = 2$ for 10,000 randomly selected perturbations in $T_f \mathbb{A}_{\boldsymbol{a}} \cap B_{S_{f,4}}$. We see that the ratios range between 4.4 and 9.4, indicating that $\mathscr{M}^{-1}$, restricted to perturbations of this type, is indeed an Hölder-1/2 map, as discussed in Remark 3.4. Figure 3.1(b) shows a similar histogram with $q = 1$ and 10,000 randomly selected perturbations with support in $S_{f,4}$. Note here that the ratios lie between 0.8 and 0.95. This is not a contradiction with the results shown in Figure 3.1(a) as $T_f \mathbb{A}_{\boldsymbol{a}} \cap B_{S_{f,4}}$ is a set of very high codimension in $B_{S_{f,4}}$. Because the dimension of $B_{S_{f,4}}$ is quite large, the phenomenon of concentration of measure makes it unlikely that a randomly selected perturbation

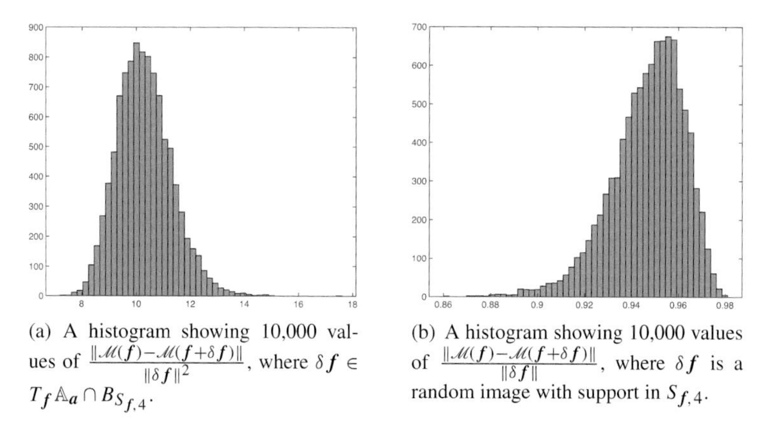

(a) A histogram showing 10,000 values of $\frac{\|\mathscr{M}(f)-\mathscr{M}(f+\delta f)\|}{\|\delta f\|^2}$, where $\delta f \in T_f \mathbb{A}_{\boldsymbol{a}} \cap B_{S_{f,4}}$.

(b) A histogram showing 10,000 values of $\frac{\|\mathscr{M}(f)-\mathscr{M}(f+\delta f)\|}{\|\delta f\|}$, where δf is a random image with support in $S_{f,4}$.

Figure 3.1 Histograms of $\frac{\|\mathscr{M}(f)-\mathscr{M}(f+\delta f)\|}{\|\delta f\|^q}$ for $q = 1, 2$.

will have a substantial projection into $T_f \mathbb{A}_a \cap B_{S_{f,4}}$. These examples illustrate the extent of the anisotropy in the continuity properties of $\widetilde{\mathcal{M}}^{-1}$ when the intersection is nontransversal. Empirically the directions where $\widetilde{\mathcal{M}}^{-1}$ is only Hölder-1/2 seem to play a central role in practical reconstruction problems, though there are circumstances where their influence is less evident. See Example 8.13.

3.2 Examples of Ill-Posedness

The well-posedness of an inverse problem is related, on the one hand, to the uniqueness of the solution and, on the other hand, to the continuity properties of the local inverse. In Section 3.1 we explored the relationship between the continuity properties of the local inverse to $\mathcal{M} \upharpoonright_{B_S}$ and the transversality of the intersection between an amplitude torus and the linear subspace defined by a support condition. As discussed in the Introduction, the solution to the phase retrieval problem, with support as the auxiliary condition is, at best, unique up to trivial associates. From the work of Bruck and Sodin and its elaboration by Hayes (see Bruck and Sodin 1979; Hayes 1982), it is also well understood that there is a subset of data for which the solution to the phase retrieval problem is genuinely nonunique. This failure of uniqueness leads inevitably to other forms of non-well-posedness, even for generic data for which the solution is unique, up to trivial associates.

In this section we explore various ways in which the phase retrieval problem can fail to have a unique solution to a given precision $\epsilon > 0$. Suppose that there are two images $\boldsymbol{f}_1, \boldsymbol{f}_2$, and a subset $S \subset J$, adequate for generic uniqueness, such that

(i) $\|\boldsymbol{f}_1\|_2 = \|\boldsymbol{f}_2\|_2$, but the minimum distance between trivial associates of $\boldsymbol{f}_1$ and $\boldsymbol{f}_2$ is much larger than $\epsilon\|\boldsymbol{f}_1\|_2$.
(ii) $S \supset \{\boldsymbol{j} : \epsilon\|\boldsymbol{f}_i\|_2 < |f_{i\boldsymbol{j}}|\}$ for $i = 1,2$.
(iii) $\|\mathcal{M}(\boldsymbol{f}_1) - \mathcal{M}(\boldsymbol{f}_2)\|_2 < \epsilon\|\boldsymbol{f}_1\|_2$.

We then say that the solution to the phase retrieval problem defined by the data $(\mathcal{M}(\boldsymbol{f}_1), S)$ is ϵ-nonunique. In the remainder of this section, we describe two different scenarios that lead to ϵ-nonuniqueness. A similar concept appears in Seldin and Fienup (1990).

3.2.1 Consequences of Genuine Nonuniqueness

The proof that a discrete image, with sufficiently small support, is generically determined by the magnitude DFT data is a consequence of the fact that polynomials in two or more variables are generically irreducible, see Appendix 1.A. If $(f_j : j \in J)$ is a d-dimensional image, then its Z-transform is given by

$$\boldsymbol{F}(z) = \sum_{j \in J} f_j z^{-j}, \tag{3.42}$$

where $z^{-j} = z_1^{-j_1} \dots z_d^{-j_d}$. There is a minimal integer vector $\boldsymbol{m} \in \mathbb{Z}^d$ so that $z^{\boldsymbol{m}} \boldsymbol{F}(z)$ is a polynomial. This polynomial is the Z-transform of the translate $\boldsymbol{f}^{(-\boldsymbol{m})}$.

Suppose that $\boldsymbol{f}$ is an image whose Z-transform, $\boldsymbol{F}(z)$ is reducible, in the sense that there are nonconstant polynomials, $\boldsymbol{F}_1, \boldsymbol{F}_2$ such that

$$\boldsymbol{F}(z) = z^{\boldsymbol{n}} \boldsymbol{F}_1(z) \boldsymbol{F}_2(z), \tag{3.43}$$

for some integer vector $\boldsymbol{n}$. If $\boldsymbol{f}_1$ and $\boldsymbol{f}_2$ are images with Z-transforms $\boldsymbol{F}_1$, $\boldsymbol{F}_2$, (up to a factor of $z^{\boldsymbol{m}_i}$ for some $\boldsymbol{m}_i$) then, there is a translation, $\boldsymbol{v}$, so that $\boldsymbol{f}^{(\boldsymbol{v})} = \boldsymbol{f}_1 * \boldsymbol{f}_2$, where $*$ denotes discrete convolution. If no trivial associate of either $\boldsymbol{f}_1$ or $\boldsymbol{f}_2$ is inversion symmetric, then the image $\boldsymbol{f}' = \boldsymbol{f}_1 * \check{\boldsymbol{f}}_2$ is not a trivial associate of $\boldsymbol{f}$ and, typically, the minimum distance between the trivial associates of $\boldsymbol{f}$ and $\boldsymbol{f}'$ is relatively large. If $\boldsymbol{f}_1$ and $\boldsymbol{f}_2$ are nonnegative, then so are $\boldsymbol{f}$ and $\boldsymbol{f}'$, and the smallest rectangles containing each image coincide.

Suppose that $\boldsymbol{f}$ and $\boldsymbol{f}'$ both have small support contained in a set S, which is small enough to generically imply uniqueness, up to trivial associates, and let $0 < \epsilon$ be much smaller than $\|\boldsymbol{f} - \boldsymbol{f}'\|_2$. Because uniqueness is generic we can modify these two images to obtain generic images $\boldsymbol{g}$ and $\boldsymbol{g}'$, that satisfy the estimates $\|\boldsymbol{g} - \boldsymbol{f}\|_2 < \epsilon/2$, $\|\boldsymbol{g}' - \boldsymbol{f}'\|_2 < \epsilon/2$, and whose supports satisfy $S_{\boldsymbol{g}} = S_{\boldsymbol{f}}, S_{\boldsymbol{g}'} = S_{\boldsymbol{f}'}$. The data $(\mathscr{M}(\boldsymbol{g}), S)$ defines both a phase retrieval problem with a unique solution, up to trivial associates, and an ϵ-nonunique problem: $\boldsymbol{g}$ and $\boldsymbol{g}'$ have support in S, and

(i) $\|\boldsymbol{g} - \boldsymbol{g}'\|_2 \geq \|\boldsymbol{f} - \boldsymbol{f}'\|_2 - \epsilon$.
(ii) $\|\mathscr{M}(\boldsymbol{g}) - \mathscr{M}(\boldsymbol{g}')\| \leq \epsilon$.

We conclude this section with a example of a pair of nonnegative images, $\boldsymbol{f}, \boldsymbol{f}'$ with exactly the *same* support and magnitude DFT data such that the distance between them, $\|\boldsymbol{f} - \boldsymbol{f}'\|_2 \approx 0.48\|\boldsymbol{f}\|_2$. The minimum distance between trivial associates is about $0.18\|\boldsymbol{f}\|_2$, but the closest trivial associates have rather different supports. These images are obtained as described above with $\boldsymbol{f}_1$ and $\boldsymbol{f}_2$ nonnegative images whose *supports* are inversion symmetric, but the images themselves are not. The left and middle images in Figure 3.2(a) show the pair of images, and the right image shows their common support. The left image in Figure 3.2(b) is the $\log_{10}$-magnitude DFT of $\boldsymbol{f}$. Finally, the right image is the difference of the two images, normalized by the maximum of $\boldsymbol{f}$.

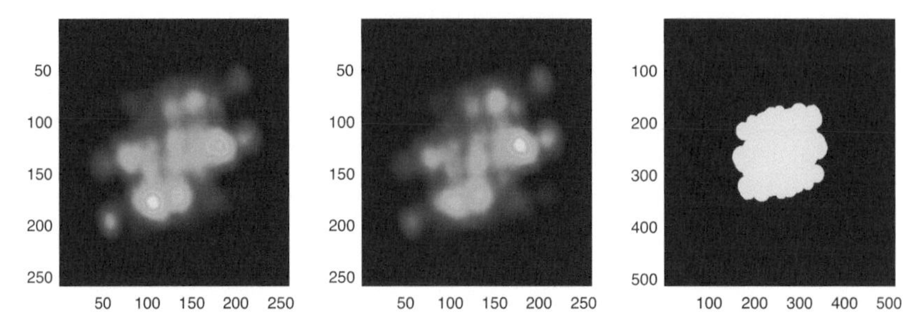

(a) The left and center images show the central 256×256 portion of a pair of 512×512 images with identical magnitude DFT data that are not trivial associates; the yellow region in the right image is their common support.

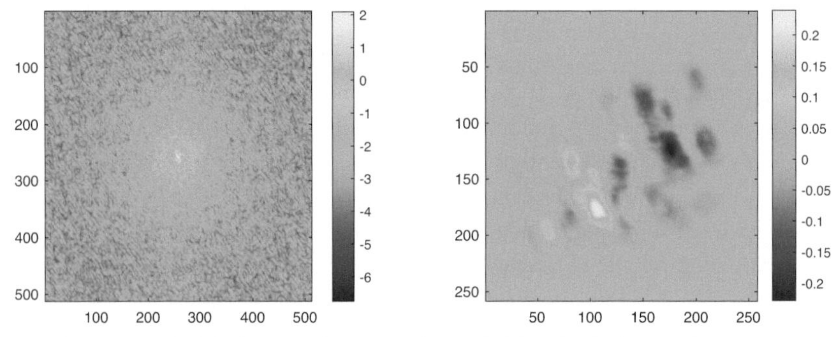

(b) The panel on the left shows the $\log_{10}$-magnitude DFT data for the images shown above; the right panel shows the normalized difference, $(f - f')/\|f\|_\infty$, between the images themselves.

Figure 3.2 An illustration of true nonuniqueness in the phase retrieval problem.

What is striking about this example is how perfectly ordinary the images and their magnitude DFT data look. The only criterion that we know of (in the continuum model) to exclude this phenomenon is that it cannot occur in an image with jump discontinuities, because the convolution of two bounded measurable functions is continuous. For the discrete model it is difficult to make this statement precise, as there are images $f = f_1 * f_2$, where, say, f_1 is a "sum of δ-functions," which provide counterexamples. Note that, for discrete images, a sum of δ-functions is modeled by an image with support in a set of isolated pixels. In fact, such examples of nonuniqueness can be found in Seldin and Fienup (1990).

As a practical matter it is very difficult to say if a given high degree polynomial in more than one variable is reducible, or "almost reducible." Ultimately the prevalence of true nonuniqueness reduces to the question in algebraic geometry of how dense the subset of reducible polynomials in

$1 < n$-variables, of degree d, is within the set of all n-variable polynomials of this degree, as d tends to infinity. There does not seem to be too much known about the answer to this question.

3.2.2 Microlocal Nonuniqueness

We now describe a second construction for building collections of distinct images whose magnitude DFT data and support are almost identical. This source of ϵ-nonuniqueness is, in some sense, a microlocal version of the translational symmetry exploited above. A similar phenomenon is explored in Alaifari et al. (2019). In this context the inversion symmetry, (1.13), also plays an important role.

Suppose that $\boldsymbol{f}$ is an image with the property that it can be decomposed as the sum of two images $\boldsymbol{f} = \boldsymbol{f}_1 + \boldsymbol{f}_2$ for which $S_{\widehat{\boldsymbol{f}}_1} \cap S_{\widehat{\boldsymbol{f}}_2} = \emptyset$. This cannot be done exactly, but the notion of ϵ-support, defined in (3.5), applies equally well to the DFT of an image:

$$S^{\epsilon}_{\widehat{\boldsymbol{f}}} = \{\boldsymbol{j} : |\widehat{f}_{\boldsymbol{j}}| > \epsilon\}. \tag{3.44}$$

As we see next, one can sometimes arrange to have $S^{\epsilon}_{\widehat{\boldsymbol{f}}_1} \cap S^{\epsilon}_{\widehat{\boldsymbol{f}}_2} = \emptyset$.

If we *could* arrange for $S_{\widehat{\boldsymbol{f}}_1} \cap S_{\widehat{\boldsymbol{f}}_2} = \emptyset$, then the collection of images

$$\begin{gathered} \pm \boldsymbol{f}_1 \pm \boldsymbol{f}_2^{(\boldsymbol{v})} \quad \text{for } \boldsymbol{v} \in J \\ \pm \check{\boldsymbol{f}}_1^{(\boldsymbol{v})} \pm \boldsymbol{f}_2^{(\boldsymbol{v}')}, \quad \pm \boldsymbol{f}_1^{(\boldsymbol{v})} \pm \check{\boldsymbol{f}}_2^{(\boldsymbol{v}')} \quad \text{for } \boldsymbol{v}, \boldsymbol{v}' \in J, \\ \pm \check{\boldsymbol{f}}_1^{(\boldsymbol{v})} \pm \check{\boldsymbol{f}}_2^{(\boldsymbol{v}')} \quad \text{for } \boldsymbol{v}, \boldsymbol{v}' \in J, \end{gathered} \tag{3.45}$$

would all have exactly the same magnitude DFT data. This is a much larger collection of images than the trivial associates of a single image and contains images that are quite different from one another. Of course, with $\epsilon = 0$, this is not strictly possible, however, if $S^{\epsilon}_{\widehat{\boldsymbol{f}}_1} \cap S^{\epsilon}_{\widehat{\boldsymbol{f}}_2} = \emptyset$, then the magnitude of any given DFT coefficient for an image, $\boldsymbol{f}'$, from the set described in (3.45), differs from that of $\boldsymbol{f}$ by at most ϵ. That is, for all $\boldsymbol{k} \in J$, $\big||\widehat{f}_{\boldsymbol{k}}| - |\widehat{f'_{\boldsymbol{k}}}|\big| < \epsilon$; we say that the magnitude DFT data agree to *precision* ϵ.

Let us, for example, define $\boldsymbol{f}' = \boldsymbol{f}_1 - \boldsymbol{f}_2$. Evidently, if ϵ is small enough we can make the difference $\||\widehat{\boldsymbol{f}}| - |\widehat{\boldsymbol{f}'}|\|_2$ as small as we like. In that case $\boldsymbol{f}'$ would have essentially the same DFT data and support as $\boldsymbol{f}$, but $\|\boldsymbol{f} - \boldsymbol{f}'\|_2 = 2\|\boldsymbol{f}_2\|_2$, could be comparable to $\|\boldsymbol{f}\|_2$.

More generally, suppose that f_0 is an object that satisfies a reasonable support condition, $S_{f_0} \subset S$, which can be decomposed as a sum of objects

$$f_0 = \boldsymbol{f}_1 + \cdots + \boldsymbol{f}_k,$$

with the following property:

$$\text{For each pair } 1 \leq p \neq q \leq k, S^{\epsilon}_{\hat{f}_p} \cap S^{\epsilon}_{\hat{f}_q} = \emptyset. \tag{3.46}$$

That is, to precision ϵ, the DFTs of the images $\{f_1, \ldots, f_k\}$ have disjoint supports. For an image of this type, the collection of images

$$f^{(\pm, V)} = \pm f_1^{(v_1)} \pm \cdots \pm f_k^{(v_k)}, \text{ with arbitrary signs and translations } V = (v_1, \ldots, v_k), \tag{3.47}$$

all have the same magnitude DFT data, to precision $k\epsilon$. It is also allowable to replace some (or all) of these components with their inversions $\{\check{f}_i\}$.

Suppose that the component images $\{f_1, \ldots, f_k\}$ also satisfy the support condition, to precision ϵ, that is

$$S^{\epsilon}_{f_j} \subset S \quad \text{for } j = 1, \ldots, k. \tag{3.48}$$

For a subset of small translations, $V \in \mathcal{V}_{\text{sm}}$, the objects $f^{\pm, V}$ satisfy the support condition specified by S, at least to precision $k\epsilon$. We call this *microlocal nonuniqueness* as it entails a decomposition essentially in the image-DFT domain. It is clear that the data $(\mathscr{M}(f_0), S)$ specified by such an image defines an $m\epsilon$ nonunique phase retrieval problem for some reasonable value of m.

A typical situation where this arises is that f_1 is a smooth, nonnegative function; the other components $\{f_2, \ldots, f_k\}$ are also smooth, and localized, but each oscillates at a different spatial frequency. These sorts of examples can be seen as a caricature of the internal components of a cell. This nonuniqueness phenomenon implies that, at a reasonable degree of precision, the magnitude DFT data and support information *do not* specify the relative locations of the component parts of these images.

In order for the DFT of an image with small (effective) support to also be well localized in the DFT domain, it seems to be necessary that the image itself be smooth. In the continuous domain, examples of this type are provided by Gaussian wave packets, $e^{-\sigma^2|x-y|^2}\cos(k \cdot x)$. Sampling these functions yields examples in the discrete setting. The degree of smoothness that is needed to achieve the conditions in (3.46) and (3.48) simultaneously, seems to depend on ϵ : a smaller ϵ requires smoother objects. The DFT of an object with sharp discontinuities decays slowly in k-space in essentially all directions. In light of the discussion in Section 1.4, the requirement for greater smoothness in the discrete image entails a finer grid spacing in the image domain, and therefore, the need for a larger discrete image.

Example 3.13 To illustrate the phenomenon of microlocal ϵ-nonuniqueness we construct an object $f = f_1 + f_2 + f_3 + f_4$, with a decomposition as

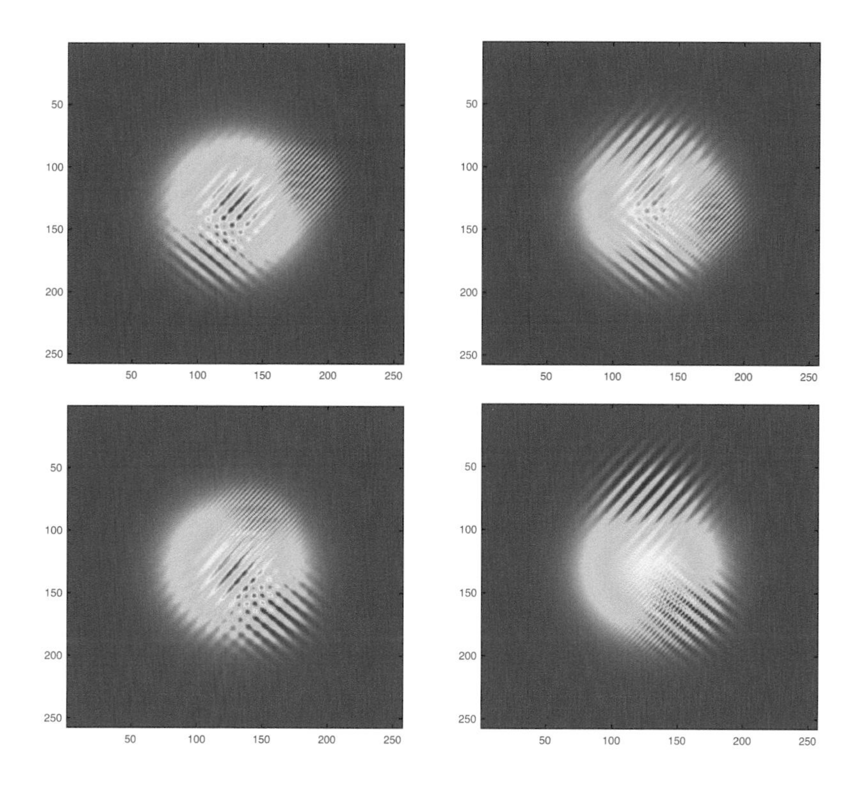

Figure 3.3 Plots of four different objects of the form $\boldsymbol{f}_1 + \boldsymbol{f}_2^{(\boldsymbol{v}_2^i)} + \boldsymbol{f}_3^{(\boldsymbol{v}_3^i)} + \boldsymbol{f}_4^{(\boldsymbol{v}_4^i)}$, for $i = 1, 2, 3, 4$. These plots show the central 256×256 portion of the 1024×1024 object used in these examples.

above. The different objects are samples of Gaussian wave packets:

$$f_{i\,\boldsymbol{j}} = e^{-\sigma_i^2 |\boldsymbol{j} - \boldsymbol{l}_i|^2} \cos(\langle \boldsymbol{k}_i, \boldsymbol{j} - \boldsymbol{l}_i \rangle), \quad i = 1, 2, 3, 4.$$

The parameters employed here are

$$\sigma_1 = 8, \qquad \sigma_2 = 20, \quad \sigma_3 = 30, \qquad \sigma_4 = 30, \tag{3.49}$$

$$\boldsymbol{k}_1 = (0, 0), \quad \boldsymbol{k}_2 = (60, 70), \quad \boldsymbol{k}_3 = (-70, 60), \quad \boldsymbol{k}_4 = (180, 200), \tag{3.50}$$

$$\boldsymbol{l}_1 = (0, 0), \qquad \boldsymbol{l}_2 = (0, 0), \quad \boldsymbol{l}_3 = (7.5, 3), \qquad \boldsymbol{l}_4 = (-4.5, 6). \tag{3.51}$$

The whole image is 1024×1024 pixels, and the spatial offsets are in pixels. The size of the image was dictated, in part, by our desire for the image to have four components with DFTs having ϵ-disjoint supports, and for (3.48) to hold with $\epsilon = 10^{-12}$.

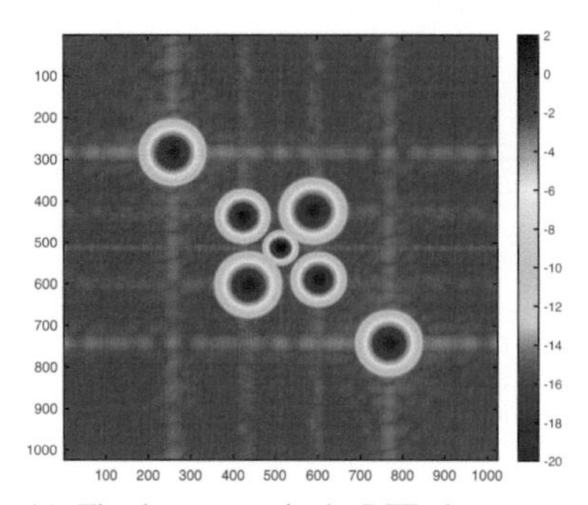

(a) The $\log_{10}$-magnitude DFT data, centered on $\boldsymbol{k} = (0,0)$, for the images shown in Figure 3.3.

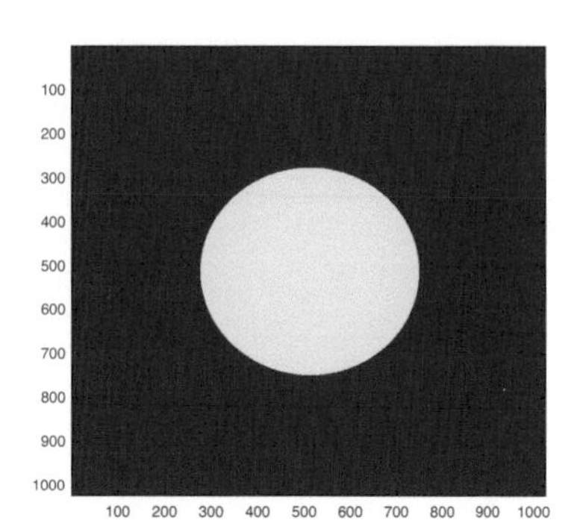

(b) A common support mask, S_{tot}, for the images shown in Figure 3.3.

Figure 3.4 The DFT data, which agrees to 15 digits, and a common support mask, S_{tot}, for the images in Figure 3.3.

In Figure 3.3 we show four objects built out of translates of these components, that is, of the form

$$\boldsymbol{f}_1 + \boldsymbol{f}_2^{(\boldsymbol{v}_2^i)} + \boldsymbol{f}_3^{(\boldsymbol{v}_3^i)} + \boldsymbol{f}_4^{(\boldsymbol{v}_4^i)}, \tag{3.52}$$

for four choices, $\{(\boldsymbol{v}_2^i, \boldsymbol{v}_3^i, \boldsymbol{v}_4^i) : i = 1,2,3,4\}$, of triplets of translation vectors. The DFT data for all of these objects agrees to 15 decimal places, as is shown in Figure 3.4(a). We define the supports as the sets

$$S_i = \left\{ \boldsymbol{j} : \left| \boldsymbol{f}_{1\boldsymbol{j}} + \boldsymbol{f}_{2\boldsymbol{j}}^{(\boldsymbol{v}_2^i)} + \boldsymbol{f}_{3\boldsymbol{j}}^{(\boldsymbol{v}_3^i)} + \boldsymbol{f}_{4\boldsymbol{j}}^{(\boldsymbol{v}_4^i)} \right| > 10^{-12} \right\} \quad i = 1,2,3,4.$$

For the four images in Figure 3.3 these sets differ by less than 220 pixels out of a total of 1024 $\times$ 1024. In Figure 3.4(b) we show the single set

$$S_{\text{tot}} = S_1 \cup S_2 \cup S_3 \cup S_4, \tag{3.53}$$

which contains the supports, at this precision, of all four images. It fits into a square with side lengths equal to 475 pixels within a field of view that is 1024 pixels on a side.

As noted, the magnitude DFT data of all of the possible combinations of translates and inversions of these 4 components, as in (3.45), agree to 15 decimal places! Small translations have their supports in essentially the same set as the support of $\boldsymbol{f}$ itself. This indicates that, as a practical matter, the phase retrieval problem is not solvable for objects of this type, even with very precise support information, in that the available data does not suffice, in finite precision arithmetic, to determine the relative positions of the objects that comprise f.

This example and the foregoing discussion show that there are several classes of images for which the solution to the phase retrieval problem is, to any desired degree of accuracy, not unique, even up to trivial associates. An important aspect of microlocal nonuniqueness is that it can usually be predicted from the magnitude DFT data itself: this phenomenon can occur when, at precision ϵ, the support of $|\widehat{f}|$ splits into a collection of well-separated sets, corresponding to the ϵ-supports of the $\{\widehat{f}_j\}$. Since the components themselves, $\{f_j\}$, must have small supports in the image domain, their DFTs must also be smooth, i.e., they are well localized and tend to 0 smoothly. As a consequence of the smoothness of the component parts, and therefore the overall image, this failure of uniqueness only occurs when the linear subspace defined by the support condition and magnitude torus meet at very small angles over a large dimensional subspace. This renders the phase retrieval problem, even at exact intersection points, locally ill-conditioned. In the nonmicrolocal setting, ϵ-nonuniqueness is not easily detectable from the magnitude DFT data itself.

The larger message of this section is that there *exist* nonpathological, formally "generic" data for which a practical statement of the discrete, classical, phase retrieval problem does *not*, in any useful sense, have a unique solution. As yet, there are no practical criteria to decide whether or not a particular set of measurements defines an ϵ-nonunique phase retrieval problem, for a reasonable value of ϵ. It is unclear how prevalent these phenomena are. In the many experiments that we have performed to date, which use samples of random sums of radial functions, as described in Section 1.4, we have not encountered examples where the phase retrieval problem seemed to be ϵ-nonunique, for a small value of ϵ. This may provide support for the widespread belief that, in principle, the phase retrieval problem is "usually" uniquely solvable, at least to within the accuracy of the available measurements. Whether it is practically solvable to this accuracy is a different question.

In Section 3.1 we have shown that the underlying abstract inverse problem often fails to be locally well conditioned because the intersections of the magnitude torus and the linear subspace defined by a support constraint are often nontransversal. In Part II we examine a variety of algorithms for approximating $\widetilde{\mathscr{M}}^{-1}$. All preexisting approaches to this problem begin by randomly selecting phases and iterating a map. These algorithms display enormous instability in their initial global phases as they search for an attracting basin. Some of these basins are defined by transversal intersections, but these do not appear to be especially attractive. For any particular initial guess, the basin in which an algorithm terminates seems not to be predictable in any useful way. Beyond that, small, but nonzero, angles between B_S and $T_{f_0}\mathbb{A}_{\boldsymbol{a}}$ lead these algorithms to stagnate or converge extremely slowly. In Part III we discuss approaches to ameliorate both the local ill conditioning, discussed in Section 3.1, and global nonuniqueness, discussed in Section 3.2, by altering the experimental protocol used to collect the data.

4

Uniqueness and the Nonnegativity Constraint

In X-ray coherent diffraction imaging (CDI) the diffraction pattern is produced essentially by the deviation, Δn, of an object's refractive index from that of the vacuum. We write $\Delta n = (\beta + i\gamma)$, where β quantifies elastic scattering of photons, and γ quantifies absorption (see Jacobsen, 2019, §3.3). In the soft X-ray range Δn lies in the positive orthant of the complex plane, and it can often be assumed that $\gamma \ll \beta$. In this case the measurement in the CDI experiment is well approximated as $C_d|\widehat{\beta}(\boldsymbol{k})|^2$, that is, a constant times the modulus squared of the discrete Fourier transform (DFT) of a nonnegative function. For this reason, one is sometimes justified in using as auxiliary information the assertion that the unknown object is described by a real-valued, nonnegative function. In astronomical applications of phase retrieval this assumption is also commonly used.

In this chapter we turn our attention to the study of the nonnegativity constraint in the discrete approximation. Let

$$B_+ = \{\boldsymbol{f} \in \mathbb{R}^J : f_{\boldsymbol{j}} \geq 0 \quad \text{for all } \boldsymbol{j} \in J\} \tag{4.1}$$

denote the orthant in $\mathbb{R}^J$ consisting of nonnegative images. We suppose that $\mathbb{A}_{\boldsymbol{a}}$ is the magnitude torus defined by the magnitudes of the DFT coefficients of a *nonnegative* image $\boldsymbol{f}_0$. The first question to consider is whether the magnitude DFT data along with the nonnegativity constraint alone define an image, up to trivial associates. It is immediately clear that it cannot. If $\boldsymbol{f}_0$ is a strictly positive image, then any sufficiently small perturbation of the phases of $\widehat{\boldsymbol{f}}_0$, with the appropriate symmetries, is the DFT data of another strictly positive image: Suppose that $\widehat{\boldsymbol{f}}_{\boldsymbol{k}} = a_{\boldsymbol{k}} e^{i\theta_{\boldsymbol{k}}}$, $\boldsymbol{k} \in J$, is the DFT data of a strictly positive image. Let $\{e^{i\phi_{\boldsymbol{k}}} : \boldsymbol{k} \in J\}$ be a collection of phases that satisfy the symmetries for the DFT of a real image: $e^{i\phi_{\boldsymbol{k}'}} = e^{-i\phi_{\boldsymbol{k}}}$, where $\boldsymbol{k}'$ is the conjugate index, see (2.17)–(2.18). Since the inverse DFT is a continuous map, there is a $\delta_0 > 0$, so that if $0 < \delta < \delta_0$, then $(a_{\boldsymbol{k}} e^{i(\theta_{\boldsymbol{k}} + \delta\phi_{\boldsymbol{k}})} : \boldsymbol{k} \in J)$ is the DFT data of another strictly positive image. That is, $\mathbb{A}_{\boldsymbol{a}} \cap B_+$ contains an open neighborhood of f_0 in $\mathbb{A}_{\boldsymbol{a}}$. Thus, some further information about the image, beyond nonnegativity, is required for the magnitude DFT data to generically determine an image, up to trivial associates.

The autocorrelation image $f_0 \star f_0$, defined in (1.21), is uniquely determined by the magnitude DFT data, and hence is the same for all points in $\mathbb{A}_{\boldsymbol{a}}$. It is, therefore, something that can, in principle, be determined directly from the measurements. If an image, f_0, is nonnegative, and the support of the autocorrelation image is sufficiently small (compared to the side lengths of the rectangular index set J), then the support of $f_0 \star f_0$ gives an upper bound for the support of f_0. This, in turn, allows us to use the fact that $f_0 \in B_+$, along with a knowledge of $S_{f_0 \star f_0}$, to prove that the points in the intersection $\mathbb{A}_{\boldsymbol{a}} \cap B_+$ are generically unique, up to trivial associates.

It should be noted, that contrary to the continuum case (the Titchmarsh Convolution Theorem), the support of $f \star f$ does not generally give a useful estimate on the support f. The support of $f \star f$ is always a subset of $S_f \ominus S_f$, see (4.6), but it is possible for $S_{f \star f}$ to be quite small, with $S_f = J$. Let $\mathbb{A}_{\boldsymbol{a}}$ be the magnitude torus defined by an image, f_0, with small support, for which $f_0 \star f_0$ also has small support. There is an open dense subset $U \subset \mathbb{A}_{\boldsymbol{a}}$ with the property that if $f \in U$, then $S_f = J$. On the other hand, $f \star f = f_0 \star f_0$, and therefore, $S_{f \star f} = S_{f_0 \star f_0}$. Hence, the a priori knowledge that f_0 is nonnegative is essential to *derive* a bound on S_{f_0} from a knowledge of $S_{f_0 \star f_0}$.

Our analysis in Chapters 2 and 3 suggests that the difficulty of the problem of finding points in $\mathbb{A}_{\boldsymbol{a}} \cap B_+$ should be directly related to the "transversality" of this intersection. In the present circumstance, the image f_0 is nonnegative with many entries equal to zero, and it therefore lies in ∂B_+. The boundary of B_+ is not smooth, and hence does not have a tangent bundle, which complicates the concept of transversality. It is instead a stratified space, where the strata of the boundary are defined by the subset of coordinates that vanish: The strata of ∂B_+ are orthants in subspaces of $\mathbb{R}^J$: if $L \subset J$, then the set

$$\partial B_+^L = \{f : f_{\boldsymbol{j}} = 0 \quad \text{for } \boldsymbol{j} \in L, \text{ and } f_{\boldsymbol{j}} > 0 \quad \text{for } \boldsymbol{j} \in J \setminus L\}, \tag{4.2}$$

is a stratum of ∂B_+. For each point $f \in \partial B_+^L$, there is an $\epsilon > 0$ and an $r > 0$, so that

$$\begin{aligned} B_+ \cap B_r(f) = \{\boldsymbol{x} \in \mathbb{R}^J : x_{\boldsymbol{j}} \geq 0 \quad &\text{for } \boldsymbol{j} \in L \\ \text{and } -\epsilon < f_{\boldsymbol{j}} - x_{\boldsymbol{j}} < \epsilon \quad &\text{for } \boldsymbol{j} \in J \setminus L\}. \end{aligned} \tag{4.3}$$

That is, a local model for B_+ near to a point $f \in \partial B_+^L$ is $[0,1)^{|L|} \times (-\epsilon, \epsilon)^{|J|-|L|}$.

For the magnitude DFT data along with the nonnegativity hypothesis to generically determine the unknown image, up to trivial associates, it is necessary that $\mathbb{A}_{\boldsymbol{a}} \cap B_+ \subset \partial B_+$. As ∂B_+ is piecewise linear, the intersection $T_f \mathbb{A}_{\boldsymbol{a}} \cap \partial B_+$ is a union of subsets lying in orthants. In Section 4.3 we analyze these intersections, showing that they are convex, conic subsets of orthants. The dimensions and sizes of these components therefore provide a good measure of the failure of transversality in this case. In generic cases these intersections are trivial. Nonetheless, "small angles" between $T_{f_0} \mathbb{A}_{\boldsymbol{a}}$ and ∂B_+ can again produce slow convergence or even stagnation.

4.1 Support and the Autocorrelation Image

For a real, d-dimensional, image f the periodic autocorrelation image is defined to be

$$[f \star f]_j = \sum_{l \in J} f_l f_{l-j} = \sum_{l \in J} f_l f_{l+j}, \tag{4.4}$$

where f is extended J-periodically to $\mathbb{Z}^d$. An elementary calculation shows that the DFT (as defined in (1.7)) of $f \star f$ is related to that of f by

$$\widehat{f \star f}_j = |\widehat{f}_j|^2. \tag{4.5}$$

Hence, $f \star f$ is the same image for all $f \in \mathbb{A}_{\boldsymbol{a}}$. Our uniqueness result for nonnegative images in $\mathbb{A}_{\boldsymbol{a}}$ requires a hypothesis about the support of $f \star f$, which is, in principle, verifiable from the magnitude DFT measurements themselves.

As noted above, for generic $\boldsymbol{a}$ and generic points $f \in \mathbb{A}_{\boldsymbol{a}}$ it is clear that $S_f = J$. As $f \star f = f_0 \star f_0$ for every $f \in \mathbb{A}_{\boldsymbol{a}}$ it is evident that, in general, the support of the *periodic* autocorrelation image gives *no* information about the support of f. This is in sharp contrast to the continuum case. In 1D, the Titchmarsh Convolution Theorem states that if I_f and I_g are the smallest intervals containing $\operatorname{supp} f$ and $\operatorname{supp} g$ respectively, then $I_f + I_g$ is the smallest interval containing $\operatorname{supp} f * g$. In higher dimensions a similar theorem holds relating the convex hulls of $\operatorname{supp} f$, $\operatorname{supp} g$, and $\operatorname{supp} f * g$.

We begin our analysis by demonstrating that if f_0 is nonnegative and the support of the autocorrelation image $f_0 \star f_0$, is a sufficiently small subset of J, then the support of $f_0 \star f_0$ bounds the support of f_0. The J-periodicity of the images under consideration makes this statement somewhat delicate to prove. For simplicity we restrict our analysis to the case that

$$J = [0 : N] \times \cdots \times [0 : N].$$

From (4.4) we can relate $S_{f\star f}$ to S_f by observing that

$$S_{f\star f} \subset S_f \ominus S_f = \{\boldsymbol{j} - \boldsymbol{l} : \boldsymbol{j}, \boldsymbol{l} \in S_f\}. \tag{4.6}$$

This must be understood in the J-periodic sense. Equation (4.4) implies that $\boldsymbol{0} \in S_{f\star f}$ and that if $\boldsymbol{j} \in S_{f\star f}$, then so is $-\boldsymbol{j}$. In light of this it is more natural to think of $S_{f\star f}$ as a subset of the centrally symmetric index set

$$J^{\text{sym}} = \left[-\left\lfloor \frac{N}{2} \right\rfloor : \left\lfloor \frac{N+1}{2} \right\rfloor\right]^d,$$

than as a subset of J.

If f is nonnegative, then the sums defining $f \star f$ are sums of nonnegative terms. If an index $\boldsymbol{j} \in S_f \ominus S_f$, then there are indices $\boldsymbol{j}_1, \boldsymbol{j}_2 \in S_f$ so that $\boldsymbol{j} = \boldsymbol{j}_1 - \boldsymbol{j}_2$, and thus the term $f_{\boldsymbol{j}_1} f_{\boldsymbol{j}_2} > 0$ appears in the sum defining $[f \star f]_{\boldsymbol{j}}$, which is, therefore, positive. Hence, for a nonnegative image, we have the opposite inclusion as well, so that

$$S_{f\star f} = S_f \ominus S_f. \tag{4.7}$$

If f is a nonnegative image with $S_f = \left[0 : \left\lfloor \frac{N+1}{2} \right\rfloor\right]^d$, then $S_{f\star f}$ is essentially all of J^{sym}, which indicates why we need to assume that $S_{f\star f}$ is a relatively small subset of J^{sym} in order to get a nontrivial bound on S_f.

For an integer vector $\boldsymbol{l} \in \mathbb{Z}^d$ recall that the translated image $f_{\boldsymbol{j}}^{(\boldsymbol{l})} \overset{d}{=} f_{\boldsymbol{j}-\boldsymbol{l}}$. It is a simple observation that $f \star f = f^{(\boldsymbol{l})} \star f^{(\boldsymbol{l})}$ Hence, when trying to estimate the size of the support of an image in terms of $S_{f\star f}$, we are free to replace f by any of its translates.

We begin with the one-dimensional case. Let f be a nonnegative sequence indexed by $J = [0 : N]$, which we regard as $(N+1)$-periodic.

Lemma 4.1 *Suppose the* $J = [0 : 4M + q]$, *for* $q \in \{-1, 0, 1, 2\}$, *and* f *is a nonnegative sequence indexed by* J, *which we regard as a* $(4M + q + 1)$-*periodic sequence. If* $0 \leq p < M$, *and we know that*

$$S_{f\star f} = S_f \ominus S_f \subset [-(M + p) : (M + p)] \mod 4M + q, \tag{4.8}$$

then there is a translate, $\widetilde{f}$ *of* f *with*

$$S_{\widetilde{f}} \subset [0 : 2(M + p)]. \tag{4.9}$$

If $p < \frac{M+1+q}{3}$ *then, in fact, there is a translate* $\widetilde{f}$ *with*

$$S_{\widetilde{f}} \subset [0 : (M + p)]. \tag{4.10}$$

Remark 4.2 In one-dimension this lemma does not imply that a signal satisfying the hypotheses of the lemma is generically uniquely determined by its magnitude DFT data. It shows only that such a signal also satisfies a support constraint, which in one-dimension, does not suffice for generic uniqueness. This case is analyzed in Beinert (2017).

Proof The proof is essentially the same in all four cases ($q = -1, 0, 1, 2$). We give the proof for the case $q = -1$. Without loss of generality, we can assume that $0 \in S_f$, and represent S_f as a subset of $[0 : 4M - 1]$. This means that $j \in S_f$ is also in $S_{f \star f}$, which is equivalent to the inequalities

$$-(M + p) \leq j \leq (M + p) \mod 4M. \tag{4.11}$$

This holds if and only if either

$$0 \leq j \leq M + p \text{ or } -(M + p) \leq j - 4M < 0. \tag{4.12}$$

These inequalities imply that $S_f \cap [0 : 4M - 1] \subset [0 : M + p] \cup [3M - p : 4M - 1]$. Taking account of the $4M$-periodicity of $\boldsymbol{f}$, and the fact that $p < M$, this implies that there is a translate, $\widetilde{\boldsymbol{f}}$ of $\boldsymbol{f}$ with $S_{\widetilde{f}} \subset [0 : 2(M + p)]$, verifying the first claim of the lemma.

To prove the second claim, we now assume that $p < \frac{M}{3}$. We can assume that $0 \in S_{\widetilde{f}}$, and therefore the hypotheses of the lemma do not exclude any part of the interval $[0 : M + p]$ from belonging to $S_{\widetilde{f}}$. On the other hand the interval $[M + p + 1 : 2(M + p)]$ is equivalent $\mod 4M$ to the interval $[p + 1 - 3M : 2p - 2M]$. If $p < \frac{M}{3}$, then this interval is disjoint from $[-(M + p) : -1]$ and therefore $S_{\widetilde{f}} \subset [0 : M + p]$. □

The higher dimensional case is essentially identical, since each index can be translated independently of the others.

Lemma 4.3 *Suppose that* $J = [0 : 4M + q]^d$, *for* $q \in \{-1, 0, 1, 2\}$, *and* $\boldsymbol{f}$ *is a nonnegative image indexed by* J, *which we regard as* $(4M + q + 1)$*-periodic in each index. If* $0 \leq p < M$, *and we know that*

$$S_{f \star f} = S_f \ominus S_f \subset [-(M + p) : (M + p)]^d \mod 4M + q, \tag{4.13}$$

then there is a translate, $\widetilde{\boldsymbol{f}}$ *of* $\boldsymbol{f}$ *with*

$$S_{\widetilde{f}} \subset [0 : 2(M + p)]^d. \tag{4.14}$$

If $p < \frac{M+1+q}{3}$ *then, in fact, there is a translate* $\widetilde{\boldsymbol{f}}$ *with*

$$S_{\widetilde{f}} \subset [0 : (M + p)]^d. \tag{4.15}$$

Proof The proof is much like the 1D-case: we can assume that S_f is represented by a subset of J. Again, we only give the details for the $q = -1$ case.

While it may not be the case that $\mathbf{0} \in S_f$, we can replace f by a translate for which S_f intersects every coordinate hyperplane. That is, for each $1 \leq i \leq d$, there is a element $\boldsymbol{j} = (j_1, \ldots, j_d) \in S_f$ such that $j_i = 0$. The hypothesis in (4.13) then implies that if $\boldsymbol{j}, \boldsymbol{l} \in S_f$, then for $1 \leq i \leq d$,

$$-(M + p) \leq j_i - l_i \leq M + p \mod 4M. \tag{4.16}$$

For each i we take for $\boldsymbol{l}$ an index in S_f with $l_i = 0$; for this choice this inequality is equivalent to either

$$0 \leq j_i \leq M + p \text{ or } -(M + p) \leq j_i - 4M < 0. \tag{4.17}$$

From this point we argue as before, to conclude that there is a translate $\widetilde{f}$ so that $S_{\widetilde{f}}$ satisfies (4.14). This is possible as we can translate independently in each index. If $p < \frac{M+q+1}{3}$, then as before, we conclude that $\widetilde{f}$ actually satisfies (4.15). □

Remark 4.4 In Fienup et al. (1982, 1990), Fienup, Crimmins, Holsztynski, and Thelen show how, in certain circumstances, the support of the autocorrelation function can be used to obtain more precise information about the support of an image.

4.2 Uniqueness for Nonnegative Images

The relationship between $S_{f\star f}$ and S_f for nonnegative images leads easily to the following uniqueness result:

Theorem 4.5 *Let f_0 be a nonnegative image indexed by $J \subset \mathbb{Z}^d$, where $J = [0 : 4M + q]^d$, for $q \in \{-1, 0, 1, 2\}$. Let $\mathbb{A}_{\boldsymbol{a}}$ be the magnitude torus defined by f_0, and suppose that support of the autocorrelation image satisfies*

$$S_{f_0 \star f_0} \subset [-(M + p) : (M + p)]^d \mod 4M + q, \tag{4.18}$$

where $p < \frac{M+q}{3}$. For a Zariski dense subset of f_0, these conditions imply that the only nonnegative images in $\mathbb{A}_{\boldsymbol{a}}$ are trivial associates of f_0.

Remark 4.6 It is important to note that, the condition in (4.18) is verifiable from the data $\{|\widehat{f}_{\boldsymbol{j}}|^2 : \boldsymbol{j} \in J\}$. This result and a little algebra show that for nonnegativity to imply uniqueness one must triple "oversample" in each dimension (given the actual support of the image), while the support condition itself requires only double "oversampling."

Proof Lemma 4.3 implies that if $\boldsymbol{f} \in \mathbb{A}_{\boldsymbol{a}}$ is nonnegative, then, after translation we can assume that

$$S_f \subset [0 : (M + p)]^d. \tag{4.19}$$

The Z-transform of the image $\boldsymbol{f}$ is given by

$$X(z) = \sum_{j \in J} f_j z_1^{-j_1} \dots z_d^{-j_d}. \tag{4.20}$$

For a Zariski dense subset of $\boldsymbol{f}$, the Z-transform, $X(z)$, is an irreducible polynomial in the reciprocal variables, $(z_1^{-1}, \dots, z_d^{-1})$, of degrees $(M + p, \dots, M + p)$ times a monomial $z_1^{m_1} \dots z_d^{m_d}$. Since $p < (M + q)/3 < M$, the standard uniqueness result from Hayes (1982) applies to show that the nonnegative points on $\mathbb{A}_{\boldsymbol{a}}$ are unique up to trivial associates. □

Remark 4.7 Here we have implicitly used the fact $\mathbb{R}_+^N$ is a Zariski dense subset of $\mathbb{C}^N$, for any N.

If we have information about the support of $\boldsymbol{f}_0$, then the support of the autocorrelation image can be used to find the "smallest" rectangle containing a translate of $\boldsymbol{f}_0$ (see Fienup et al. 1982, 1990).

Proposition 4.8 *Suppose that $\boldsymbol{f}_0$ is a nonnegative image indexed by $J = [0 : 4N - 1]^d$. If a translate of $\boldsymbol{f}_0$ is known to have support in the rectangle $R_0 = [0 : 2N - 1]^d$, and the autocorrelation image has support in*

$$R_a = [-(N + p_1) : N + p_1] \times \dots \times [-(N + p_d) : N + p_d], \tag{4.21}$$

for integers $0 \le p_j \le N - 1$, then a translate of $\boldsymbol{f}_0$ is supported in

$$R_1 = [0 : N + p_1] \times \dots \times [0 : N + p_d]. \tag{4.22}$$

Remark 4.9 Given that the support of $\boldsymbol{f}_0$ is contained in R_0, this result says that the "width" of the support of $S_{\boldsymbol{f}_0 \star \boldsymbol{f}_0}$ is twice the width of the support of $\boldsymbol{f}_0$ itself. Similar results hold with $4N - 1$ replaced by $4N + q$ for $q = 0, 1, 2$.

Proof The proof of this proposition is similar to, but easier than that of Lemma 4.3. We can assume that $\boldsymbol{f}_0$ is replaced by a translate (also denoted $\boldsymbol{f}_0$), with $S_{\boldsymbol{f}_0} \subset R_0$, and for each $1 \le i \le d$, there is an index $\boldsymbol{j}^0 \in S_{\boldsymbol{f}_0}$ so that $j_i^0 = 0$. If $\boldsymbol{l}$ is any other point in $S_{\boldsymbol{f}_0}$, then $0 \le l_j \le 2N - 1$, for $1 \le j \le d$, and

$$0 \le l_i - j_i^0 = l_i \le N + p_i. \tag{4.23}$$

Because $0 \leq p_i \leq N-1$, we do not need to consider the intervals $[-(N+p_i) : -1]$. As such an inequality holds for each $1 \leq i \leq d$, this proves the proposition. □

Remark 4.10 These considerations can be extended to complex images, with the autocorrelation image defined as

$$[f \star f]_j = \sum_{l \in J} f_l \overline{f_{j-l}}. \tag{4.24}$$

As in the real case, the DFT of $f \star f$ is $(|\widehat{f}_k|^2 : k \in J)$. The support of $f \star f$ always satisfies:

$$S_{f \star f} \subset S_f \ominus S_f, \tag{4.25}$$

and generally speaking one cannot infer a bound on S_f from a knowledge of $S_{f\star f}$. The complex analogue of a nonnegative image is an image f that satisfies an estimate either of the form

$$\theta_0 \leq \arg f_j < \theta_0 + \frac{\pi}{2} \quad \text{for all } j \in J, \tag{4.26}$$

or of the form

$$\theta_0 < \arg f_j \leq \theta_0 + \frac{\pi}{2} \quad \text{for all } j \in J. \tag{4.27}$$

Note that the strict inequality on one side is necessary. With either of these hypotheses, the nonzero terms appearing in the sum defining $f \star f$ satisfy

$$-\frac{\pi}{2} < \arg(f_l \overline{f_{j-l}}) < \frac{\pi}{2}. \tag{4.28}$$

That is, all the nonzero terms appearing in these sums lie strictly in the right half plane, and therefore cannot sum to zero. With either hypothesis we can again conclude that $S_{f\star f} = S_f \ominus S_f$. Using this observation, the uniqueness results proved above can easily be extended to complex images satisfying either (4.26) or (4.27). We leave the details to the interested reader. An adequate constraint for phase retrieval with complex images is to assume an a priori bound on $S_{f\star f}$, and that the coordinates of the image satisfy either (4.26) or (4.27).

These results show that nonnegativity, in conjunction with control on the support of the autocorrelation image, gives generic uniqueness in the phase retrieval problem. As remarked above, for such an image, the intersection of the magnitude torus $\mathbb{A}_a$ with B_+ lies in ∂B_+. In the next section we study the ℓ_1-norm as a function on the tangent space to the magnitude torus at a nonnegative image. This leads to a computational method for determining

$T_{f_0}\mathbb{A}_a \cap \partial B_+$. The size of this intersection is a measure of the extent of the failure of transversality in this case.

4.3 Nonnegative Images and the 1-Norm

In this section we analyze the ℓ_1-norm as a function on the magnitude torus $\mathbb{A}_a$ defined by a nonnegative image, f_0, and as a function on the fiber of the tangent bundle at f_0, $T_{f_0}\mathbb{A}_a$. The magnitude torus is defined by the conditions

$$|\widehat{f}_k| = |\widehat{f}_{0k}| \stackrel{d}{=} a_k \quad \text{for all } k \in J. \tag{4.29}$$

The zero-DFT coefficient is given by

$$\widehat{f}_0 = \sum_{j \in J} f_j; \tag{4.30}$$

for a nonnegative image, f, this equals $\|f\|_1$. From this it is immediate that for any $f \in \mathbb{A}_a$ we have

$$|\widehat{f}_0| \leq \sum_{j \in J} |f_j| = \|f\|_1. \tag{4.31}$$

The inequality is strict unless the image is of a single sign, in which case

$$|\widehat{f}_0| = \|f\|_1, \tag{4.32}$$

and therefore, the ℓ_1-norm assumes its minimum value at points on $\mathbb{A}_a$ with all coordinates of a single sign, and, at such points, the global minimum is assumed.

This, therefore, provides a new constraint that is essentially dual to the nonnegativity constraint, which can be used as auxiliary information in the phase retrieval problem. For $f_0 \in \mathbb{A}_a \cap B_+$, set $r_1 = \|f_0\|_1$. We can then insist that $f \in \mathbb{A}_a$ satisfies the additional constraint

$$\|f\|_1 = r_1 \text{ that is } f \in \mathbb{A}_a \cap B^1_{r_1}, \tag{4.33}$$

where $B^1_{r_1} = \{f : \|f\|_1 \leq r_1\}$. A priori, this intersection may not consist entirely of trivial associates of a single such point, but all points in the intersection have coordinates of a single sign. With sufficient sampling in the DFT domain, Theorem 4.5 implies that, generically all points in the intersection are trivial associates of $\pm f_0$.

Minimizing the ℓ_1-norm also provides a new source of maps for algorithms to solve the phase retrieval problem for nonnegative images. In analyzing the ℓ_1-norm as a function on $T_{f_0}\mathbb{A}_a$ we discover the surprising fact that

$$T_{f_0}\mathbb{A}_a \cap \partial B_+ = T_{f_0}\mathbb{A}_a \cap \partial B^1_{r_1}, \tag{4.34}$$

which, in turn, leads to an effective method for computing this intersection.

Remark 4.11 It is tempting to try to use an ℓ_1-constraint as a proxy for small support, even for images without the nonnegativity constraint. Simple examples show that, without nonnegativity, the minimum of the ℓ_1-norm on the magnitude torus may *fail* to coincide with the images that have small support. Indeed, these images may not even be local minima. Work in 1D of Shechtman, Eldar, Szameit, Segev, et al. has shown that with sufficient *sparsity*, a condition much stronger than *small support*, this approach can be made to work (see Shechtman et al. 2011; Sidorenko et al. 2015). In 1D the phase retrieval problem with a support constraint does not have a unique solution, up to trivial associates. A main point of this work is that using sparsity one can often unambiguously solve the 1D phase retrieval problem. In 1D an image with small support has 50% of its pixels not equal to zero, whereas in Shechtman et al. (2011) their algorithm is successful on 1D images that have 5% of their pixels not equal to zero.

4.4 The 1-Norm on the Tangent Space

The analysis in Section 3.1 shows that it is very useful to understand the transversality properties of the intersections, $\mathbb{A}_a \cap B$, where B is a set defined by adequate auxiliary information. For the cases $B = B_+$ or $B^1_{r_1}$, analyzing the ℓ_1-norm on the tangent space, $T_{f_0}\mathbb{A}_a$, to the torus at f_0 provides considerable insight into this question. As before, we write

$$T_f\mathbb{A}_a = f + T^0_f\mathbb{A}_a, \text{ and } N_f\mathbb{A}_a = f + N^0_f\mathbb{A}_a, \tag{4.35}$$

where $T^0_f\mathbb{A}_a$ and $N^0_f\mathbb{A}_a$ are subspaces of $\mathbb{R}^J$.

Recall that if f is an image indexed by J, which is regarded as a J-periodic image indexed by $\mathbb{Z}^d$, then for $v \in \mathbb{Z}^d$ we define the translate of f by v to be

$$f^{(v)}_j \stackrel{d}{=} f_{j-v}. \tag{4.36}$$

In the proof of Lemma 3.8 we showed that if f_0 is an image such that

$$\widehat{f}_{0j} \neq 0 \quad \text{for any } j \in J, \tag{4.37}$$

then, for any $f \in \mathbb{A}_a$, the vectors,

$$\tau^{(v)} = f^{(v)} - f^{(-v)}, \quad v \in J, \tag{4.38}$$

contain a spanning set for the tangent space $T_f^0\mathbb{A}_a$, and the vectors

$$\boldsymbol{\nu}^{(v)} = \boldsymbol{f}^{(v)} + \boldsymbol{f}^{(-v)}, \quad \boldsymbol{v} \in J, \tag{4.39}$$

contain a spanning set for the normal space, $N_f^0\mathbb{A}_a$. This result does not require f_0 to be nonnegative. The condition in (4.37) is again generic.

For $\boldsymbol{f} \in \mathbb{A}_a$ let $J_f^t, J_f^n \subset J$, be chosen so that $\{\boldsymbol{\tau}^{(v)} : \boldsymbol{v} \in J_f^t\}$ is a basis for $T_f^0\mathbb{A}_a$, and $\{\boldsymbol{\nu}^{(v)} : \boldsymbol{v} \in J_f^n\}$ is a basis for $N_f^0\mathbb{A}_a$. We begin with a lemma about the ℓ_1-norm as a function on $T_{f_0}\mathbb{A}_a$. To that end we define the function $L : T_{f_0}\mathbb{A}_a \to [0,\infty)$ by

$$L(\boldsymbol{\epsilon}) = \left\| \boldsymbol{f}_0 + \sum_{l \in J_{f_0}^t} \epsilon_l \boldsymbol{\tau}^{(l)} \right\|_1, \tag{4.40}$$

here, $\boldsymbol{\epsilon} = (\epsilon_l : \boldsymbol{l} \in J_{f_0}^t)$.

Lemma 4.12 *Let $\boldsymbol{f}_0$ be a nonnegative image that satisfies* (4.37) *and let $\mathbb{A}_a$ be the magnitude torus it defines. As a function on $T_{f_0}\mathbb{A}_a$, the ℓ_1-norm assumes its minimum value at $\boldsymbol{f}_0$.*

Proof Let $\boldsymbol{\epsilon} = \{\epsilon_l : \boldsymbol{l} \in J_{f_0}^t\}$ be a collection of real numbers indexed by $J_{f_0}^t$. We need to show that the function,

$$L(\boldsymbol{\epsilon}) = \sum_{j \in J} \left| f_{0j} + \sum_{l \in J_{f_0}^t} \epsilon_l \tau_j^{(l)} \right|, \tag{4.41}$$

assumes its minimum value at $\boldsymbol{\epsilon} = \boldsymbol{0}$. Note that if $\boldsymbol{j} \in S_{f_0}$, then $f_{0j} > 0$, hence, there exists a $\delta > 0$, so that if $\|\boldsymbol{\epsilon}\|_1 < \delta$, then

$$0 \le f_{0j} + \sum_{l \in J_{f_0}^t} \epsilon_l \tau_j^{(l)} \quad \text{for all } \boldsymbol{j} \in S_{f_0}. \tag{4.42}$$

For such $\boldsymbol{\epsilon}$ we can rewrite $L(\boldsymbol{\epsilon})$ as

$$L(\boldsymbol{\epsilon}) = \sum_{j \in S_{f_0}} \left[f_{0j} + \sum_{l \in J_{f_0}^t} \epsilon_l \tau_j^{(l)} \right] + \sum_{j \in S_{f_0}^c} \left| \sum_{l \in J_{f_0}^t} \epsilon_l \tau_j^{(l)} \right|. \tag{4.43}$$

The proof of the lemma follows from the fact that

$$\sum_{j \in J} \sum_{l \in J_{f_0}^t} \epsilon_l \tau_j^{(l)} = \sum_{l \in J_{f_0}^t} \epsilon_l \sum_{j \in J} [f_j^{(l)} - f_j^{(-l)}] = 0. \tag{4.44}$$

The last equality follows as the image $\boldsymbol{f}$ is J-periodic, and therefore, for each $\boldsymbol{l}$,

$$\sum_{j \in J} f_j^{(l)} = \sum_{j \in J} f_j. \tag{4.45}$$

Hence, the sums over $\boldsymbol{j} \in J$ on the right-hand side of (4.44) vanish for every $\boldsymbol{l} \in J^t_{f_0}$.

From the triangle inequality and (4.43) it is clear that

$$L(\boldsymbol{\epsilon}) \geq \sum_{j \in S_{f_0}} \left[f_{0j} + \sum_{l \in J^t_{f_0}} \epsilon_l \tau_j^{(l)} \right] + \sum_{j \in S^c_{f_0}} \left[\sum_{l \in J^t_{f_0}} \epsilon_l \tau_j^{(l)} \right] = L(0), \tag{4.46}$$

where the last equality follows from (4.44). This shows that L has a local minimum at $\boldsymbol{\epsilon} = \boldsymbol{0}$. As L is a convex function a local minimum is necessarily a global minimum, which completes the proof of the lemma. □

The function $\|\boldsymbol{f}\|_1$ has strict local minima on $\mathbb{A}_{\boldsymbol{a}}$ at nonnegative images. At points where L also has a strict local minimum, the ℓ_1-norm satisfies an interesting lower bound. Such an estimate is only possible because the ℓ_1-norm is *not* a differentiable function.

Proposition 4.13 *Let $\boldsymbol{f}_0$ be a nonnegative image that satisfies the hypotheses of Lemma 4.12 and for which L has a strict local minimum at $\boldsymbol{\epsilon} = \boldsymbol{0}$. There is* $\mu > 0$, *and a non-empty, open neighborhood $U \subset \mathbb{A}_{\boldsymbol{a}}$ of $\boldsymbol{f}_0$ so that*

$$\|\boldsymbol{f}\|_1 \geq \|\boldsymbol{f}_0\|_1 + \mu \|\boldsymbol{f} - \boldsymbol{f}_0\|_1 \quad \textit{for } \boldsymbol{f} \in U. \tag{4.47}$$

Proof If we let $L_0(\boldsymbol{\epsilon}) = L(\boldsymbol{\epsilon}) - L(0)$, then equation 4.43 implies that there exists a $0 < \delta$ so that if $\|\boldsymbol{\epsilon}\|_1 \leq \delta$, then

$$L(\lambda \boldsymbol{\epsilon}) = L(0) + \lambda L_0(\boldsymbol{\epsilon}) \quad \text{for } \lambda \in [0,1]. \tag{4.48}$$

Since L has a strict local minimum at $\boldsymbol{0}$ it follows that there is a $0 < \eta$ so that

$$\min_{\{\boldsymbol{\epsilon} :\, \|\boldsymbol{\epsilon}\|_1 = \delta\}} L_0(\boldsymbol{\epsilon}) = \eta. \tag{4.49}$$

Combining this with (4.48) shows that, for $\{\boldsymbol{\epsilon} : \|\boldsymbol{\epsilon}\|_1 \leq \delta\}$, we have

$$L(\boldsymbol{\epsilon}) \geq L(0) + \frac{\|\boldsymbol{\epsilon}\|_1}{\delta} \eta. \tag{4.50}$$

Let

$$\boldsymbol{v}_{\epsilon} = \sum_{l \in J^t_{f_0}} \epsilon_l \boldsymbol{\tau}^{(l)}, \tag{4.51}$$

and $\pi_A : T_{f_0}\mathbb{A}_{\boldsymbol{a}} \to \mathbb{A}_{\boldsymbol{a}}$ be the nearest point map. For $0 < \delta$, let

$$V_\delta = \{\boldsymbol{v}_\epsilon : \|\boldsymbol{\epsilon}\|_1 < \delta\}, \tag{4.52}$$

and $U_\delta = \pi_A(V_\delta)$. Since the $\{\boldsymbol{v}_\epsilon\}$ are vectors tangent to $\mathbb{A}_{\boldsymbol{a}}$, there is a constant $0 < C$, so that the estimate

$$\|f_0 + \lambda \boldsymbol{v}_\epsilon - \pi_A(f_0 + \lambda \boldsymbol{v}_\epsilon)\|_1 \leq C\lambda^2 \tag{4.53}$$

holds uniformly for $\boldsymbol{v}_\epsilon \in V_\delta$ and $\lambda \in [0,1]$. The triangle inequality, (4.50), and (4.53) then imply that there is another constant $0 < C_1$ so that

$$\begin{aligned} \|\pi_A(f_0 + \lambda \boldsymbol{v}_\epsilon)\|_1 &\geq \|f_0 + \lambda \boldsymbol{v}_\epsilon\|_1 - \|f_0 + \lambda \boldsymbol{v}_\epsilon - \pi_A(f_0 + \lambda \boldsymbol{v}_\epsilon)\|_1 \\ &\geq \|f_0\|_1 + C_1\lambda\eta - C\lambda^2. \end{aligned} \tag{4.54}$$

This estimate implies that there are positive constants δ' and C_2 so that if $\boldsymbol{v}_\epsilon \in V_{\delta'}$, then

$$\|\pi_A(f_0 + \boldsymbol{v}_\epsilon)\|_1 \geq \|f_0\|_1 + C_2\|\boldsymbol{v}_\epsilon\|_1. \tag{4.55}$$

Since $U_{\delta'} = \pi_A(V_{\delta'})$ contains a non-empty, open neighborhood of f_0, and the ℓ_1-distance $\|\pi_A(f_0 + \boldsymbol{v}_\epsilon) - f_0\|_1$ is uniformly comparable to $\|\boldsymbol{v}_\epsilon\|_1$, this completes the proof of the proposition. □

4.5 Transversality of $\mathbb{A}_{\boldsymbol{a}} \cap \partial B_+$ and $\mathbb{A}_{\boldsymbol{a}} \cap \partial B^1_{r_1}$

Using the results from the previous section we can analyze the transversality of the intersections $\mathbb{A}_{\boldsymbol{a}} \cap \partial B_+$ and $\mathbb{A}_{\boldsymbol{a}} \cap \partial B^1_{r_1}$. The boundaries of B_+ and $B^1_{r_1}$ are not smooth submanifolds, but instead stratified spaces with affine-components of different codimensions. For both cases the codimension of a stratum is equal to the number of coordinates that vanish along it. A sparse, nonnegative image $f_0 \in \mathbb{A}_{\boldsymbol{a}} \cap B_S$ therefore lies on strata of both ∂B_+ and $\partial B^1_{r_1}$ of very high codimension. In 2D examples the image f_0 vanishes at no less than $\frac{3}{4}$ of the indices in J.

If f_0 is an isolated point in $\mathbb{A}_{\boldsymbol{a}} \cap B_+$, then the triangle inequality implies that it is also an isolated point in $\mathbb{A}_{\boldsymbol{a}} \cap B^1_{r_1}$. Examining the proof of Lemma 4.12 we see that $L(\boldsymbol{\epsilon})$ only *fails* to have a strict minimum at $\boldsymbol{\epsilon} = \boldsymbol{0}$ if there exists a $\boldsymbol{\epsilon} \neq \boldsymbol{0}$ so that,

$$0 \leq \sum_{l \in J^t_{f_0}} \epsilon_l \tau_j^{(l)} \quad \text{for every } j \in S^c_{f_0}. \tag{4.56}$$

In this case there is a $\lambda_0 > 0$ so that, if $0 < \lambda < \lambda_0$, then

$$0 \leq f_{0j} + \lambda \sum_{l \in J^t_{f_0}} \epsilon_l \tau^{(l)}_j \quad \text{for all } \boldsymbol{j} \in J, \tag{4.57}$$

and $L(\lambda\boldsymbol{\epsilon}) = L(0)$. Note that for $-\lambda_0 < \lambda < 0$, we have $L(\lambda\boldsymbol{\epsilon}) > L(0)$ unless equality holds in (4.56), for all $\boldsymbol{j} \in S^c_{f_0}$. As we see below, this is unlikely to occur. Thus, the failure to be a strict minimum occurs on rays, of which an interval lies in $\partial B_+ \cap \partial B^1_{r_1}$.

Hence, if $T_{f_0}\mathbb{A}_{\boldsymbol{a}} \cap \partial B^1_{r_1}$ does not consist of $\boldsymbol{f}_0$ alone, then there are directions $\boldsymbol{v} \in T^0_{f_0}\mathbb{A}_{\boldsymbol{a}}$ so that

$$\|\boldsymbol{f}_0 + \lambda\boldsymbol{v}\|_1 = \|\boldsymbol{f}_0\|_1 \quad \text{for } 0 \leq \lambda \leq 1. \tag{4.58}$$

As noted, this happens if and only if $\boldsymbol{f}_0 + \lambda\boldsymbol{v}$ is a nonnegative vector, for $0 \leq \lambda \leq 1$. If $\boldsymbol{v}_1 \neq \boldsymbol{v}_2$ are two nonzero vectors for which (4.58) holds, then it is also true for the convex combinations

$$\lambda(\boldsymbol{f}_0 + \boldsymbol{v}_1) + (1-\lambda)(\boldsymbol{f}_0 + \boldsymbol{v}_2) = \boldsymbol{f}_0 + \lambda\boldsymbol{v}_1 + (1-\lambda)\boldsymbol{v}_2 \quad \text{for } \lambda \in [0,1]. \tag{4.59}$$

Thus, the set $T_{f_0}\mathbb{A}_{\boldsymbol{a}} \cap \partial B^1_{r_1}$ is a convex, conic subset of an orthant. Moreover, we have shown that the intersections $T_{f_0}\mathbb{A}_{\boldsymbol{a}} \cap \partial B^1_{r_1}$ and $T_{f_0}\mathbb{A}_{\boldsymbol{a}} \cap \partial B_+$ actually coincide. We summarize these observations in a theorem:

Theorem 4.14 *The intersections $\mathbb{A}_{\boldsymbol{a}} \cap \partial B_+$ and $\mathbb{A}_{\boldsymbol{a}} \cap \partial B^1_{r_1}$ are transversal at $\boldsymbol{f}_0$ if and only if L has a strict minimum at $\boldsymbol{0}$. In all cases, the connected component of*

$$T_{f_0}\mathbb{A}_{\boldsymbol{a}} \cap \partial B^1_{r_1} = T_{f_0}\mathbb{A}_{\boldsymbol{a}} \cap \partial B_+,$$

containing $\boldsymbol{f}_0$, is a convex conic subset of an orthant.

As a counterpoint to Proposition 4.13 we have the following result, which further justifies our notion that a nontrivial intersection of $T_{f_0}\mathbb{A}_{\boldsymbol{a}}$ with ∂B_+ or $\partial B^1_{r_1}$ is a failure of transversality.

Proposition 4.15 *Let $\boldsymbol{f}_0$ be an image that satisfies the hypotheses of Lemma 4.12 and for which L does* **not** *have a strict local minimum at $\boldsymbol{\epsilon} = \boldsymbol{0}$. Let $\boldsymbol{\epsilon} \neq \boldsymbol{0}$ be a vector such that* (4.57) *holds for $\lambda \in [0,1]$. There is a constant C so that*

$$\|\pi_A(\boldsymbol{f}_0 + \lambda\boldsymbol{v}_\epsilon)\|_1 \leq \|\boldsymbol{f}_0\|_1 + C\lambda^2, \tag{4.60}$$

here

$$\boldsymbol{v}_\epsilon = \sum_{l \in J^t_{f_0}} \epsilon_l \boldsymbol{\tau}^{(l)}. \tag{4.61}$$

Note that $\|\boldsymbol{f}_0 - \pi_A(\boldsymbol{f}_0 + \lambda\boldsymbol{v}_\epsilon)\|_1 \approx \lambda\|\boldsymbol{v}_\epsilon\|_1$.

Proof The proof is essentially the same as that of Proposition 4.13, but with the inequality reversed

$$\begin{aligned}\|\pi_A(\boldsymbol{f}_0 + \lambda \boldsymbol{v}_\epsilon)\|_1 &\leq \|\boldsymbol{f}_0 + \lambda \boldsymbol{v}_\epsilon\|_1 + \|\pi_A(\boldsymbol{f}_0 + \lambda \boldsymbol{v}_\epsilon) - (\boldsymbol{f}_0 + \lambda \boldsymbol{v}_\epsilon)\|_1 \\ &\leq \|\boldsymbol{f}_0\|_1 + C\lambda^2.\end{aligned} \tag{4.62}$$

Here we use the triangle inequality, equation (4.53), and the fact that $\|\boldsymbol{f}_0 + \lambda \boldsymbol{v}_\epsilon\|_1 = \|\boldsymbol{f}_0\|_1$, for $\lambda \in [0, 1]$. The existence of a constant C so that

$$\|\boldsymbol{f}_0 + \lambda \boldsymbol{v}_\epsilon - \pi_A(\boldsymbol{f}_0 + \lambda \boldsymbol{v}_\epsilon)\|_1 \leq C\lambda^2 \tag{4.63}$$

follows from the fact that the ℓ_1 and ℓ_2 norms are equivalent in finite dimensional spaces. □

Using these results, we can quantify the failure of transversality of the intersection of $\mathbb{A}_{\boldsymbol{a}}$ with ∂B_+ near to $\boldsymbol{f}_0 \in \mathbb{A}_{\boldsymbol{a}} \cap B_+$, as well as with $\partial B^1_{r_1}$. This aspect of the geometry of these subsets is expected to have a decisive effect on the behavior of iterative algorithms for finding points in the intersection.

Definition 4.16 For a subset $K \subset J$ we define the linear map from $\mathbb{E}_K : \mathbb{R}^{J^t_{f_0}}$ to $\mathbb{R}^{K^c}$ as the restriction

$$\mathbb{E}_K : \boldsymbol{\epsilon} \longrightarrow \sum_{l \in J^t_{f_0}} \epsilon_l \tau^{(l)}_j \restriction_{j \in K^c} . \tag{4.64}$$

Without a subscript, $\mathbb{E} : \mathbb{R}^{J^t_{f_0}} \to T^0_{\boldsymbol{f}_0} \mathbb{A}_{\boldsymbol{a}} \subset \mathbb{R}^J$.

As noted above, the exact intersection of $T_{\boldsymbol{f}_0}\mathbb{A}_{\boldsymbol{a}}$ with ∂B_+ is defined by vectors $\boldsymbol{\epsilon} \in \mathbb{R}^{J^t_{f_0}}$ so that $\mathbb{E}_{S_{f_0}}(\boldsymbol{\epsilon})$ is a nonnegative vector. For small enough $0 < \lambda$, the vectors $\boldsymbol{f}_0 + \lambda\mathbb{E}(\boldsymbol{\epsilon})$ then have nonnegative coefficients, and therefore, belong to $T_{\boldsymbol{f}_0}\mathbb{A}_{\boldsymbol{a}} \cap \partial B_+$. With

$$\mathscr{I}_{\boldsymbol{f}_0} \overset{d}{=} \{\boldsymbol{\epsilon} \in \mathbb{R}^{J^t_{f_0}} : 0 \leq \boldsymbol{f}_0 + \mathbb{E}(\boldsymbol{\epsilon})\}, \tag{4.65}$$

the sets,

$$T_{\boldsymbol{f}_0}\mathbb{A}_{\boldsymbol{a}} \cap \partial B_+ = T_{\boldsymbol{f}_0}\mathbb{A}_{\boldsymbol{a}} \cap \partial B^1_{r_1} = \{\boldsymbol{f}_0 + \mathbb{E}(\boldsymbol{\epsilon}) : \boldsymbol{\epsilon} \in \mathscr{I}_{\boldsymbol{f}_0}\}. \tag{4.66}$$

The set $\mathscr{I}_{\boldsymbol{f}_0}$ is a convex, conic set with vertex at 0, the image $\{\mathbb{E}_{S_{f_0}}(\boldsymbol{\epsilon}) : \boldsymbol{\epsilon} \in \mathscr{I}_{\boldsymbol{f}_0}\}$ lies in an orthant in $\mathbb{R}^N$, where $N = |J| - |S_{f_0}|$.

The dimension of the vector space underlying $\mathscr{I}_{\boldsymbol{f}_0}$ and the $\max\{\|\boldsymbol{\epsilon}\|_1 : \boldsymbol{\epsilon} \in \mathscr{I}_{\boldsymbol{f}_0}\}$ are measures of the failure of transversality of the intersection $\mathbb{A}_{\boldsymbol{a}} \cap \partial B_+$ near to $\boldsymbol{f}_0$. The function $L(\boldsymbol{\epsilon})$ is constant in the set $\mathscr{I}_{\boldsymbol{f}_0}$; as indicated by Proposition 4.15, to leading order, the points in $T_{\boldsymbol{f}_0}\mathbb{A}_{\boldsymbol{a}} \cap \partial B_+$ belong to $\mathbb{A}_{\boldsymbol{a}} \cap \partial B_+$.

In real applications the exact support of f_0 is unknown. We now show how a good estimate for S_{f_0} leads to effectively computable sufficient conditions for the intersection to be transversal. It is often the case that there is no $\boldsymbol{\epsilon} \neq \mathbf{0}$ so that (4.56) holds. To see why, we note that for each $\boldsymbol{l} \in J^n_{f_0}$ and $\boldsymbol{m} \in J^t_{f_0}$ the orthogonality of $T_{f_0}\mathbb{A}_{\boldsymbol{a}}$ and $N_{f_0}\mathbb{A}_{\boldsymbol{a}}$ implies that

$$\sum_{j \in J} \nu_j^{(l)} \tau_j^{(m)} = 0 \tag{4.67}$$

Suppose that $R_{f_0} \supset S_{f_0}$, is a rectangle, whose sides lengths are at most half those of J. A moment's thought shows that for many choices of $\boldsymbol{l} \in J$ the supports of $f_0^{\pm l}$, and therefore also of $\boldsymbol{\nu}^{(l)}$, are disjoint from R_{f_0}. We let $J^{n0}_{f_0} \subset J^n_{f_0}$ denote the indices for which this is true. If $\boldsymbol{l} \in J^{n0}_{f_0}$, then (4.67) implies that

$$\sum_{j \in R^c_{f_0}} \nu_j^{(l)} \tau_j^{(m)} = 0 \quad \text{for all } \boldsymbol{m} \in J^t_{f_0}. \tag{4.68}$$

If we let

$$\boldsymbol{\nu}^{\text{tot}} = \sum_{l \in J^{n0}_{f_0}} \boldsymbol{\nu}^{(l)}, \tag{4.69}$$

then, for any choice of $\boldsymbol{\epsilon}$

$$\sum_{j \in R^c_{f_0}} \nu_j^{\text{tot}} \left[\sum_{l \in J^t_{f_0}} \epsilon_l \tau_j^{(l)} \right] = 0. \tag{4.70}$$

If R_{f_0} is a tight estimate for S_{f_0}, then it is often the case that

$$\nu_j^{\text{tot}} > 0 \quad \text{for all } \boldsymbol{j} \in R^c_{f_0} \subset S^c_{f_0}. \tag{4.71}$$

In this case (4.56), (4.70), and (4.71) imply that

$$\sum_{l \in J^t_{f_0}} \epsilon_l \tau_j^{(l)} = 0 \quad \text{for every } \boldsymbol{j} \in R^c_{f_0}. \tag{4.72}$$

If the null-space of $\mathbb{E}_{R_{f_0}}$ is trivial, this would imply that $\boldsymbol{\epsilon} = \mathbf{0}$, and therefore, the strict inequality $L(\boldsymbol{\epsilon}) > L(0)$ holds for any $\boldsymbol{\epsilon} \neq \mathbf{0}$. In two or more dimensions $|R^c_{f_0}| >> |J^t_{f_0}|$, hence dimensional considerations suggest that $\ker \mathbb{E}_{R_{f_0}}$ is often trivial, and generically this appears to be the case.

Even if (4.71) holds, it is possible that $\mathbb{E}_{R_{f_0}}$ has a nontrivial null-space. If $\ker \mathbb{E}_{R_{f_0}}$ is nontrivial, then the question: "does L have a strict minimum at $\mathbf{0}$?" reduces to that of the existence of a nonzero vector $\boldsymbol{\epsilon} \in \ker \mathbb{E}_{R_{f_0}}$ for which

$$\sum_{l \in J^t_{f_0}} \epsilon_l \tau_j^{(l)} \geq 0 \quad \text{for } \boldsymbol{j} \in R_{f_0} \setminus S_{f_0}. \tag{4.73}$$

In particular examples one can compute $\ker \mathbb{E}_{R_{f_0}}$ and attempt to answer this question, though in practice it is a rather difficult computational question.

If the image is a convolution $f_0 = \boldsymbol{g} * \boldsymbol{\varphi}$, with $\boldsymbol{\varphi}$ inversion symmetric, then Proposition 2.15 shows that $\ker \mathbb{E}_{S_{f_0}} \neq \{\mathbf{0}\}$: If $S_{\boldsymbol{\varphi}}$ is the support of $\boldsymbol{\varphi}$, then there is a subspace of $T^0_{\boldsymbol{\varphi}} \mathbb{A}_{\boldsymbol{a}_{\boldsymbol{\varphi}}}$, with dimension about $|S_{\boldsymbol{\varphi}}|/2$, consisting of tangent vectors supported in $S_{\boldsymbol{\varphi}}$. Convolving with $\boldsymbol{g}$ leads to vectors in $T^0_{f_0} \mathbb{A}_{\boldsymbol{a}_{f_0}}$ supported in $S_{f_0} = S_{\boldsymbol{\varphi}} + S_{\boldsymbol{g}}$ (assuming that $\boldsymbol{g}$ and $\boldsymbol{\varphi}$ are both non-negative). It therefore follows from (4.56)–(4.57) that the intersection between $\mathbb{A}_{\boldsymbol{a}}$ and ∂B_+ is nontransversal.

For example, Figure 4.1 shows a nonnegative 128×128 image obtained by convolving $(1 - \|\boldsymbol{j}\|^2/M^2)\chi_{[0,1]}(\|\boldsymbol{j}\|/M)$ with a piecewise constant image. The vector $\boldsymbol{\nu}^{\text{tot}}$, defined in (4.69), is positive throughout $R^c_{f_0}$; its minimum value is $4.5 \times 10^{-5}\|\boldsymbol{\nu}^{\text{tot}}\|_\infty$. Nonetheless, the dimension of $\ker \mathbb{E}_{R_{f_0}} = 40$ and there is a 34-dimensional subspace of $\ker \mathbb{E}_{R_{f_0}}$ that intersects ∂B_+ in a convex cone. For images defined by convolution with an inversion symmetric image, $\boldsymbol{\varphi}$, the dimension of this intersection depends on $|S_{\boldsymbol{\varphi}}|$. Other than these sorts of cases, we have *not* seen an example where the $\ker \mathbb{E}_{R_{f_0}}$ is nontrivial, producing nontransverse intersections.

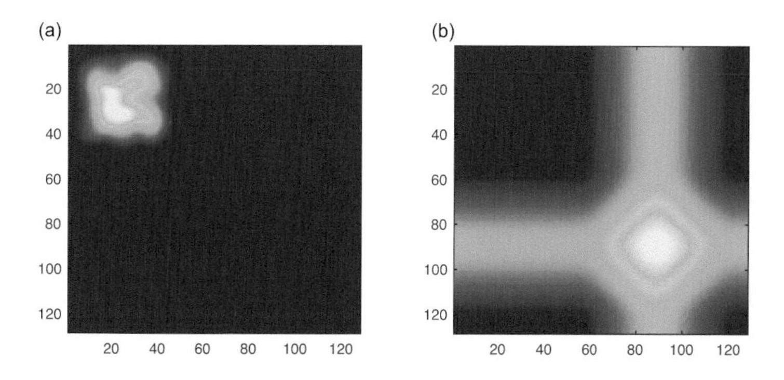

Figure 4.1 The object, defined as a convolution, is shown in part (a) and the vector $\boldsymbol{\nu}^{\text{tot}}$ is shown in part (b).

Table 4.1. *The dimensions of* $T_{f_0}\mathbb{A}_{\boldsymbol{a}} \cap \partial B_+$ *for* $k = 2,4,6$, *for images obtained by convolving a piecewise constant image with Gaussians of the form* $e^{-\frac{\sigma^2\|x\|^2}{k^2}}$, *where* σ *is set to be the inverse of the pixel size*

k	2	4	6
$\dim T_{f_0}\mathbb{A}_{\boldsymbol{a}} \cap \partial B_+$	4	18	34

Table 4.1 summarizes the results of computing $\dim T_{f_0}\mathbb{A}_{\boldsymbol{a}_{f_0}} \cap \partial B_+$, where f_0 is the result of convolving a piecewise constant image with samples of Gaussians of various widths. The nontransversality of these intersections has a markedly deleterious effect on the performance of algorithms for finding $\mathbb{A}_{\boldsymbol{a}} \cap B_+$.

When the intersections $\mathbb{A}_{\boldsymbol{a}} \cap \partial B_+$ and $\mathbb{A}_{\boldsymbol{a}} \cap \partial B^1_{r_1}$ are transversal at f_0, one expects the phase retrieval problems, with these auxiliary conditions, to be "well conditioned." It is, of course, still possible for $T_{f_0}\mathbb{A}_{\boldsymbol{a}}$ to make a very shallow angle with ∂B_+ over a large subspace, which can again lead to very slow convergence, or stagnation of iterative methods for finding points in $\mathbb{A}_{\boldsymbol{a}} \cap \partial B_+$, or $\mathbb{A}_{\boldsymbol{a}} \cap \partial B^1_{r_1}$. This is a rather difficult thing to assess analytically. For a vector $\boldsymbol{\epsilon}$, one needs to locate the indices $\{\boldsymbol{j} \in S^c_{f_0}\}$ such that

$$\sum_{l \in J^t_{f_0}} \epsilon_l \tau^{(l)}_j < 0. \tag{4.74}$$

Denote this subset of $S^c_{f_0}$ by $S^{c-}_{f_0}(\boldsymbol{\epsilon})$. The ratio that measures the "ℓ_1-angle" is

$$\frac{L(\boldsymbol{\epsilon}) - L(\mathbf{0})}{\|\boldsymbol{v}_{\epsilon}\|_1} = \frac{2\sum_{j \in S^{c-}_{f_0}(\boldsymbol{\epsilon})} \left|\sum_{l \in J^t_{f_0}} \epsilon_l \tau^{(l)}_j\right|}{\|\boldsymbol{v}_{\epsilon}\|_1}, \tag{4.75}$$

where

$$\boldsymbol{v}_{\epsilon} = \sum_{l \in J^t_{f_0}} \epsilon_l \boldsymbol{\tau}^{(l)}. \tag{4.76}$$

One would then seek vectors $\boldsymbol{\epsilon}$ that minimize this ratio. Even numerically this is difficult to do. In any case, a small minimum value for this ratio would be reflected in a small value for the coefficient μ appearing in the estimate (4.47), indicating a near tangency between $T_{f_0}\mathbb{A}_{\boldsymbol{a}}$ and ∂B_+. This near tangency would likely lead to very slow convergence of algorithms for finding $\mathbb{A}_{\boldsymbol{a}} \cap B_+$, which, empirically, is the rule for smoother images.

5
Some Preliminary Conclusions

To conclude our theoretical analysis of the discrete, classical, phase retrieval problem, we summarize some of the things that we have learned thus far. We have largely adopted the perspective that phase retrieval is the problem of finding intersections of two sets $A,\ B \subset \mathbb{R}^J$, where $A = \mathbb{A}_{\boldsymbol{a}}$ is a magnitude torus determined by directly measurable data and B is defined by auxiliary information. We have considered $B = B_S$, the set of images with support in S, B_+, nonnegative images, and $B^1_{r_1}$, nonnegative images with ℓ_1-norm bounded by r_1. Our main results show that the inherent difficulty of finding isolated points in $\mathbb{A}_{\boldsymbol{a}} \cap B$ hinges on the local geometry near to intersection points, and in particular, whether or not such an intersection is transversal.

These results are largely applications of Lemma 2.3, which gives a very explicit description for the tangent and normal spaces to a magnitude torus. If the base point $\boldsymbol{f}_0 \in \mathbb{A}_{\boldsymbol{a}}$ satisfies a support condition, then this is reflected in support properties of the respective bases $\{\boldsymbol{\tau}^{(\boldsymbol{v})}\}, \{\boldsymbol{\nu}^{(\boldsymbol{v})}\}$. Using this description, we have shown that, for reasonable support conditions, the intersections lying in $\mathbb{A}_{\boldsymbol{a}} \cap B_S$ are frequently not transversal. The failure of transversality is a consequence of the inevitable nonuniqueness of the solution to the phase retrieval problem caused by the existence of trivial associates, and in particular, the translation invariance of magnitude discrete Fourier transform (DFT) data. This failure of transversality, as measured by $\dim T_{\boldsymbol{f}_0}\mathbb{A}_{\boldsymbol{a}} \cap B_S$, is worse for looser support constraints. Beyond the exact intersections, there are often large subspaces where $T_{\boldsymbol{f}_0}\mathbb{A}_{\boldsymbol{a}}$ and B_S make very small angles. This problem becomes more severe as the images become smoother. As we see in Part II, such subspaces lead to very slow convergence for iterative algorithms.

In Chapter 3 we examined the extent to which the phase retrieval problem with support as the auxiliary condition is locally well posed. We first examined the map, $\mathscr{M}$, from images $\boldsymbol{f}$ to their magnitude DFT data, $|\widehat{\boldsymbol{f}}|$, among images with support in a given set S. Our main result was that this mapping has a local Lipschitz inverse near to a point $\boldsymbol{f}_0$ if and only if the intersection $\mathbb{A}_{\boldsymbol{a}_{f_0}} \cap B_S$ is transversal at $\boldsymbol{f}_0$. Regardless of the algorithm that is used, the phase retrieval problem is ill posed near to any image where this intersection fails to be transversal.

As shown by the formula in (2.42), without a very precise support constraint, the intersection, $\mathbb{A}_{\boldsymbol{a}_{f_0}} \cap B_S$, is usually not transversal. In finite precision arithmetic, a non-Lipschitz inverse can be expected to lead to a substantial loss of accuracy. Theorem 3.3 implies that the inverse is at most Hölder-continuous of order $\frac{1}{2}$, and therefore the number of accurate digits in the solution is expected to be about half the number of accurate digits in the data, though a precise accounting would depend on the constants in the continuity estimate. In examples in Part II, we see that the failure of transversality often leads to problems much worse than those predicted by a Hölder-$\frac{1}{2}$ local inverse. In Example 3.12 we also showed that the continuity properties of the local inverse at a nontransversal intersection can be highly anisotropic. This leaves open the possibility of algorithms that, in certain situations, might avoid some of the difficulties produced by nontransversality. In Example 8.13 we examine a situation where this is, to some extent, accomplished.

We also introduced the notion of ϵ-nonuniqueness and showed that it can arise in several distinct ways. If the data $(\boldsymbol{a}, S)$ is ϵ-nonunique, then there will be at least two markedly different images that satisfy the requirements to be a solution of the phase retrieval problem to within a tolerance of ϵ. This is clearly quite problematic for a scientific method. Uniqueness is generic, but there is also data for which the phase retrieval problem does not have a unique solution. These facts lead inevitably to the existence of data for which the phase retrieval problem has a formally unique solution, but which is also ϵ-nonunique for any $\epsilon > 0$. While a precise characterization is not yet available, we conjecture that images displaying ϵ-nonuniqueness are precluded from have sharp edges or jump discontinuities.

We also explored a second source of ϵ-nonunique data, which is essentially a microlocal version of the translation symmetry explored earlier. There are images that can be decomposed as a sum $\boldsymbol{f} = \boldsymbol{f}_1 + \cdots + \boldsymbol{f}_k$ where the ϵ-supports of the components all lie in a fixed set S and the ϵ-supports of their DFT data are disjoint. When this occurs one can independently replace each component in the sum with either $\pm \boldsymbol{f}_j^{(\boldsymbol{v}_j)}$ or $\pm \check{\boldsymbol{f}}_j^{(\boldsymbol{v}_j)}$ without changing the magnitude DFT data, at precision $k\epsilon$. Many images obtained this way will also satisfy the support constraint, at precision $k\epsilon$.

Fortunately, the possibility of such a decomposition can be detected in the magnitude DFT data and seems to require that the component parts are smooth objects. If ϵ is taken to be machine precision ($\approx 10^{-16}$ for double precision), then for images of this type, the magnitude DFT data and a reasonable support constraint do *not* suffice to determine an image, up to trivial associates. One can easily imagine, however, that such objects could arise in images of biological cell structures at the nanometer scale. The component substructures could correspond to convoluted substructures including the endoplasmic reticulum or mitochondria.

We then turned to the study of the nonnegativity constraint. For nonnegative images it is natural to use the orthant, B_+ as the auxiliary set, and look for intersections, $\mathbb{A}_{\boldsymbol{a}} \cap B_+$. We showed that generically a nonnegative image is uniquely determined, up to trivial associates, by its magnitude DFT data, if the autocorrelation image has sufficiently small support. Once again this is a condition that can be checked from the magnitude DFT data itself. In this case the image also has small support and therefore lies in the boundary of the orthant. This boundary is not a smooth space, but instead a stratified space made up of orthants of various dimensions.

The intersection of $\mathbb{A}_{\boldsymbol{a}}$ and ∂B_+ is transversal at $\boldsymbol{f}_0$ if there is a $0 < \delta$ so that

$$T_{\boldsymbol{f}_0}\mathbb{A}_{\boldsymbol{a}} \cap \partial B_+ \cap \{\boldsymbol{f} : \|\boldsymbol{f} - \boldsymbol{f}_0\| < \delta\} = \{\boldsymbol{f}_0\}.$$

As the ∂B_+ is not a smooth submanifold, we needed to provide a way to quantify the failure of transversality. To that end we studied the ℓ_1-norm on $T_{\boldsymbol{f}_0}\mathbb{A}_{\boldsymbol{a}}$. Our analysis began with the observation that $\|\boldsymbol{f}\|_1$ assumes its global minima at nonnegative images lying on the torus. Let $\boldsymbol{f}_0$ be such an image and let $r_1 = \|\boldsymbol{f}_0\|_1$. If $B^1_{r_1}$ is the ℓ_1-ball with this radius, then generically $\mathbb{A}_{\boldsymbol{a}} \cap B^1_{r_1}$ consists of trivial associates of $\pm\boldsymbol{f}_0$.

The ℓ_1-norm on $T_{\boldsymbol{f}_0}\mathbb{A}_{\boldsymbol{a}}$ also has a global minimum at $\boldsymbol{f}_0$, but this minimum need not be strict. In fact, there may be a convex, conic subset of $T_{\boldsymbol{f}_0}\mathbb{A}_{\boldsymbol{a}}$ where the ℓ_1-norm is constant. This set is equal to $T_{\boldsymbol{f}_0}\mathbb{A}_{\boldsymbol{a}} \cap \partial B_+$, and, by definition, it also equals $T_{\boldsymbol{f}_0}\mathbb{A}_{\boldsymbol{a}} \cap \partial B^1_{r_1}$. The dimension and diameter of these convex sets provide a way to quantify the failure of transversality.

Along a high codimensional stratum, like the one containing $\boldsymbol{f}_0$, the boundaries of both B_+ and $B^1_{r_1}$ are strictly convex in many directions, and therefore one might expect it to be easier for these intersections to be transversal. In fact, our analysis shows that, in generic cases, the intersection can be expected to be transversal, and so the problem of finding the intersections $\mathbb{A}_{\boldsymbol{a}} \cap B_+$ and $\mathbb{A}_{\boldsymbol{a}} \cap B^1_{r_1}$ should be easier than finding $\mathbb{A}_{\boldsymbol{a}} \cap B_S$. This expectation is borne out by examples in Part II. We also provide a computationally effective method for finding at least a subspace of $T_{\boldsymbol{f}_0}\mathbb{A}_{\boldsymbol{a}} \cap \partial B_+$, if it is nontrivial.

Images defined by convolving an image $\boldsymbol{g}$ with an inversion symmetric image $\boldsymbol{\varphi}$ are nongeneric in several ways. Let $f_0 = \boldsymbol{g} * \boldsymbol{\varphi}$. From the perspective of phase retrieval, the most significant issue is the fact that $\dim T_{f_0}\mathbb{A}_{\boldsymbol{a}} \cap B_S$ is essentially never zero; even if $S = S_{f_0}$. The symmetry of $\boldsymbol{\varphi}$ forces the intersection $T^0_{\varphi}\mathbb{A}_{\boldsymbol{a}} \cap B_{S_{\varphi}}$ to be nontrivial. Because

$$T^0_{g*\varphi}\mathbb{A}_{\boldsymbol{a}_{g*\varphi}} = T^0_{\varphi}\mathbb{A}_{\boldsymbol{a}_{\varphi}} * \boldsymbol{g} \text{ and } S_{g*\varphi} \subset S_{g} + S_{\varphi}, \tag{5.1}$$

this failure of transversality infects $\mathbb{A}_{\boldsymbol{a}_{g*\varphi}} \cap B_{S_{g*\varphi}}$. It renders the rather common practice of apodizing data with a Gaussian inadvisable in the context of phase retrieval.

Our theoretical analyses of the phase retrieval problem provide cogent explanations for the oft-seen failure of iterative algorithms. The lack of transversality at the intersections of $\mathbb{A}_{\boldsymbol{a}}$ and B makes the problem ill posed. As we see in the next part of the book, the lack of transversality also causes the maps, used to find the intersections, to be weakly contracting, or even neutrally stable, at some, or all, of their fixed points. The distance function between $\mathbb{A}_{\boldsymbol{a}}$ and B, $d_{\mathbb{A}_{a}B}(\boldsymbol{x}, \boldsymbol{y}) = \|\boldsymbol{x} - \boldsymbol{y}\|_2$, defines a landscape between these two subsets. The critical points of this function produce multiple attracting basins for essentially all known iterative algorithms, which can, in turn, lead to very complicated, nonconvergent dynamics.

Understanding the failure mechanisms for standard approaches is certainly the first step toward improving the behavior of these algorithms. As shown by examples in Part II, when the infinitesimal translation symmetry is not present, standard algorithms can converge to give very accurate reconstructions. In Part III we present several additional proposals for how this might be done. The issues presented by ϵ-nonuniqueness might appear even more intractable, but, in fact, similar modifications of experimental protocols can obviate these difficulties as well.

PART II

Analysis of Algorithms for Phase Retrieval

6
Introduction to Part II

We turn now to the analysis of a representative choice from among the many algorithms that are currently used for practical phase retrieval. We assume that we have magnitude discrete Fourier transform (DFT) data, $\boldsymbol{a} \in \mathbb{R}_+^J$, which defines a magnitude torus $\mathbb{A}_{\boldsymbol{a}}$, and we also have some auxiliary information about the image, which defines a second subset B of $\mathbb{R}^J$. The auxiliary information is assumed to be adequate so that the set $\mathbb{A}_{\boldsymbol{a}} \cap B$ consists of finitely many points, which, generically, are trivial associates of a single image. The algorithms attempt to find points in this intersection. We largely restrict our attention to real images, though much of what we say remains true for complex images.

The building blocks for most of these algorithms are nearest point maps and reflections defined by subsets $M \subset \mathbb{R}^N$. The nearest point map, P_M is defined by

$$\begin{aligned} P_M(\boldsymbol{f}) &= \text{the point in } M \text{ closest to } \boldsymbol{f}, \text{ with respect to Euclidean distance} \\ &= \underset{\boldsymbol{x} \in M}{\arg\min} \|\boldsymbol{x} - \boldsymbol{f}\|_2. \end{aligned} \tag{6.1}$$

If M is not convex, then $P_M(\boldsymbol{f})$ is not defined everywhere, and cannot be extended to be continuous throughout all of $\mathbb{R}^N$. It is not defined at points $\boldsymbol{f}$ for which the closest point on M is nonunique. This is, generically, a set of codimension one and, in practice, does not pose a serious problem, though it may contribute to the extreme sensitivity to initial conditions that these algorithms can display. The reflection across M is defined by

$$R_M(\boldsymbol{f}) = 2P_M(\boldsymbol{f}) - \boldsymbol{f}; \tag{6.2}$$

it is defined where P_M is defined. If M is a linear subspace, then P_M, is just the orthogonal projection onto M, and R_M is the usual orthogonal reflection fixing

M. If M is not linear, then M is still the fixed-point set of R_M, but the map can be quite different from a linear reflection. For example, if $M = B_+$, then

$$[R_{B_+}(\boldsymbol{f})]_{\boldsymbol{j}} = |f_{\boldsymbol{j}}|. \tag{6.3}$$

In this part of the book, we examine the behavior of various types of algorithms in light of the results from the previous part. These algorithms are defined by iterating maps, which have two general forms: alternating projection, and hybrid iterative maps. Alternating projection is based on a map of the type

$$AP(\boldsymbol{f}) = P_B \circ P_A(\boldsymbol{f}).$$

We concentrate here on a class of maps suggested by the *hybrid input-output (HIO) maps*, introduced by Fienup in Fienup (1978, 1982).

For a subset B that defines adequate auxiliary information, and $\beta \in (0, 1]$, we define the map

$$D^{\beta}_{BA}(\boldsymbol{f}) = \boldsymbol{f} + P_B \circ [(1+\beta)P_A(\boldsymbol{f}) - \boldsymbol{f}] - \beta P_A(\boldsymbol{f}). \tag{6.4}$$

The map P_B is positive homogeneous of degree 1, if

$$\lambda P_B(\boldsymbol{f}) = P_B(\lambda \boldsymbol{f}) \text{ for all } \lambda \in [0, \infty). \tag{6.5}$$

With $A = \mathbb{A}_{\boldsymbol{a}}$ and P_B satisfying (6.5), any map of the form given in (6.4) is a viable candidate for defining an iterative algorithm to solve the phase retrieval problem. To see this, we observe that $\boldsymbol{f}^*$ is a fixed point of $D^{\beta}_{B\mathbb{A}_{\boldsymbol{a}}}$ if and only if

$$P_{\mathbb{A}_{\boldsymbol{a}}}(\boldsymbol{f}^*) = P_B(\beta^{-1}[(1+\beta)P_{\mathbb{A}_{\boldsymbol{a}}}(\boldsymbol{f}^*) - \boldsymbol{f}^*]) \in \mathbb{A}_{\boldsymbol{a}} \cap B. \tag{6.6}$$

If $\beta = 1$, then it is not necessary to assume that P_B is homogeneous.

Fienup's HIO maps are not defined directly in terms of the maps P_A, P_B, but instead, drawing inspiration from control theory and the concept of negative feedback, in terms of the subset where the image space constraint fails to be satisfied. For $B = B_S$ the HIO map is given by $D^{\beta}_{B_S A}$, but if $B = B_+$, then Fienup's HIO map is given by a very different expression, for $\boldsymbol{j} \in J$,

$$[D^{\beta}_{\text{HIO}}(\boldsymbol{f})]_{\boldsymbol{j}} = \chi_{(-\infty,0)}(P_A(\boldsymbol{f})_{\boldsymbol{j}})\boldsymbol{f}_{\boldsymbol{j}} + [(1+\beta)P_{B_+} \circ P_A(\boldsymbol{f}) + \beta P_A(\boldsymbol{f})]_{\boldsymbol{j}}. \tag{6.7}$$

The algorithm defined by iterating this map is quite different from that defined by $D^{\beta}_{B_+A}$. Fienup's HIO maps have had, and continue to have, a profound impact on many problems in optics, among them phase retrieval.

Another generalization of HIO maps is referred to as *difference maps*. These have also had a major impact in phase retrieval as well as in other fields (Elser 2003; Elser et al. 2007; Gravel and Elser 2008). They take the form

$$D^{\beta}_{BA\Delta}(f) = f + \beta[P_B(g_A(f)) - P_A(g_B(f))], \tag{6.8}$$

$$g_B(f) = P_B(f) - \frac{1}{\beta}[P_B(f) - f], \tag{6.9}$$

$$g_A(f) = P_A(f) + \frac{1}{\beta}[P_A(f) - f]. \tag{6.10}$$

Many of these maps provide examples of Douglas–Rachford splittings.

When $\beta = 1$, the map $D^1_{BA\Delta} = D^1_{BA}$; for this case we introduce the simplified notation:

$$D_{BA}(f) = f + P_B \circ R_A(f) - P_A(f), \tag{6.11}$$

and refer to the map in (6.11) as a *hybrid iterative* map. For $\beta = 1$ the roles of A and B may be interchanged; we define

$$D_{AB}(f) = f + P_A \circ R_B(f) - P_B(f). \tag{6.12}$$

If $A = \mathbb{A}_{\boldsymbol{a}}$ and B is adequate for phase retrieval, then the fixed points of $D_{\mathbb{A}_{\boldsymbol{a}}B}$ again define points in $\mathbb{A}_{\boldsymbol{a}} \cap B$, as in (6.6). In the sequel we consider both classes of maps. Figure 1.4 shows simple examples of the action of hybrid iterative maps.

While it is common in applications to use a value of $\beta < 1$, we use the maps D_{AB}, D_{BA}, to define model algorithms. Since we have not explored all parameter ranges for either hybrid maps or difference maps, we do not claim to have thoroughly investigated all choices of β, which could conceivably yield different outcomes. Experimentally, we have not found that taking $\beta < 1$ significantly changes the behavior of algorithms based on iterating D^{β}_{BA} or D^{β}_{AB}.

Analyses of algorithms like HIO, and Douglas–Rachford, and related nonconvex optimization problems appear in Hesse and Luke (2013), Levin and Bendory (2019), Li and Pong (2016), and Phan (2016). A very thorough review of research on the Douglas–Rachford splitting, with an extensive bibliography, can be found in Lindstrom and Sims (2021). An analysis of the convergence properties of algorithms used in phase retrieval, as inconsistent and nonconvex feasibility problems, is presented in Luke and Martins (2020).

We first carefully examine how various iterative algorithms behave when A and B are affine subspaces of $\mathbb{R}^J$. This is important in its own right, and a necessary step when linearizing these maps in case A or B is nonlinear. If a pair of linear subspaces meet at a point, then the rates of convergence of algorithms defined by AP, D_{AB}, or D_{BA} depend on the angles between the subspaces. As described in the introduction to Chapter 2, to quantify these relationships we let Q and R be matrices whose columns are orthonormal bases for A and

B, respectively. The cosines of the angles between A and B are the singular values of the matrix $H = R^t Q$. If $\sigma_1 < 1$ is the largest singular value, then these algorithms converge at a rate proportional to σ_1^n. We also consider the case that $A \cap B = F$ is a positive dimensional subspace. This is the linear model for a nontransversal intersection. In this case, $\sigma_1 = 1$, and the number of singular values equal to 1 is the $\dim A \cap B$.

If A, B are linear subspaces, then

$$A + B = \{a + b : a \in A, b \in B\} \tag{6.13}$$

is again a subspace. We let $(A + B)^\perp$ denote the orthogonal complement of $A + B$; it is easily seen to equal $A^\perp \cap B^\perp$. For the case of hybrid iterative maps the subspace, $C = (A + B)^\perp$, is usually of large dimension; the maps D_{AB}, D_{BA} act as the identity on C. We call this subspace the *center manifold* for the map. The hybrid iterative maps converge to points on the center manifold. The intersection point is then found by applying a projection to this limit point. If $F = A \cap B$ is positive dimensional, then this analysis needs to be slightly modified; as before, C is $(A + B)^\perp$. The hybrid iterative map now acts as the identity on $C \oplus F$ and converges to points on this larger set.

With the linear analysis complete we consider what happens when one of the subspaces is replaced by a nonlinear submanifold of $\mathbb{R}^J$. The behavior of alternating projection is quite different in the nonlinear case, as the map $P_A \circ P_B$ can have many fixed points that are not intersection points. In applications to phase retrieval it seems that the iterates of alternating projection usually converge to such uninteresting fixed points.

All fixed points of hybrid iterative maps are connected to intersection points, via the appropriate generalization of the center manifold. The linearization of a hybrid iterative map at an intersection point is the identity in the center manifold directions and hence *noncontracting*. If the intersection at f_0 is nontransversal, then the subspace $T_{f_0} A \cap T_{f_0} B$ plays the role of F in the linearization of the map at f_0. These are directions normal to the center manifold where the linearized map is also the identity and therefore noncontracting.

Experimentally, when the iterates of a hybrid map converge, it is always to points on the center manifold that are far from the intersection point that defines it. The linearization of the hybrid iterative map at points along the center manifold is not obviously contracting. Examining it carefully at points to which the iterates converge, it is often found to be a nonnormal operator, which has singular values that are *larger* than 1. Nonetheless, its eigenvalues are complex with modulus less than 1, which allows the iterates to converge.

Such linear maps have complicated dynamics, spiraling along very eccentric orbits toward their fixed points, a phenomenon which is also seen in phase retrieval using hybrid iterative maps. Because many points on the (positive dimensional) center manifold are attracting, it is not possible to reliably predict to which point the iterates converge. This fact along with the existence of multiple attracting basins seem to largely explain the extreme sensitivity these iterations display to initial conditions.

While this part of the book focuses on the analysis of algorithms used in practical phase retrieval, we also introduce and study some simpler model geometries where the features of the maps underlying these algorithms are easier to appreciate. In particular, we demonstrate clearly that hybrid iterative maps can have attracting subsets that are not center manifolds defined by a true intersection point. Such subsets do not contain fixed points, but nonetheless, seem to play an important role in the dynamics of this class of maps. For two subsets $A,\ B \subset \mathbb{R}^N$ we let $d_{AB} : A \times B \to [0, \infty)$ be defined by

$$d_{AB}(\boldsymbol{x}, \boldsymbol{y}) = \|\boldsymbol{x} - \boldsymbol{y}\|_2. \tag{6.14}$$

If $(\boldsymbol{x}_0, \boldsymbol{y}_0)$ is a critical point of this map, then the line segment, $\ell \boldsymbol{x}_0 \boldsymbol{y}_0$, joining these points is perpendicular to both A and B. This line segment belongs to the intersection of the fibers of the normal bundles, $N_{\boldsymbol{x}_0} A \cap N_{\boldsymbol{y}_0} B$; this intersection defines a generalized center manifold. *As a set* it is fixed by the hybrid iterative map and may be attracting. We show, in examples, that it is not even necessary for the critical point to be a local minimum for this to be true. The function d_{AB} defines the *landscape* between the two sets A and B. The complexity of this landscape appears to be a principal determinant of the dynamics of hybrid iterative maps, alternating projection, and likely all of the maps defined above.

The weak contraction properties of hybrid iterative maps are significantly exacerbated by the fact that nonzero critical points of the function $d_{\mathbb{A}_a B}$ can also define attracting basins for the difference maps. These appear to exist in great profusion and seem to play a significant role in the stagnation behavior of these hybrid iterative maps. The alternating projection map can be used as a tool to probe the distribution of nonzero minima of $d_{\mathbb{A}_a B}$, which is, in turn, an indication of the complexity of the landscape defined by this function. It is plausible that these maps have strange attractors to which the iterates are entrained. While it is very difficult to prove the existence of a strange attractor, we provide evidence for these assertions with numerical examples. This possibility has been suggested several times in the literature (see Elser et al. 2007).

In most of our experiments we see that the iterates of a hybrid iterative map either converge, if the intersection is transversal and the angles between $T_{f_0}\mathbb{A}_{\boldsymbol{a}}$ and B_S are not too small or stagnate. In the latter case the iterates remain in a small neighborhood of a center manifold, or manifolds, but the successive differences between iterates remain essentially constant in size. There is a third possibility, which is that the iterates of a hybrid iterative map could slowly converge to a point on the center manifold defined by a nontransversal intersection that is distant from $\boldsymbol{f}_0 \in \mathbb{A}_{\boldsymbol{a}} \cap B_S$. Because the limit point is not at an intersection, it may be possible to "circumvent" the fact that the linearization at $\boldsymbol{f}_0$ reduces to Id in the$T_{f_0}\mathbb{A}_{\boldsymbol{a}} \cap B_S$-directions. In Sections 8.1.1 and 8.1.2 we give several simple (nonphase retrieval) examples where this is seen to occur. While it does not seem to occur in phase retrieval for most realistic examples, we have found a class of objects for which it does happen, see Example 8.13.

Remark 6.1 As stated earlier, in numerical examples the support of an image, $\boldsymbol{f}$, is taken to the set $\{\boldsymbol{j} : |f_{\boldsymbol{j}}| \geq 10^{-14}\|\boldsymbol{f}\|\}$. To simplify the notation, we often write $S_{\boldsymbol{f}}$ instead of the more accurate $S_{\boldsymbol{f}}^{10^{-14}}$.

7

Algorithms for Phase Retrieval

In this chapter we introduce the basic types of algorithms used in phase retrieval. Among other things we analyze their behavior on pairs of linear subspaces. This analysis shows that, when two linear subspaces meet at a very shallow angle, the known algorithms can be expected to converge very slowly. We begin with the classical *alternating projection* algorithm (AP algorithm), and then consider algorithms based on *hybrid iterative maps*, which are motivated by the hybrid input output (HIO) algorithms introduced by Fienup (1982) (see also Elser 2003). We also include a brief analysis of the relaxed averaged alternating reflection (RAAR) algorithm. We have found that the nonorthogonal splittings introduced in this chapter to analyze the case of linear subspaces, $\mathbb{R}^N = A \oplus B \oplus C,\ C = (A + B)^{\perp}$, are of great general applicability.

Some of the analysis in this chapter is anticipated by results in Elser (2003), however, Elser does not consider the effects of the angles between the two subspaces on the behavior of the iterates of these types of maps. This aspect of the problem has been considered in the math literature (see, for example Bauschke and Borwein 1993). As the results of Part I show, this can be expected to be a crucial determinant in applications to phase retrieval. In a final section we outline a new, noniterative method for phase retrieval that uses the Hilbert transform to directly determine the unknown phase from the measured magnitude of the Fourier transform. This approach requires a modification to the standard experimental protocol, which we describe. Reconstruction methods using the Hilbert transform, but without modifications to the experimental protocol, were explored in Nakajima and Asakura (1985, 1986) and Nakajima (1995).

7.1 Classical Alternating Projection

We begin by analyzing algorithms related to classical AP. The idea for this algorithm, in a Banach space setting, goes back to the early work of John von Neumann (1950, Theorem 13.7) and Bauschke and Borwein (1993). It was introduced into optics in the work of Gerchberg and Saxton (1972), and significantly improved by Fienup (1978). In the phase retrieval community this algorithm is often referred to as *Error Reduction*. While the algorithm of Gerchberg and Saxton works well for its original intended purpose, it is not generally a reliable method for locating points in $\mathbb{A}_{\boldsymbol{a}} \cap B_S$. Nonetheless, it represents a good starting point for the discussion of practical algorithms. In practice, AP is often used in combination with other iterative algorithms.

For two subsets, A and B of $\mathbb{R}^N$, an AP algorithm is defined by the iteration

$$\boldsymbol{x}^{(n)} = P_B \circ P_A(\boldsymbol{x}^{(n-1)}). \tag{7.1}$$

In this section we begin by considering the case where A and B are linear subspaces of $\mathbb{R}^N$, which meet at 0. In Appendix 7.A we analyze a more general case and show the connection between the AP algorithm and gradient flow. This connection has also been investigated in Fienup (1982) and Marchesini (2007). Other relationships between phase retrieval algorithms and gradient descent on an objective function have been considered in Bauschke et al. (2002) and Levin and Bendory (2019).

Let A, B be linear subspaces of $\mathbb{R}^N$ such that $C = (A+B)^{\perp} \neq \{\mathbf{0}\}$. In this case $\mathbb{R}^N = A \oplus B \oplus C$; with this splitting $\boldsymbol{x} = Q\boldsymbol{x}_1 + R\boldsymbol{x}_2 + V\boldsymbol{x}_3$, where Q, R, and V are matrices with orthogonal columns spanning A, B, and C, respectively. The Euclidean inner product then takes the form

$$\begin{aligned} &((\boldsymbol{x}_1,\boldsymbol{x}_2,\boldsymbol{x}_3),(\boldsymbol{y}_1,\boldsymbol{y}_2,\boldsymbol{y}_3))_H \\ &\quad = \langle \boldsymbol{x}_1,\boldsymbol{y}_1\rangle + \langle H\boldsymbol{x}_1,\boldsymbol{y}_2\rangle + \langle \boldsymbol{x}_2, H\boldsymbol{y}_1\rangle + \langle \boldsymbol{x}_2,\boldsymbol{y}_2\rangle + \langle \boldsymbol{x}_3,\boldsymbol{y}_3\rangle, \end{aligned} \tag{7.2}$$

where the matrix

$$H = R^t Q. \tag{7.3}$$

To proceed we find expressions for the maps P_A and P_B with respect to this splitting of $\mathbb{R}^N$. Simple calculations show that

$$P_A(\boldsymbol{x}_1,\boldsymbol{x}_2,\boldsymbol{x}_3) = (\boldsymbol{x}_1 + H^t\boldsymbol{x}_2, 0, 0) \text{ and } P_B(\boldsymbol{x}_1,\boldsymbol{x}_2,\boldsymbol{x}_3) = (0, \boldsymbol{x}_2 + H\boldsymbol{x}_1, 0), \tag{7.4}$$

and therefore, the matrix representing $P_B \circ P_A$ is

$$P_B \circ P_A \boldsymbol{x} = \begin{pmatrix} 0 & 0 & 0 \\ H & HH^t & 0 \\ 0 & 0 & 0 \end{pmatrix} \begin{pmatrix} \boldsymbol{x}_1 \\ \boldsymbol{x}_2 \\ \boldsymbol{x}_3 \end{pmatrix}, \tag{7.5}$$

and

$$[P_B \circ P_A]^n = \begin{pmatrix} 0 & 0 & 0 \\ (HH^t)^{n-1}H & (HH^t)^n & 0 \\ 0 & 0 & 0 \end{pmatrix}. \tag{7.6}$$

If we let

$$P_B \circ P_A(\boldsymbol{x}) = (\mathbf{0}, H\boldsymbol{x}_1 + HH^t\boldsymbol{x}_2, \mathbf{0})^t = (\mathbf{0}, \boldsymbol{y}_0, \mathbf{0})^t,$$

then

$$[P_B \circ P_A]^{n+1}\boldsymbol{x} = (\mathbf{0}, (HH^t)^n \boldsymbol{y}_0, \mathbf{0})^t.$$

As described in the introduction to Chapter 2, the singular values of H are the cosines of the angles between special extremal orthonormal bases for A and B respectively. The eigenvalues of HH^t, which are the squares of the singular values of H, are nonnegative and less than 1; hence, $(HH^t)^n \boldsymbol{y}_0$ converges geometrically to 0. Note that the *rate* of convergence depends on the size of the maximum singular value of H. If some of these angles are very small, then some of these singular values are very close to 1, leading to very slow convergence of the algorithm. In the linear case, the zero vector is the entire fixed-point set of $P_B \circ P_A$. Similar results can be found in Deutsch (1985) and Bauschke and Borwein (1993).

If $A \cap B = F$ is a positive dimensional subspace, then we decompose $\mathbb{R}^N = A_0 \oplus B_0 \oplus F \oplus C$, where $A_0 = A \cap F^\perp$, $B_0 = B \cap F^\perp$, and $C = (A+B)^\perp$. With this decomposition

$$P_B \circ P_A = \begin{pmatrix} 0 & 0 & 0 & 0 \\ H & HH^t & 0 & 0 \\ 0 & 0 & \mathrm{Id} & 0 \\ 0 & 0 & 0 & 0 \end{pmatrix}, \tag{7.7}$$

and the behavior of the algorithm is essentially the same, with

$$\lim_{n\to\infty} [P_B \circ P_A]^n (\boldsymbol{x}_1, \boldsymbol{x}_2, \boldsymbol{x}_3, \boldsymbol{x}_4) = (0, 0, \boldsymbol{x}_3, 0). \tag{7.8}$$

It should be noted that when one, or both, of the subsets A, B is nonlinear, the AP map can have attracting fixed points that are not related to $A \cap B$, see Appendix 7.A. When F is nontrivial this map has a subspace on which the linearization is noncontracting. Hence, nonlinear effects can alter the behavior of iterates in ways that cannot be predicted by this analysis.

Remark 7.1 The AP algorithm is the first successful algorithm used to attack the phase retrieval problem. It is successful provided that the magnitudes of both the Fourier and image space data are known. It also has an interesting application to the problem of finding intersections of subspaces of very high-dimensional Euclidean spaces. Suppose that U and V are subspaces of $\mathbb{R}^N$, and that P_U and P_V denote the respective orthogonal projections. One iterate of AP for this case is $\boldsymbol{x}^{(1)} = P_V \circ P_U(\boldsymbol{x}^{(0)})$. Assume that $W = U \cap V \neq \{\boldsymbol{0}\}$, and let P_W denote the orthogonal projection onto this subspace. If $\boldsymbol{x}^{(0)}$ is the initial vector, then our analysis in Chapter 7.1, and in particular equation (7.8), shows that

$$\lim_{n\to\infty} [P_V \circ P_U]^n \boldsymbol{x}^{(0)} = P_W \boldsymbol{x}^{(0)}. \tag{7.9}$$

The rate of convergence is determined by the largest singular value of H that is less than 1. This method is, of course, nothing more, nor less than the classical power method; it already appears in von Neumann (1950).

In applications to phase retrieval $V = B_S$, the set of images with support in S and $U = T_f \mathbb{A}_{\boldsymbol{a}}$, where $\boldsymbol{f} \in B_S \cap \mathbb{A}_{\boldsymbol{a}}$, with $\mathbb{A}_{\boldsymbol{a}}$ the discrete Fourier transform (DFT)-magnitude torus defined by the point $\boldsymbol{f} \in B_S$. The novelty here is that, as shown in Appendix 2.B, we can find extremely fast algorithms for the projections $P_{T_f \mathbb{A}_a}$ and P_{B_S}, even when the dimensions of these subspaces are quite large, on the order of millions. This AP-based algorithm has proven to be quite effective for computing the intersections, $T_f \mathbb{A}_{\boldsymbol{a}} \cap B_S$, for images up to about 1024×1024.

7.2 Hybrid Iterative Maps

In phase retrieval, AP algorithms have been largely supplanted by algorithms obtained by iterating an HIO or difference map. As noted in Chapter 6, these maps were originally introduced by Fienup, in the form of the HIO method (6.4) (for the support constraint), and (6.7) (for the nonnegativity constraint). A related class of maps was introduced by Elser et al. (2007) taking the form (6.8). We restrict our attention to the case $\beta = 1$ in which case the hybrid iterative maps take the form (6.11) (or (6.12)); the iteration is then given by

$$\boldsymbol{x}^{(n+1)} = D_{BA}(\boldsymbol{x}^{(n)}) = \boldsymbol{x}^{(n)} + [P_B(R_A(\boldsymbol{x}^{(n)})) - P_A(\boldsymbol{x}^{(n)})]. \tag{7.10}$$

7.2.1 Attracting Basins for the Hybrid Iterative Map: The Center Manifold

For the moment we let g_A, g_B denote maps that reduce to the identity on A, B, respectively. With Id denoting the identity map, $\boldsymbol{x}^*$ is a fixed point of $D = \mathrm{Id} + P_B \circ g_A - P_A \circ g_B$, if and only if $P_B(g_A(\boldsymbol{x}^*)) - P_A(g_B(\boldsymbol{x}^*)) = 0$. The point

$$\boldsymbol{x}^{**} = P_B(g_A(\boldsymbol{x}^*)) = P_A(g_B(\boldsymbol{x}^*)), \tag{7.11}$$

evidently belongs to $A \cap B$. It is worth noting that the fixed-point set of D is generally *not* contained in $A \cap B$, but rather, points in $A \cap B$ are easily found once a fixed point is located, via (7.11). An important difference with AP is that all fixed points of hybrid iterative maps *are* simply related to $A \cap B$.

Let L_A, $L_B \subset X$ be the subsets such that $P_A(L_A) = \boldsymbol{x}^{**}$ and $P_B(L_B) = \boldsymbol{x}^{**}$. A necessary condition for $\boldsymbol{x}^*$ to be a fixed point of D is that

$$\boldsymbol{x}^* \in g_A^{-1}(L_B) \cap g_B^{-1}(L_A) \overset{d}{=} \mathscr{C}_{AB}^{\boldsymbol{x}^{**}}. \tag{7.12}$$

We call this set, $\mathscr{C}_{AB}^{\boldsymbol{x}^{**}}$, the *center manifold* defined by $\boldsymbol{x}^{**}$ for this map. Even if there is a unique point $\boldsymbol{x}^{**} \in A \cap B$ in the intersection, the center manifold, $\mathscr{C}_{AB}^{\boldsymbol{x}^{**}}$, can be high dimensional and contain points quite distant from $\boldsymbol{x}^{**}$ itself. If A or B is nonlinear, then some points of $\mathscr{C}_{AB}^{\boldsymbol{x}^{**}}$ may fail to be fixed points; additional conditions may be required. Identifying which points of $\mathscr{C}_{AB}^{\boldsymbol{x}^{**}}$ are *attracting* fixed points is far from settled in any nontrivial case.

Let A and B be affine subspaces of $\mathbb{R}^N$ that intersect in a point, $\boldsymbol{x}^{**}$. Let $g_A = \mathrm{Id}$ and $g_B = R_B$, the linear reflection; hence, $R_B^{-1} = R_B$. The set $L_A = A^\perp + \boldsymbol{x}^{**}$, the affine subspace orthogonal to A at $\boldsymbol{x}^{**}$, and similarly $L_B = B^\perp + \boldsymbol{x}^{**}$. The intersection satisfies

$$R_B^{-1}(L_A) \cap L_B = L_A \cap L_B = \boldsymbol{x}^{**} + A^\perp \cap B^\perp;$$

it is the center manifold. The $\dim L_A \cap L_B \geq N - (\dim A + \dim B)$, and therefore, if $\dim A + \dim B < N$, then $\mathscr{C}_{AB}^{\boldsymbol{x}^{**}}$ is positive dimensional.

Definition 7.2 In the sequel we use the notation

$$\begin{aligned} D_{AB}(\boldsymbol{x}) &= \boldsymbol{x} + P_A(R_B(\boldsymbol{x})) - P_B(\boldsymbol{x}), \\ D_{BA}(\boldsymbol{x}) &= \boldsymbol{x} + P_B(R_A(\boldsymbol{x})) - P_A(\boldsymbol{x}). \end{aligned} \tag{7.13}$$

Remark 7.3 If A and B are smooth submanifolds that have an isolated, transverse intersection at $\boldsymbol{x}^{**}$, then the descriptions of L_A and L_B, near to $\boldsymbol{x}^{**}$, are similar: L_A is locally the fiber of the normal bundle to A at $\boldsymbol{x}^{**}$, which we can think of as an affine linear subspace of $\mathbb{R}^N$. See Appendix 2.A. Similarly, near to $\boldsymbol{x}^{**}$, L_B is the fiber of the normal bundle to B at $\boldsymbol{x}^{**}$. If B is a smooth submanifold, then the action of R_B on $L_A \cap L_B$ is linear; in general, it is not. These are affine subspaces, and the dimensional calculation is exactly as before. In case A, for example, is nonlinear, the set L_A may not include the entire fiber of the normal bundle through $\boldsymbol{x}^{**}$, but may also have components therein, which are distant from the base point. It depends on how A is embedded into $\mathbb{R}^N$. As noted above, the center manifold defined in (7.12) may contain points that are, in fact, not fixed by the hybrid iterative map.

Algorithms to find points in an intersection $A \cap B$ typically exhibit two distinct phases:

(i) A global phase – during which the algorithm searches for an attracting basin.
(ii) A local phase – during which the iterates remain in the attracting basin defined by a particular fixed-point set.

Heuristically, this explains why hybrid iterative map algorithms might work better than AP-based algorithms in the global phase: if $\dim \mathscr{C}_{AB}^{\boldsymbol{x}^{**}} > 0$, then the attracting basins for hybrid iterative maps are neighborhoods of positive dimensional subsets, hence, potentially easier to find than the neighborhood of a point. This intuition can be found in Fienup (1982) and in Elser (2003), Elser et al. (2007), and Gravel and Elser (2008). While most work on algorithms, including this book, focus on the local phase, it would be very useful if methods could be found to influence the global phase. For example, it would very likely improve the final result, if attracting basins defined by transverse intersections of A and B could be made especially attractive.

There are circumstances where A and B do not intersect, but a hybrid iterative map has a positive dimensional invariant set, $\widetilde{\mathscr{C}}$, which may be transversely attracting. The dynamics of D_{AB} on $\widetilde{\mathscr{C}}$ are then nonconvergent. If there is a pair of distinct points $\boldsymbol{x}_A \in A$ and $\boldsymbol{x}_B \in B$ that define a critical point of the function $d_{AB} : A \times B \to [0, \infty)$, defined in (6.14), then $\widetilde{\mathscr{C}} = N_{\boldsymbol{x}_A} A \cap N_{\boldsymbol{x}_B} B \neq \emptyset$. This intersection is an invariant set for the map D_{AB}, which may be an attracting set. Note that the pair $(\boldsymbol{x}_A, \boldsymbol{x}_B)$ is *not* required to be a local minimum of d_{AB}. This phenomenon is illustrated in the following example:

Example 7.4 We show, in a simple two-dimensional example, that a generalized center manifold, defined by a saddle point of d_{AB}, may be transversely attracting, The A-curve (green curve in Figure 7.1) is the unit circle $\{(x, y) : x^2 + y^2 = 1\}$, and the B-curve (blue curve in Figure 7.1) is given parametrically as

$$B_{\alpha,\beta,p} = \{(\alpha + \beta \cos pt)(\cos t, \sin t) : t \in [0, 2\pi)\},$$

where α, $\beta > 0$ and $p \in \mathbb{N}$. If $\alpha + \beta < 1$, then the intersection $A \cap B_{\alpha,\beta,p}$ is empty, if $\alpha + \beta = 1$, then this intersection consists of p nontransversal intersections and if $\alpha + \beta$ is a little larger than 1, then this set consists of $2p$ transversal intersections.

In Figure 7.1 we show the results of running an iterative algorithm, based on the hybrid iterative map $D_{AB_{\alpha,\beta,p}}$, for 200 iterates with $\alpha = 0.95$, $\beta = 0.06$, and

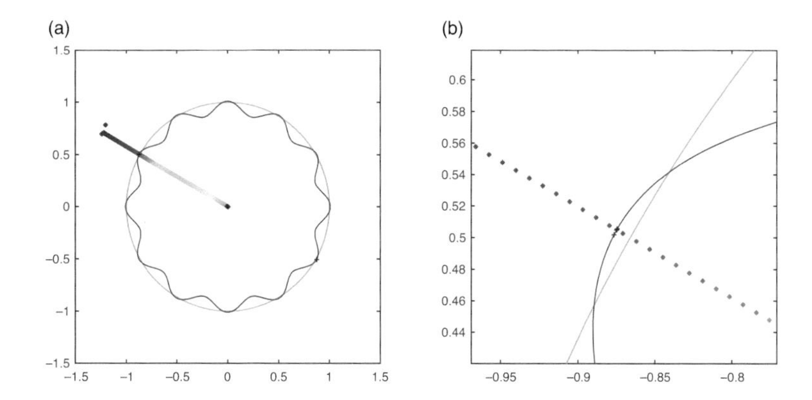

Figure 7.1 In part (a), we see 200 iterates of $D_{AB_{\alpha,\beta,p}}$ with $\alpha = 0.95$, $\beta = 0.06$, $p = 12$. Points are ordered from blue to red. Part (b) is a magnification of the portion of the trajectory near to the closest intersections of A and $B_{0.95,0.06,12}$. The iterates are attracted to a generalized center manifold defined by a saddle point of $d_{AB_{0.95,0.06,12}}$.

$p = 12$. As $\alpha + \beta = 1.01$ the set $A \cap B_{0.95,0.06,12}$ consists of 12 pairs of very nearby transverse intersections. We let $\boldsymbol{x}^{(n+1)} = D_{AB_{0.95,0.06,12}}(\boldsymbol{x}^{(n)})$, denote the iterates and $\boldsymbol{r}^{(n)} = P_{B_{0.95,0.06,12}}(\boldsymbol{x}^{(n)})$ the approximate reconstructions. Part (a) shows the full trajectory, with the earlier points in blue and the later ones in red. Part (b) is a blowup showing that the trajectory has accumulated on the generalized center manifold defined by a pair of points that define a saddle point for $d_{AB_{0.95,0.06,12}}$. After the first few iterates, the distances between successive iterates equal the constant value 0.01, which is the distance between A and $B_{0.95,0.06,12}$ along the generalized center manifold.

Phenomena quite similar to this are observed when using hybrid iterative maps to try to find intersections of $\mathbb{A}_{\boldsymbol{a}}$ and B_S. This shows that, while the true fixed points of these maps are connected to actual intersections of the subsets, hybrid iterative maps may also have (quasi-)attracting basins that are *not* associated to such intersections, but instead to critical points of $d_{\mathbb{A}_{\boldsymbol{a}}B_S}$. From elementary Morse theory (see Milnor 1963), it follows that the topology of $\mathbb{A}_{\boldsymbol{a}}$ forces this function to have a great profusion of critical points other than minima and maxima.

7.2.2 Hybrid Iterative Maps for Pairs of Linear Subspaces

Before we analyze hybrid iterative maps in the context of phase retrieval, we consider the case of a pair of linear subspaces. Let $A, B \subset \mathbb{R}^N$ be affine subspaces, which intersect at a single point that can be taken to be 0. The nearest point maps P_A, P_B are linear orthogonal projections, and the subsets L_A, L_B are simply $A^\perp, B^\perp$. As $R_A(A^\perp) = A^\perp$, it follows that

$$R_A^{-1}(L_B) \cap L_A = R_A(B^\perp) \cap A^\perp = R_A(B^\perp \cap A^\perp) = B^\perp \cap A^\perp. \tag{7.14}$$

Thus, the dimension of this intersection is

$$\dim R_A^{-1}(L_B) \cap L_A = \dim B^\perp + \dim A^\perp - N = N - (\dim A + \dim B). \tag{7.15}$$

This gives the hybrid iterative maps an advantage in the global phase if this dimension is greater than $0 = \dim A \cap B$; henceforth we assume that $\dim A + \dim B < N$. In applications to phase retrieval such a hypothesis is automatically satisfied.

To do the analysis we let $C = (A+B)^\perp$, which is easily seen to agree with the intersection $A^\perp \cap B^\perp = L_A \cap L_B$. In the terminology above this is the center manifold for D_{BA} defined by $\mathbf{0}$, which would be denoted by $\mathscr{C}_{BA}^{\mathbf{0}}$. We have the (nonorthogonal) decomposition

$$\mathbb{R}^N = A \oplus B \oplus C. \tag{7.16}$$

If Q, R, and V are matrices with orthonormal columns so that the columns span A, B, and C, respectively, then we can write

$$\boldsymbol{x} = Q\boldsymbol{x}_1 + R\boldsymbol{x}_2 + V\boldsymbol{x}_3. \tag{7.17}$$

The Euclidean inner product is then given by (7.2); as before $H = R^t Q$. In the sequel $\langle \cdot, \cdot \rangle$ is the Euclidean inner product in spaces of various dimensions. The specific space should be clear from the context. Using (7.4) we obtain

$$\begin{aligned} R_A(\boldsymbol{x}_1, \boldsymbol{x}_2, \boldsymbol{x}_3) &= (\boldsymbol{x}_1 + 2H^t\boldsymbol{x}_2, \ -\boldsymbol{x}_2, \ -\boldsymbol{x}_3) \text{ and} \\ R_B(\boldsymbol{x}_1, \boldsymbol{x}_2, \boldsymbol{x}_3) &= (-\boldsymbol{x}_1, \ \boldsymbol{x}_2 + 2H\boldsymbol{x}_1, \ -\boldsymbol{x}_3), \end{aligned} \tag{7.18}$$

which, along with (7.4), imply that

$$\begin{aligned} D_{AB}(\boldsymbol{x}) &= \boldsymbol{x} + P_A(R_B(\boldsymbol{x})) - P_B(\boldsymbol{x}) = (H^t\boldsymbol{x}_2 + 2H^tH\boldsymbol{x}_1, -H\boldsymbol{x}_1, \ \boldsymbol{x}_3), \\ D_{BA}(\boldsymbol{x}) &= \boldsymbol{x} + P_B(R_A(\boldsymbol{x})) - P_A(\boldsymbol{x}) = (-H^t\boldsymbol{x}_2, \ H\boldsymbol{x}_1 + 2HH^t\boldsymbol{x}_2, \ \boldsymbol{x}_3). \end{aligned} \tag{7.19}$$

To show that these maps contract onto the center manifold we use the following lemma:

Lemma 7.5 *The linear transformations on the space $A+B$ defined by*

$$A_H : \begin{pmatrix} \boldsymbol{x}_1 \\ \boldsymbol{x}_2 \end{pmatrix} \rightarrow \begin{pmatrix} 2H^t H & H^t \\ -H & 0 \end{pmatrix} \begin{pmatrix} \boldsymbol{x}_1 \\ \boldsymbol{x}_2 \end{pmatrix} \text{ and } B_H : \begin{pmatrix} \boldsymbol{x}_1 \\ \boldsymbol{x}_2 \end{pmatrix} \rightarrow \begin{pmatrix} 0 & -H^t \\ H & 2HH^t \end{pmatrix} \begin{pmatrix} \boldsymbol{x}_1 \\ \boldsymbol{x}_2 \end{pmatrix} \tag{7.20}$$

have norm less than one with respect to the inner product

$$((\boldsymbol{x}_1, \boldsymbol{x}_2), (\boldsymbol{y}_1, \boldsymbol{y}_2))_H = \langle \boldsymbol{x}_1, \boldsymbol{y}_1 \rangle + \langle H\boldsymbol{x}_1, \boldsymbol{y}_2 \rangle + \langle \boldsymbol{x}_2, H\boldsymbol{y}_1 \rangle + \langle \boldsymbol{x}_2, \boldsymbol{y}_2 \rangle. \tag{7.21}$$

Remark 7.6 Recall that $H = R^t Q$; the inner product in (7.21) is the standard Euclidean inner product for $\mathbb{R}^N$ restricted to $A \oplus B$, expressed in terms of the basis Q, R, see (7.2).

Proof Let

$$G_H = \begin{pmatrix} \mathrm{Id} & H^t \\ H & \mathrm{Id} \end{pmatrix}$$

be the matrix defining the inner product, $(\cdot,\cdot)_H$. We give the proof for B_H; the other case is quite similar. We need to show that

$$\langle G_H B_H \boldsymbol{x}, B_H \boldsymbol{x} \rangle < \langle G_H \boldsymbol{x}, \boldsymbol{x} \rangle. \tag{7.22}$$

Expanding and simplifying, we see that this is equivalent to

$$\langle B'_H \boldsymbol{x}, \boldsymbol{x} \rangle < \langle G_H \boldsymbol{x}, \boldsymbol{x} \rangle, \tag{7.23}$$

where

$$B'_H = \begin{pmatrix} H^t H & H^t H H^t \\ H H^t H & H H^t \end{pmatrix} = \begin{pmatrix} 0 & H^t \\ H & 0 \end{pmatrix} G_H \begin{pmatrix} 0 & H^t \\ H & 0 \end{pmatrix}. \tag{7.24}$$

We let $L = \begin{pmatrix} 0 & H^t \\ H & 0 \end{pmatrix}$ and J_H denote the positive square root of G_H. Since $L = G_H - \mathrm{Id}$, the matrix J_H commutes with L. Hence, (7.23) is equivalent to showing that

$$\langle L\boldsymbol{y}, L\boldsymbol{y} \rangle < \langle \boldsymbol{y}, \boldsymbol{y} \rangle, \tag{7.25}$$

where $\boldsymbol{y} = J_H \boldsymbol{x}$. Since $H = R^t Q$, it is immediate that

$$\|L(\boldsymbol{y}_1, \boldsymbol{y}_2)\|_2^2 = \|(H^t \boldsymbol{y}_2, H\boldsymbol{y}_1)\|_2^2 = \|H\boldsymbol{y}_1\|_2^2 + \|H^t \boldsymbol{y}_2\|_2^2 < \|\boldsymbol{y}_1\|_2^2 + \|\boldsymbol{y}_2\|_2^2. \tag{7.26}$$

The inequality is strict because $A \cap B = \{0\}$, and therefore, the singular values of H (and H^t) are strictly less than 1. Thus, the operator norm $\|B_H\|_H < 1$. The argument for A_H is essentially identical. □

Remark 7.7 Examining the proof, we see that the singular values, $\{\sigma_n\}$, of B_H relative to the inner product defined by G_H equal the square roots of the eigenvalues of the matrix $\begin{pmatrix} H^t H & 0 \\ 0 & HH^t \end{pmatrix}$. These are, in turn, given by

$$\sigma_j = \cos^2\theta_j \approx 1 - \theta_j^2, \tag{7.27}$$

where $\{\theta_j\}$ are the angles between A and B defined in (2.8). The $\{\sigma_n\}$ are, of course, the singular values of H.

Using this matrix notation we can express the action of the two hybrid iterative maps, in the linear case, as

$$\begin{aligned} D_{AB}(\boldsymbol{x}_1, \boldsymbol{x}_2, \boldsymbol{x}_3) &= \begin{pmatrix} A_H & 0 \\ 0 & \mathrm{Id} \end{pmatrix} \begin{pmatrix} \begin{bmatrix} \boldsymbol{x}_1 \\ \boldsymbol{x}_2 \end{bmatrix} \\ \boldsymbol{x}_3 \end{pmatrix} \\ D_{BA}(\boldsymbol{x}_1, \boldsymbol{x}_2, \boldsymbol{x}_3) &= \begin{pmatrix} B_H & 0 \\ 0 & \mathrm{Id} \end{pmatrix} \begin{pmatrix} \begin{bmatrix} \boldsymbol{x}_1 \\ \boldsymbol{x}_2 \end{bmatrix} \\ \boldsymbol{x}_3 \end{pmatrix}, \end{aligned} \tag{7.28}$$

which shows that the iterates of these maps leave the $\boldsymbol{x}_3$-component invariant and

$$\lim_{n\to\infty} D_{AB}^n(\boldsymbol{x}_1, \boldsymbol{x}_2, \boldsymbol{x}_3) = (0, 0, \boldsymbol{x}_3) \text{ and } \lim_{n\to\infty} D_{BA}^n(\boldsymbol{x}_1, \boldsymbol{x}_2, \boldsymbol{x}_3) = (0, 0, \boldsymbol{x}_3). \tag{7.29}$$

From (7.28) it is clear that the behavior of the iterates of D_{AB} and D_{BA} is determined by the singular values $\{\sigma_n\}$, of H. The "more orthogonal" A is to B, i.e., the smaller σ_1 is, the faster the convergence to the center manifold. Whereas, singular values close to 1 lead to slow convergence for both of these algorithms.

Remark 7.8 In case the auxiliary information is a support constraint, the HIO algorithm uses the map given in (6.4), with $A = \mathbb{A}_{\boldsymbol{a}}$, $B = B_s$, and $0 < \beta < 1$. We can analyze this map in the case that A and B are linear subspaces of $\mathbb{R}^N$, with $A \cap B = \boldsymbol{0}$ and $\dim A + \dim B < N$. Using the splitting $\boldsymbol{x} = Q\boldsymbol{x}_1 + R\boldsymbol{x}_2 + V\boldsymbol{x}_3$, this map takes the form

$$D_{BA}^{\beta}(\boldsymbol{x}) = \begin{pmatrix} (1-\beta)\,\mathrm{Id} & -\beta H^t & 0 \\ \beta H & (1+\beta) H H^t & 0 \\ 0 & 0 & \mathrm{Id} \end{pmatrix} \cdot \begin{pmatrix} \boldsymbol{x}_1 \\ \boldsymbol{x}_2 \\ \boldsymbol{x}_3 \end{pmatrix}. \tag{7.30}$$

The subspace on which the matrix B_H has singular value σ_j is spanned by two vectors $\boldsymbol{u}_j$ and $\boldsymbol{v}_j$, which satisfy $H\boldsymbol{u}_j = \sigma_j \boldsymbol{v}_j$ and $H^t \boldsymbol{v}_j = \sigma_j \boldsymbol{u}_j$. If we denote the upper 2×2 block in the matrix on the right-hand side of (7.30) by B_H^{β}, then we see that the subspaces $\{(a\boldsymbol{u}_j, b\boldsymbol{v}_j) : a, b \in \mathbb{R}\}$ are invariant subspaces for B_H^{β}. On these subspaces B_H^{β} reduces to

$$B_H^{\beta} \cdot \begin{pmatrix} a\boldsymbol{u}_j \\ b\boldsymbol{v}_j \end{pmatrix} = \begin{pmatrix} (1-\beta) & -\beta\sigma_j \\ \beta\sigma_j & (1+\beta)\sigma_j^2 \end{pmatrix} \cdot \begin{pmatrix} a\boldsymbol{u}_j \\ b\boldsymbol{v}_j \end{pmatrix}. \tag{7.31}$$

We denote the singular values on this subspace by $\sigma_j^-(\beta) \le \sigma_j^+(\beta)$. A simple calculation shows that

$$\sigma_j^-(\beta) \cdot \sigma_j^+(\beta) = \sigma_j^2, \tag{7.32}$$

but $\sigma_j^-(\beta) \neq \sigma_j^+(\beta)$, unless $\beta = 1$. From this we conclude that $\sigma_1 = \sigma_1(1) < \sigma_1^+(\beta)$, for $\beta < 1$, and therefore, the sequences $\{[D_{BA}^{\beta}]^n(\boldsymbol{x}^{(0)})\}$ have the fastest rate of convergence when $\beta = 1$.

If A and B intersect in a positive dimensional subspace, $F = A \cap B$, then we further split

$$\mathbb{R}^N = A_0 \oplus B_0 \oplus F \oplus C, \tag{7.33}$$

where $A_0 = A \cap F^{\perp}$, $B_0 = B \cap F^{\perp}$, and $C = (A+B)^{\perp}$, as before. With respect to this splitting, $\boldsymbol{x} = Q\boldsymbol{x}_1 + R\boldsymbol{x}_2 + U\boldsymbol{x}_3 + V\boldsymbol{x}_4$, with Q, R, U, V orthonormal bases for the respective subspaces. The maps now take the form

$$\begin{aligned} D_{AB}(\boldsymbol{x}_1,\, \boldsymbol{x}_2,\, \boldsymbol{x}_3,\, \boldsymbol{x}_4) &= \begin{pmatrix} A_H & 0 & 0 \\ 0 & \mathrm{Id} & 0 \\ 0 & 0 & \mathrm{Id} \end{pmatrix} \left(\begin{matrix} \begin{bmatrix} \boldsymbol{x}_1 \\ \boldsymbol{x}_2 \end{bmatrix} \\ \boldsymbol{x}_3 \\ \boldsymbol{x}_4 \end{matrix} \right) \\ D_{BA}(\boldsymbol{x}_1,\, \boldsymbol{x}_2,\, \boldsymbol{x}_3,\, ,\boldsymbol{x}_4) &= \begin{pmatrix} B_H & 0 & 0 \\ 0 & \mathrm{Id} & 0 \\ 0 & 0 & \mathrm{Id} \end{pmatrix} \left(\begin{matrix} \begin{bmatrix} \boldsymbol{x}_1 \\ \boldsymbol{x}_2 \end{bmatrix} \\ \boldsymbol{x}_3 \\ \boldsymbol{x}_4 \end{matrix} \right), \end{aligned} \tag{7.34}$$

which satisfy

$$\begin{aligned} &\lim_{n\to\infty} D_{AB}^n(\boldsymbol{x}_1,\, \boldsymbol{x}_2,\, \boldsymbol{x}_3,\, \boldsymbol{x}_4) = (0,0,\, \boldsymbol{x}_3,\, \boldsymbol{x}_4) \text{ and} \\ &\qquad\qquad \lim_{n\to\infty} D_{BA}^n(\boldsymbol{x}_1,\, \boldsymbol{x}_2,\, \boldsymbol{x}_3,\, \boldsymbol{x}_4) = (0,0,\, \boldsymbol{x}_3,\, \boldsymbol{x}_4). \end{aligned} \tag{7.35}$$

Remark 7.9 In the linear case, equations (7.5) and (7.7) show that AP also converges. A crucial difference is that the hybrid iterative maps are converging to the center manifold, and not to intersection, $A \cap B$. The heuristics suggesting that the hybrid map is “looking” for a higher-dimensional target are borne out in these linear examples. On the other hand, it should be noted that the hybrid iterative maps are not contracting in the center manifold directions, and this considerably complicates the analysis of their nonlinear perturbations.

Remark 7.10 It is possible to do a more refined analysis of the linear case, which reveals additional subspaces on which the hybrid iterative maps D_{AB}, D_{BA}, converge in a single step. For this analysis we define the subspaces

$$F = A \cap B, \ K = A \cap B^{\perp}, \text{ and } L = A^{\perp} \cap B, \tag{7.36}$$

and then set

$$\widetilde{A}_0 = A \cap (F \oplus K)^{\perp} \text{ and } \widetilde{B}_0 = B \cap (F \oplus L)^{\perp}. \tag{7.37}$$

As before, $C = (A+B)^{\perp}$, and we let $\widetilde{Q}$, $\widetilde{R}$, U, V, X, Y be orthonormal bases for $\widetilde{A}_0$, $\widetilde{B}_0$, F, C, K, L, respectively; with $\mathbb{R}^N = \widetilde{A}_0 \oplus \widetilde{B}_0 \oplus F \oplus C \oplus K \oplus L$,

$$\boldsymbol{x} = \widetilde{Q}\boldsymbol{x}_1 + \widetilde{R}\boldsymbol{x}_2 + U\boldsymbol{x}_3 + V\boldsymbol{x}_4 + X\boldsymbol{x}_5 + Y\boldsymbol{x}_6, \tag{7.38}$$

the inner product is given by

$$\langle \boldsymbol{x}, \boldsymbol{y} \rangle = \langle \boldsymbol{x}_1, \boldsymbol{y}_1 \rangle + \langle \widetilde{H}\boldsymbol{x}_1, \boldsymbol{y}_2 \rangle + \langle \boldsymbol{x}_2, \widetilde{H}\boldsymbol{y}_1 \rangle + \langle \boldsymbol{x}_2, \boldsymbol{y}_2 \rangle + \\ \langle \boldsymbol{x}_3, \boldsymbol{y}_3 \rangle + \langle \boldsymbol{x}_4, \boldsymbol{y}_4 \rangle + \langle \boldsymbol{x}_5, \boldsymbol{y}_5 \rangle + \langle \boldsymbol{x}_6, \boldsymbol{y}_6 \rangle, \tag{7.39}$$

where $\widetilde{H} = \widetilde{R}^t \widetilde{Q}$. Calculations like those above show that

$$D_{AB}(\boldsymbol{x}) = (2\widetilde{H}^t \widetilde{H}\boldsymbol{x}_1 + \widetilde{H}^t \boldsymbol{x}_2, \ -\widetilde{H}\boldsymbol{x}_1, \ \boldsymbol{x}_3, \ \boldsymbol{x}_4, \ \boldsymbol{0}, \ \boldsymbol{0}). \tag{7.40}$$

The $K \oplus L$-components converge in a single iterate to $\boldsymbol{0}$; the argument used to prove Lemma 7.5 can be modified to show that

$$\lim_{n \to \infty} D_{AB}^n(\boldsymbol{x}) = (\boldsymbol{0}, \ \boldsymbol{0}, \ \boldsymbol{x}_3, \ \boldsymbol{x}_4, \ \boldsymbol{0}, \ \boldsymbol{0}). \tag{7.41}$$

In our applications, the matrix $\widetilde{H}$ has many singular values close to 1, and F itself is often nonzero. In this circumstance the existence of even a large dimensional subspace, where the linearization converges in a single step, does not seem to have much impact on the behavior of the types of nonlinear perturbations of these maps that arise in phase retrieval. In light of this, most of the analysis in subsequent sections employs the simpler decomposition $\mathbb{R}^N = A_0 \oplus B_0 \oplus F \oplus C$.

Remark 7.11 It is interesting to note that in the linear case, the pairs of subspaces (A, B) and $(A^{\perp}, B^{\perp})$ play entirely symmetric roles in the map D_{AB}. A moment's thought shows that this map can be rewritten as

$$D_{AB}(\boldsymbol{x}) = P_{A^{\perp}} \circ P_{B^{\perp}}(\boldsymbol{x}) + P_A \circ P_B(\boldsymbol{x}). \tag{7.42}$$

With this reformulation it is easy to show that the fixed-point set of D_{AB} equals $A \cap B \cup A^{\perp} \cap B^{\perp}$.

A small modification of the analysis above allows us to consider the case that A and B do not intersect. We normalize A to be a linear subspace, such that $\mathbf{0}$ is the unique point in A of minimum distance to B. (The case where a positive dimensional subspace of A is at minimum distance can be similarly analyzed; we leave this to the reader.) We set $B = \boldsymbol{v} + B^0$, where B^0 is a linear subspace and $\boldsymbol{v}$ is the unique vector realizing this shortest distance. This implies that $A \cap B^0 = \{\mathbf{0}\}$. The reduced center manifold is

$$C_0 = (A + B^0 + [\![\boldsymbol{v}]\!])^{\perp} = A^{\perp} \cap B^{0\perp} \cap [\![\boldsymbol{v}]\!]^{\perp},$$

where we use the notation $[\![W]\!]$ to denote the linear subspace spanned by the vectors in set W. The center manifold, $C = (A + B^0)^{\perp} = C_0 + [\![\boldsymbol{v}]\!]$, is an invariant set for D_{BA}, whereas C_0 is not.

With the splitting $\mathbb{R}^N = A \oplus B^0 \oplus C_0 \oplus [\![\boldsymbol{v}]\!]$, we write $\boldsymbol{x} = Q\boldsymbol{x}_1 + R\boldsymbol{x}_2 + V\boldsymbol{x}_3 + \boldsymbol{v}\boldsymbol{x}_4$. A simple calculation shows that

$$D_{BA}(\boldsymbol{x}_1, \boldsymbol{x}_2, \boldsymbol{x}_3, \boldsymbol{x}_4) = \begin{pmatrix} B_H & 0 & 0 \\ 0 & \mathrm{Id} & 0 \\ 0 & 0 & \mathrm{Id} \end{pmatrix} \begin{pmatrix} \begin{bmatrix} \boldsymbol{x}_1 \\ \boldsymbol{x}_2 \end{bmatrix} \\ \boldsymbol{x}_3 \\ \boldsymbol{x}_4 \end{pmatrix} + \begin{pmatrix} 0 \\ 0 \\ 0 \\ 1 \end{pmatrix}, \tag{7.43}$$

and so

$$D^n_{BA}(\boldsymbol{x}_1, \boldsymbol{x}_2, \boldsymbol{x}_3, \boldsymbol{x}_4) = \begin{pmatrix} B^n_H & 0 & 0 \\ 0 & \mathrm{Id} & 0 \\ 0 & 0 & \mathrm{Id} \end{pmatrix} \begin{pmatrix} \begin{bmatrix} \boldsymbol{x}_1 \\ \boldsymbol{x}_2 \end{bmatrix} \\ \boldsymbol{x}_3 \\ \boldsymbol{x}_4 \end{pmatrix} + \begin{pmatrix} 0 \\ 0 \\ 0 \\ n \end{pmatrix}. \tag{7.44}$$

The first three components of these trajectories spiral in toward $(0, 0, \boldsymbol{x}_3)$. The iterates themselves move linearly toward infinity along the direction $\boldsymbol{v}$. The analysis of D_{AB} is essentially identical. When $\|\boldsymbol{v}\| \ll 1$ the initial iterates produced by a near miss look quite similar to those that would have been produced by an exact intersection.

Remark 7.12 Perhaps the most interesting feature of the analysis when $A \cap B = \emptyset$ is that the center manifold, $(A + B^0)^{\perp}$, is very stable under perturbations in A and B, and has a good definition even when A and B do not quite intersect. This is important for the application of this analysis to nonlinear perturbations of the subsets as it provides a stable target, in the global phase, for the iterates of algorithms based on hybrid iterative maps, at least for a small $\boldsymbol{v}$. This analysis of the near-miss case explains the behavior observed in Example 7.4.

In real applications, where the data consists of samples of the magnitude of the continuum Fourier transform, and is contaminated by noise, a pair of subsets that do not actually intersect is almost a certainty. Beyond that, very small, but nonzero, critical points of d_{AB} (see (1.35)) define (quasi-)invariant subsets for the hybrid iterative map, which may well be transversely attracting.

7.2.3 The RAAR Algorithm

Another algorithm that is quite popular in the experimental literature is called the *relaxed averaged alternating reflection* algorithm, or RAAR (see Luke 2005; Chapman et al. 2006). With our notation, this algorithm is defined by the map

$$RA(\boldsymbol{x}) = \frac{\beta}{2}[R_B \circ R_A(\boldsymbol{x}) + \boldsymbol{x}] + (1-\beta)P_A(\boldsymbol{x}), \tag{7.45}$$

where β, the relaxation parameter, lies in $(0,\ 1)$. In the linear case, using the decomposition $\mathbb{R}^N = A_0 \oplus B_0 \oplus F \oplus C$, we easily obtain the block matrix representation

$$RA(\boldsymbol{x}_1,\ \boldsymbol{x}_2,\ \boldsymbol{x}_3,\ \boldsymbol{x}_4) = \begin{pmatrix} (1-\beta)\begin{pmatrix} I & H^t \\ 0 & 0 \end{pmatrix} + \beta B_H & \mathbf{0} \\ \mathbf{0} & \begin{pmatrix} \mathrm{Id} & 0 \\ 0 & \beta\,\mathrm{Id} \end{pmatrix} \end{pmatrix} \begin{pmatrix} \boldsymbol{x}_1 \\ \boldsymbol{x}_2 \\ \boldsymbol{x}_3 \\ \boldsymbol{x}_4 \end{pmatrix}. \tag{7.46}$$

If $F = \{\mathbf{0}\}$, then one simply removes the third row and column (containing Id) from this matrix. We let Q_β denote the upper 2×2 block.

Proceeding as in the proof of Lemma 7.5 we see that

$$Q_\beta^t G_H Q_\beta = \begin{pmatrix} (1-\beta)\,\mathrm{Id} + \beta H^t H & (1-\beta)H^t + \beta H^t H H^t \\ (1-\beta)H + \beta H H^t H & H H^t \end{pmatrix}. \tag{7.47}$$

A calculation shows that if $\{\sigma_n^2\}$ are the nonzero singular values of $H^t H$, then the squares of the nonzero singular values of this matrix are $\{(1-\beta)+\beta\sigma_n^2\} \cup \{\beta\sigma_n^2\}$. As $(1-\beta)+\beta\sigma_n^2 > \sigma_n^2$, an algorithm based on iterating RA for $0 < \beta < 1$, can be expected to converge more slowly than those defined by D_{AB} and D_{BA}. In any case, very small angles between A and B lead to a very slow rate of convergence. If $\dim A \cap B \neq 0$, then there is a positive dimensional subspace, F, on which RA acts as the identity.

7.3 Nonlinear Submanifolds

The analysis above extends easily to the case that A is a smooth, nonlinear submanifold of $\mathbb{R}^N$ and B is an affine subspace, if we assume that $A \cap B$ consists of isolated points $\{\boldsymbol{x}_l^{**}\}$ *and* that these intersections are transversal. That is, for each l the tangent space of A at $\boldsymbol{x}_l^{**}$ meets B only at $\boldsymbol{x}_l^{**}$:

$$T_{\boldsymbol{x}_l^{**}} A \cap B = \{\boldsymbol{x}_l^{**}\}. \tag{7.48}$$

We analyze the behavior of the hybrid iterative maps in a neighborhood of such an intersection point. Since A is nonlinear, the nearest point map $P_A(\boldsymbol{x})$ may fail to be single valued or continuous. To do this analysis, we need to assume that we are close enough to an intersection point that P_A is both single valued and differentiable. We have already seen that, for the phase retrieval problem, these intersections are usually *not* transversal, and so we briefly address the linearized analysis in this case in Section 7.3.2.

The analysis in Section 7.2 shows that, in the linear case, the maps D_{AB} and D_{BA} are contracting in directions normal to the center manifold, but act as the identity in $C \oplus F$, where $F = A \cap B$. In fact, if the iterates of such a map converge, then they usually converge to a point on the center manifold quite distant from the intersection point that defines it. For the sake of completeness, we include the linearized analysis near points in $A \cap B$. In Chapter 8, we revisit this question for hybrid iterative maps in the context of phase retrieval by studying some simplified models in great detail. For the actual phase retrieval problem, we linearize the hybrid iterative maps at arbitrary points along the center manifold.

7.3.1 The Transversal Case

Let us shift the coordinates so that the point of intersection $\boldsymbol{x}^{**} = \boldsymbol{0}$, and let $A_T = T_{\boldsymbol{x}^{**}} A$, the tangent space to A. Since $\boldsymbol{x}^{**} = \boldsymbol{0}$, A_T is a subspace of $\mathbb{R}^N$, rather than an affine subspace. In a neighborhood of $\boldsymbol{x}^{**}$ the set L_A agrees with the normal bundle to A at $\boldsymbol{x}^{**}$, which we denote $A_T^{\perp}$. Let

$$C = (A_T + B)^{\perp} = A_T^{\perp} \cap B^{\perp} = L_A \cap L_B.$$

Following the linear case, we let $\mathbb{R}^N = A_T \oplus B \oplus C$. With $\boldsymbol{x} = R\boldsymbol{x}_1 + Q\boldsymbol{x}_2 + V\boldsymbol{x}_3$, the inner product is again given by (7.2). Near to $\boldsymbol{x}^{**}$, A is given as a graph over its tangent space by

$$A = \{R\boldsymbol{x}_1 + Qq_2(\boldsymbol{x}_1) + Vq_3(\boldsymbol{x}_1) + \text{h.o.t.}\}, \tag{7.49}$$

where

$$q_p(\boldsymbol{x}_1) = \langle q_p \boldsymbol{x}_1, \boldsymbol{x}_1 \rangle = (q_{p1}^{ij} x_{1i} x_{1j}, \ldots, q_{pl_p}^{ij} x_{1i} x_{1j}) \tag{7.50}$$

are vector-valued symmetric quadratic forms and "h.o.t." are additional terms that are $O(\|\boldsymbol{x}_1\|^3)$.

We want to calculate the linearizations of D_{AB} and D_{BA} along the center manifold near to the intersection point, which is represented by the set $\{(0, 0, \boldsymbol{x}_3)\}$, with $\|\boldsymbol{x}_3\|$ small. This is an elementary, but tedious, calculation using the chain rule. We just summarize the results:

$$\begin{aligned} D_{AB}(\boldsymbol{x}_1, \boldsymbol{x}_2, \boldsymbol{x}_3) &= \begin{pmatrix} 2(\mathrm{Id}+K(\boldsymbol{x}_3))H^t H - K(\boldsymbol{x}_3) & (\mathrm{Id}+K(\boldsymbol{x}_3))H^t & 0 \\ -H & 0 & 0 \\ 0 & 0 & \mathrm{Id} \end{pmatrix} \begin{pmatrix} \boldsymbol{x}_1 \\ \boldsymbol{x}_2 \\ \boldsymbol{x}_3 \end{pmatrix} \\ &\quad + \mathrm{h.o.t.} \\ D_{BA}(\boldsymbol{x}_1, \boldsymbol{x}_2, \boldsymbol{x}_3) &= \begin{pmatrix} K(\boldsymbol{x}_3) & -(\mathrm{Id}+K(\boldsymbol{x}_3))H^t & 0 \\ H(\mathrm{Id}+2K(\boldsymbol{x}_3)) & 2H(\mathrm{Id}+K(\boldsymbol{x}_3))H^t & 0 \\ 0 & 0 & \mathrm{Id} \end{pmatrix} \begin{pmatrix} \boldsymbol{x}_1 \\ \boldsymbol{x}_2 \\ \boldsymbol{x}_3 \end{pmatrix} + \mathrm{h.o.t.} \end{aligned} \tag{7.51}$$

Here, $K(\boldsymbol{x}_3)$ is a linear transformation defined by

$$\mathrm{Id}+K(\boldsymbol{x}_3) = (\mathrm{Id}-2q_3^{\dagger}(\boldsymbol{x}_3))^{-1} \tag{7.52}$$

where for $\boldsymbol{x}_1, \boldsymbol{x}_1' \in A_T$, and $\boldsymbol{x}_3 \in C$, $q_3^{\dagger}(\boldsymbol{x}_3) : A_T \to A_T$ is a linear transformation defined by

$$\langle q_3^{\dagger}(\boldsymbol{x}_3)\boldsymbol{x}_1, \boldsymbol{x}_1' \rangle = \langle \boldsymbol{x}_3, q_3(\boldsymbol{x}_1, \boldsymbol{x}_1') \rangle. \tag{7.53}$$

Note that $K(0) = 0$. The higher-order terms here are at least quadratic in $(\boldsymbol{x}_1, \boldsymbol{x}_2)$.

From Lemma 7.5 it follows immediately that, for sufficiently small $\boldsymbol{x}_3$, the maps D_{AB} and D_{BA} are contractions in the directions normal to C. Assuming that the $\boldsymbol{x}_3$-components remain small and converge, standard contraction mapping theorems would then show that the iterates converge to a point on the center manifold. As noted earlier, because the linearizations of the hybrid iterative maps reduce to the identity along the center manifold, any contraction properties in these components are genuinely nonlinear, and so would require a much more elaborate analysis to demonstrate. Moreover, one would expect that these components are contracting on some portions of the center manifold and are either neutral, or expanding along other portions, see Section 7.3.3. In the phase retrieval problem, we have found that, when the iterates of a hybrid iterative map, $D_{\mathbb{A}_a B}$ converge, it is often to points, $\boldsymbol{x}$, on the center manifold where the linearized map, $dD_{\mathbb{A}_a B}(\boldsymbol{x})$, is *not* a contraction. In fact, $dD_{\mathbb{A}_a B}(\boldsymbol{x})$ is a highly nonnormal linear map, with complex eigenvalues of modulus less than 1, and many singular values greater than 1. See Section 8.2.

7.3.2 The Nontransversal Case

We briefly discuss how the analysis needs to be modified if $F = T_{\boldsymbol{x}^{**}} A \cap T_{\boldsymbol{x}^{**}} B$ is positive dimensional. The linear analysis in this case uses a decomposition like that in (7.33), with $\mathbb{R}^N = A_0 \oplus B_0 \oplus F \oplus C$, where we let $A_0 = T_{\boldsymbol{x}^{**}} A \cap F^{\perp}$, $B_0 = B \cap F^{\perp}$, and $C = (A+B)^{\perp}$. The tangent space to A, $T_{\boldsymbol{x}^{**}} A$, is represented by $\{(\boldsymbol{x}_1, \boldsymbol{0}, \boldsymbol{x}_3, \boldsymbol{0})\}$, and B is given by $\{(\boldsymbol{0}, \boldsymbol{x}_2, \boldsymbol{x}_3, \boldsymbol{0})\}$. Locally A is represented as a graph over its tangent plane by

$$Q\boldsymbol{x}_1 + Rq_2(\boldsymbol{x}_1, \boldsymbol{x}_3) + U\boldsymbol{x}_3 + Vq_4(\boldsymbol{x}_1, \boldsymbol{x}_3), \tag{7.54}$$

where, for $j = 2, 4$ we have that

$$q_j(\boldsymbol{0}, \boldsymbol{0}) = \boldsymbol{0} \text{ and } \nabla q_j(\boldsymbol{0}, \boldsymbol{0}) = \boldsymbol{0}. \tag{7.55}$$

The center manifold is defined near to $\boldsymbol{x}^{**}$ by $\{(\boldsymbol{0}, \boldsymbol{0}, \boldsymbol{0}, \boldsymbol{x}_4)\}$.

The linearization of D_{AB} along the center manifold is given by:

$$D_{AB}(\boldsymbol{x}_1, \boldsymbol{x}_2, \boldsymbol{x}_3, \boldsymbol{x}_4) = \begin{pmatrix} 2(\mathrm{Id}+K_1(\boldsymbol{x}_4))H^t H - K_1(\boldsymbol{x}_4) & (\mathrm{Id}+K_1(\boldsymbol{x}_4))H^t & -K_2(\boldsymbol{x}_4) & 0 \\ -H & 0 & 0 & 0 \\ K_3(\boldsymbol{x}_4)(\mathrm{Id}-2H^t H) & -K_3(\boldsymbol{x}_4)H^t & \mathrm{Id}-K_4(\boldsymbol{x}_4) & 0 \\ 0 & 0 & 0 & \mathrm{Id} \end{pmatrix} \begin{pmatrix} \boldsymbol{x}_1 \\ \boldsymbol{x}_2 \\ \boldsymbol{x}_3 \\ \boldsymbol{x}_4 \end{pmatrix} + \text{h.o.t.} \tag{7.56}$$

Here, K_1, K_2, K_3, and K_4 are linear matrix-valued functions, which vanish at $\boldsymbol{x}_4 = \boldsymbol{0}$; the higher-order terms are at least quadratic in $(\boldsymbol{x}_1, \boldsymbol{x}_2, \boldsymbol{x}_3)$. There is a similar expression for D_{BA}. For small enough $\boldsymbol{x}_4$ the linearized map is contracting in the $A_0 \oplus B_0$-directions. On the other hand, at first order these maps are *not* obviously contracting in directions lying in F, which are normal to the center manifold. At the intersection point $(\boldsymbol{0}, \boldsymbol{0}, \boldsymbol{0}, \boldsymbol{0})$, these components of the map reduce to the identity. The map is truly nonlinear in these directions and its behavior cannot be predicted without considerable additional information about the nonlinear manifold A and its relationship to B.

The linearization in the C-components reduces to the identity along the subset of the center manifold that is fixed, pointwise, by the map. Experimentally, in the phase retrieval problem (where $A = \mathbb{A}_{\boldsymbol{a}}$ and $B = B_S$), the iterates in the C- and F-directions behave quite similarly. At points $\boldsymbol{x} \in \mathbb{A}_{\boldsymbol{a}}$ close enough to $\boldsymbol{x}^{**}$, the tangent space $T_{\boldsymbol{x}}\mathbb{A}_{\boldsymbol{a}}$ has a subspace that makes a very small angle with B_S. The dimension of this subspace is bounded below by $\dim F$ as $\boldsymbol{x} \to \boldsymbol{x}^{**}$ and is usually much larger. The iterates of the algorithms either converge geometrically to a point on the center manifold, or the iterates fall into an attracting basin, and then stagnate. The bulk of the (more or less constant sized) differences, $\{\boldsymbol{x}^{(n)} - \boldsymbol{x}^{(n+1)}\}$, are in the $C \oplus F$-directions. Convergence occurs when $T_{\boldsymbol{x}^{**}}\mathbb{A}_{\boldsymbol{a}}$ and B_S intersect in a very low-dimensional subspace and the nonzero angles between them are not too small. Even a one-dimensional intersection between $T_{\boldsymbol{x}^{**}}\mathbb{A}_{\boldsymbol{a}}$ and B_S can cause an algorithm based on a hybrid iterative map to stagnate. These phenomena are illustrated in Example 8.7.

7.3.3 Simple Nonlinear Examples

In the Introduction, we discussed the use of Newton's method for finding the intersections of the curves $\{y = x^p\}$, with $\{y = 0\}$, for $2 \leq p$. These examples illustrate the well-known fact that, instead of the supergeometric convergence obtained for a transversal intersection, Newton's method converges geometrically with a rate that decreases as the order of contact increases.

We conclude this section with a few simple examples in three-dimensions, using hybrid iterative maps, that illustrate the effects of a nontransversal intersection between A and B, and their dependence on the order of contact between A and B at the intersection point. For integers $n \geq 2$, let

$$A_n = \{(t, t^n, t^n) : t \in \mathbb{R}\}, \tag{7.57}$$

and $B = \{(t, 0, 0) : t \in \mathbb{R}\}$. These sets meet at $(0, 0, 0)$, and, for all n, the center manifold (near enough to $(0,\ 0,\ 0)$) is the 2-plane $\mathscr{C}^{\mathbf{0}} = \{x_1 = 0\}$. Plots of these sets along with the center manifold are shown in Figure 7.2.

The order of contact between A_n and B equals $n - 1$. For $n = 2$, we see that the Jacobian of the map along the center manifold is given by

$$dD_{A_2 B}(0,\ x_2,\ x_3) = \begin{pmatrix} \frac{1}{1+2x_2+2x_3} & 0 & 0 \\ 0 & 1 & 0 \\ 0 & 0 & 1 \end{pmatrix}. \tag{7.58}$$

This shows that the fixed point $(0,\ x_2,\ x_3)$ is not linearly stable unless $|1 + 2(x_2 + x_3)| > 1$. If one begins near a point on the center manifold where $|1 + 2(x_2 + x_3)| < 1$, then the iterates move quickly toward the part of the center manifold where it is greater than 1, and thence toward convergence. This is shown in Figure 7.3(b).

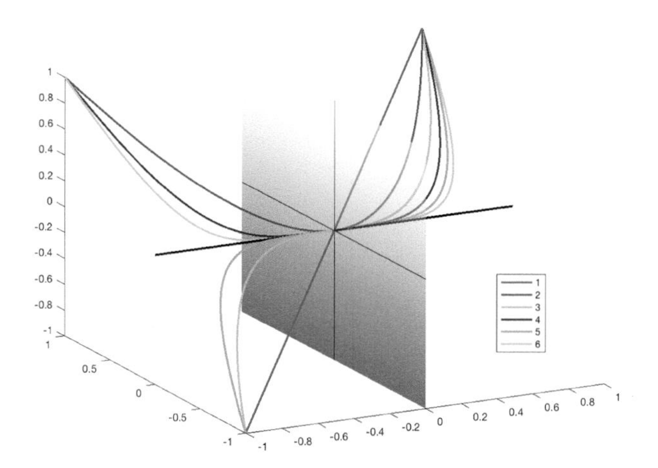

Figure 7.2 A plot showing the sets A_n, for $n = 1$ to 6, and the center manifold (shaded x_2x_3-plane). B is the x_1-axis, which is shown in black.

We have implemented the algorithms based on D_{A_nB} in MATLAB for a range of values of n, that is

$$\boldsymbol{x}_{m+1} = D_{A_nB}(\boldsymbol{x}_m), \tag{7.59}$$

with $\boldsymbol{x}_0$ a random starting point. The results of 500 iterates for each of the maps D_{A_nB} with $n = 2, 4, 6$, are shown in Figure 7.3(a). The solid-line plots show the $\log_{10}$ of the errors normal to the center manifold, $\{\log_{10} |x_{1m} - x_{1\infty}|\}$, and the dashed-line plots show the $\log_{10}$ of the errors parallel to the center manifold, $\{\log_{10}[|x_{2m} - x_{2\infty}| + |x_{3m} - x_{3\infty}|]\}$. For $n = 2$, the convergence to a point on the center manifold is geometric, but the x_2 and x_3 variables converge at twice the rate that x_1 converges. As n increases the rates of convergence decrease, with x_1 always converging at a slower rate than x_2 and x_3. In Figures 7.3(b, c) we show the limit points on $\{x_1 = 0\}$ of the algorithms based on iterating the maps D_{A_nB} with $n = 2, 3$. For each map we show the results with 1,000 randomly selected starting points. In Figure 7.3(b) we also show the lines $|1 + 2(x_2 + x_3)| = 1$, which clearly delimit the attracting portion of the center manifold.

In the following sections we see that this simple model problem gives some insight into the qualitative behavior of algorithms for solving the phase retrieval problem, when the unknown object is somewhat smooth, or the support constraint is imprecise. The much greater complexity of the phase retrieval problem, and the existence of multiple attracting basins lead to phenomena not captured by these simple examples.

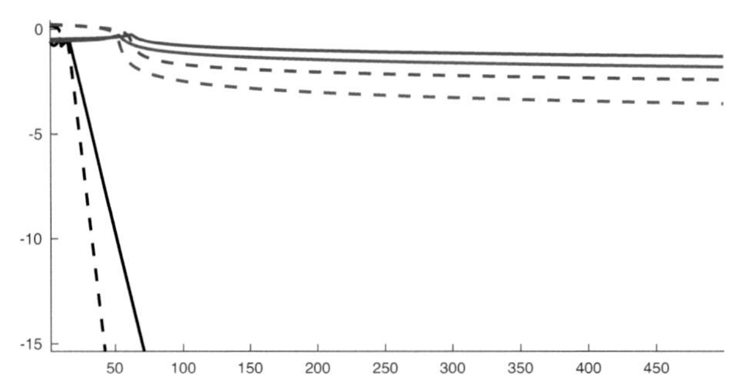

(a) Behavior of the iterates for algorithms based on the map $D_{A_n B}$ for $n = 2, 4, 6$, corresponding to the colors: black, red, and blue. The solid line, in a given color, shows $\log_{10}|x_{1m} - x_{1\infty}|$, the F-components, and the dashed line shows $\log_{10}[|x_{2m} - x_{2\infty}| + |x_{3m} - x_{3\infty}|]$, the C-components.

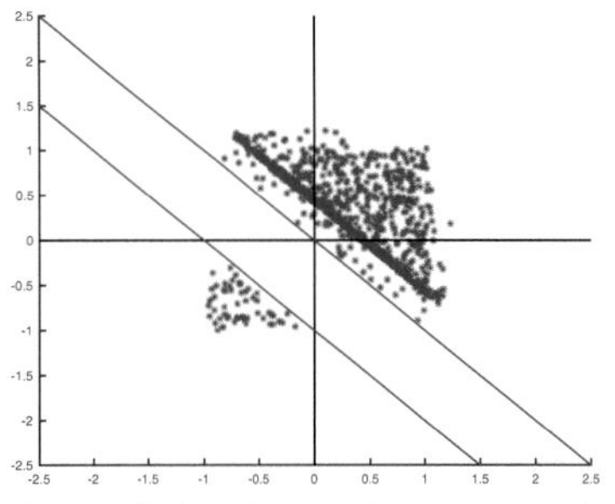

(b) The limit points on the center manifold, $\{x_1 = 0\}$, resulting from 1,000 random starting points with $n = 2$. The lines $|1 + 2(x_2 + x_3)| = 1$ are shown in red.

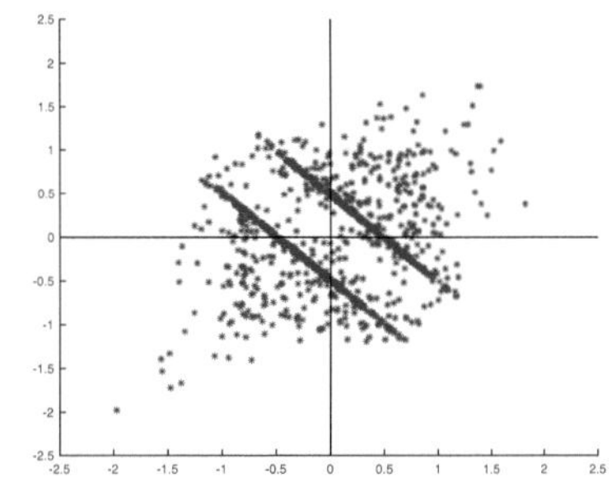

(c) The limit points on the center manifold, $\{x_1 = 0\}$, resulting from 1,000 random starting points with $n = 3$.

Figure 7.3 Plots of the limit points on the center manifolds for 1,000 random starting points for the iterates of the maps $D_{A_n B}$, with $n = 2, 3$.

7.4 A Noniterative Approach to Phase Retrieval

There is, in principle, an entirely different approach to phase retrieval that relies on the compact support of the objects we seek to image, and some elementary 1D-complex analysis. In this approach, the Hilbert transform is used to recover the unmeasured phase of the Fourier data. We close this chapter with a brief explanation of this approach. It uses a "holographic" experimental setup, which is related to, but distinct from, "external holography" methods as explained in Maretzke and Hohage (2017) and Jacobsen (2019). In Nakajima and Asakura (1985, 1986) and Nakajima (1995) an attempt was made to solve the phase retrieval problem in CDI using the Hilbert transform, though without a holographic experimental setup. This general idea for phase retrieval in 1D goes back to Kolmogorov. (See Wu et al. 1996; Huang et al. 2016).

While the measurements are again samples of the squared magnitude of the Fourier transform of a function, the approach here requires a somewhat different experimental setup. It is necessary that the object be compactly supported, though an estimate on its support is not needed. Up to this point our discussion has focused exclusively on the discrete model for the phase retrieval problem introduced in Section 1.1, and we have not discussed, in detail, how the discrete and continuum problems are related. To explain this new approach, which uses the Hilbert transform, it is simpler to work with the continuum model. We call this approach to phase retrieval the *Holographic Hilbert Transform* method or HHT. The details of implementing this method on finitely sampled data can be found in Chapter 13. This section uses some results from complex analysis in one-dimension, which can be found in Stein and Shakarchi (2003a).

7.4.1 An Idealized Model

First, recall that the Hilbert transform is defined on smooth, rapidly decaying functions, by the formula

$$\mathscr{H}f(x) = \mathscr{F}^{-1}[-i\,\text{sign}\,\xi\,\widehat{f}(\xi)], \tag{7.60}$$

where the continuum Fourier transform, and its inverse are defined by Stein and Shakarchi (2003b):

$$\begin{aligned}\widehat{f}(\xi) = \mathscr{F}(f)(\xi) &\stackrel{d}{=} \int_{-\infty}^{\infty} f(x)e^{-2\pi i x\xi}dx \\ f(x) = \mathscr{F}^{-1}(\widehat{f})(x) &\stackrel{d}{=} \int_{-\infty}^{\infty} \widehat{f}(\xi)e^{2\pi i x\xi}d\xi.\end{aligned} \tag{7.61}$$

These operators can be extended as bounded operators on other spaces, among them $L^2(\mathbb{R})$. A good treatment of the basic properties of the Hilbert transform can be found in Stein and Shakarchi (2003a) and Katznelson (1968). Suppose that $f \in L^2(\mathbb{R})$, and $\mathscr{H}(f)$ denotes its Hilbert transform; the functions $f \pm i\mathscr{H}(f)$ are the boundary values of functions, $F_\pm(z)$, that are holomorphic in the half planes, $H_\pm = \{z : 0 < \pm\text{Im}\,z\}$, and square integrable on the lines $\text{Im}\,z = c$. That is $(\partial_x + i\partial_y)F_\pm(x+iy) = 0$, where $\pm y > 0$,

$$\begin{aligned}&\lim_{y\to 0^\pm} F_\pm(x+iy) = f(x) \pm i\mathscr{H}(f)(x) \\ &\sup_{c>0}\int_{-\infty}^{\infty} |F_\pm(x \pm ic)|^2 dx < \infty.\end{aligned} \tag{7.62}$$

Indeed, these properties uniquely determine the Hilbert transform, $\mathscr{H}(f)$.

Suppose that $f(x)$ is a reasonably smooth function that is supported in the finite interval $[a,\ b]$, with $a < 0 < b$. The Fourier transform of f, has an analytic continuation to the whole complex plane, which satisfies the estimates

$$\begin{aligned} |\widehat{f}(\xi+i\tau)| &\leq \|f\|_1 e^{2\pi b\tau} \quad \text{for } 0 < \tau \\ |\widehat{f}(\xi+i\tau)| &\leq \|f\|_1 e^{2\pi a\tau} \quad \text{for } 0 > \tau. \end{aligned} \tag{7.63}$$

The function $Me^{-2\pi ic\xi} + \widehat{f}(\xi)$ is the Fourier transform of $M\delta(x-c)+f(x)$, and we observe that, on the real axis,

$$|Me^{-2\pi ic\xi} + \widehat{f}(\xi)| = M|1+M^{-1}\widehat{f}(\xi)e^{2\pi ic\xi}|. \tag{7.64}$$

If $c < a$, then the estimates in (7.63) imply that the function $\widehat{f}(\xi + i\tau)e^{2\pi ic(\xi+i\tau)}$ tends to 0 in the lower half plane as $\tau \to -\infty$. If $c > b$, then $\widehat{f}(\xi+i\tau)e^{2\pi ic(\xi+i\tau)}$ is analytic in H_+ and tends to 0 as $\tau \to \infty$. For the moment we assume that $c < a$.

If we choose $M > \|f\|_1 \geq \|\widehat{f}\|_{L^\infty}$, then the function $1+M^{-1}\widehat{f}(\xi)e^{2\pi ic\xi}$ does not vanish in the lower half plane, and therefore, with $\zeta = \xi + i\tau$,

$$F_c(\zeta) = 1 + M^{-1}\widehat{f}(\zeta)e^{2\pi ic\zeta}$$

is a nonvanishing, analytic function in H_- that tends to 1 as $|\zeta| \to \infty$ in H_-. From this, we easily conclude that $\log F_c(\zeta)$ is also an analytic function in H_- that tends to 0 as $|\zeta| \to \infty$. Indeed, the function $\log F_c(\zeta)$ is square integrable on horizontal lines in the lower half plane.

Note that the real and imaginary parts of $\log F_c$ are given by

$$\log F_c(\zeta) = \log|F_c(\zeta)| + i\arg(F_c(\zeta)). \tag{7.65}$$

The defining properties of the Hilbert transform then imply that

$$\arg(F_c(\xi)) = -\mathscr{H}[\log|F_c|](\xi). \tag{7.66}$$

Hence, a knowledge of $\{|F_c(x)| : x \in \mathbb{R}\}$ suffices to determine the $\{\arg(F_c(x)) : x \in \mathbb{R}\}$, and therefore, $\log F_c(x)$. If we also know the value of M, then we see that

$$\widehat{f}(\xi)e^{2\pi ic\xi} = M[\exp(\log(F_c(\xi))) - 1]. \tag{7.67}$$

To relate this to the 1D phase retrieval model for CDI, imagine that we can construct a highly diffracting, very small object that can be placed to the left of the object described by $f(x)$, to obtain a composite object that is well approximated by $M\delta(x-c)+f(x)$. We assume that the experimental apparatus is able to measure the squared magnitude of the Fourier transform,

$|Me^{-2\pi ic\xi} + \widehat{f}(\xi)|^2$. We call $M\delta(x-c)$ a δ-spike of size M at c. This is, in essence, a holographic method, since the measurement is the result of the interference between the Fourier transform of f and that of the "known" reference, $\delta(x-c)$. If we know M, then we can compute

$$\mathscr{H}\left[\log\left(\frac{|Me^{-2\pi ic\xi} + \widehat{f}(\xi)|}{M}\right)\right] = \mathscr{H}[\log|1 + M^{-1}e^{2\pi ic\xi}\widehat{f}(\xi)|]. \quad (7.68)$$

The relations in (7.66) and (7.67) then allow us to determine the Fourier transform of $f(x+c)$.

This method is easily extended to higher dimensions. For the 2D case, we assume the existence of a very small, highly diffracting object to create a composite object well approximated by $M\delta(x-c, y-d) + f(x,y)$.

Here (c, d) is selected to lie "to the left" of the smallest rectangle containing the supp f. The measurement is then proportional to

$$|Me^{-2\pi i(c\xi+d\eta)} + \widehat{f}(\xi,\eta)|^2 = M^2|1 + M^{-1}e^{2\pi i(c\xi+d\eta)}\widehat{f}(\xi,\eta)|^2. \quad (7.69)$$

We let $F_{c,\,d}(\xi,\eta)$ denote the function inside the absolute value bars on the right-hand side of (7.69). For each fixed $\eta \in \mathbb{R}$ this function has an analytic extension in ξ to the lower half plane.

If $\|f\|_1 < M$, then $F_{c,d}(\zeta,\eta)$ is nonvanishing for $\zeta \in H_-$ and tends to 1 as $|\zeta| \to \infty$, hence, $\log|F_{c,d}(\zeta,\eta)|$ tends to 0. We apply the one-dimensional Hilbert transform in the first argument to $\log|F_{c,d}(\xi,\eta)|$ and proceed exactly as in the 1D case to determine $\arg F_{c,d}(\xi,\eta)$ and then $M^{-1}e^{2\pi i(c\xi+d\eta)}\widehat{f}(\xi,\eta)$ itself. The δ-source can actually lie anywhere outside the smallest rectangle containing the supp f. Its location (left or right vs. above or below) determines in which variable to apply the 1D Hilbert transform, and whether to use $\pm\mathscr{H}_\xi(\log|F_{c,d}|)$ to reconstruct $\widehat{f}$.

The feasibility of this approach hinges on the possibility of creating a δ-source for diffraction with sufficient strength to attain $\|f\|_1 < M$. This requirement may restrict this approach to specific X-ray energies, and 2D imaging problems, where, for example, one could reduce $\|f\|_1$ by using a thinner sample.

7.4.2 A More Realistic Model

Of course, no real object is pointlike, and is instead described by an integrable function $\varphi(\boldsymbol{x})$. For definiteness we assume that φ is even, $\varphi(\boldsymbol{x}) = \varphi(-\boldsymbol{x})$. The model for our composite object would then be $(M\varphi(\boldsymbol{x}-\boldsymbol{c}) + f(\boldsymbol{x}))$, and the far field measurement would be modeled as

$$|Me^{-2\pi i c\cdot\boldsymbol{\xi}}\widehat{\varphi}(\boldsymbol{\xi})+\widehat{f}(\boldsymbol{\xi})|^2 = |M\widehat{\varphi}(\boldsymbol{\xi})|^2\cdot\left|1+M^{-1}e^{2\pi i c\cdot\boldsymbol{\xi}}\frac{\widehat{f}(\boldsymbol{\xi})}{\widehat{\varphi}(\boldsymbol{\xi})}\right|^2. \tag{7.70}$$

The mathematical requirements for φ are that:

(i) φ be compactly supported.
(ii) $\widehat{f}(\boldsymbol{\xi})/\widehat{\varphi}(\boldsymbol{\xi})$ has analytic extensions (in one of the variables ξ_j, for $j\in\{1,\ldots,d\}$) to a half plane, that tend to 0 at infinity.

We then need to choose M so that

$$F_{\varphi,c} = 1+M^{-1}e^{2\pi i c\cdot\boldsymbol{\xi}}\frac{\widehat{f}(\boldsymbol{\xi})}{\widehat{\varphi}(\boldsymbol{\xi})} \tag{7.71}$$

is nonvanishing for $(\xi_j+i\tau)\boldsymbol{e}_j$ in this half plane and tends to 1 as $|\xi_j+i\tau|$ goes to infinity. Here $(\xi_1,\ldots,\ \xi_{j-1},\ \xi_{j+1},\ldots,\ \xi_d)$ is fixed. There are functions φ that satisfy these conditions; one can even require that $0\le\varphi(\boldsymbol{x})$ and $0<\widehat{\varphi}(\boldsymbol{\xi})$. In Section 13.4.2 we show that these conditions can be relaxed considerably, which renders this approach a practical possibility.

We now restrict to the 1D case. Assuming that we know $|M\widehat{\varphi}(\xi)|$, we proceed as before, observing that

$$\log|Me^{-2\pi i c\xi}\widehat{\varphi}(\xi)+\widehat{f}(\xi)| - \log|M\widehat{\varphi}(\xi)| = \log|F_{\varphi,c}(\xi)|. \tag{7.72}$$

We compute the $\mathscr{H}\log|F_{\varphi,c}(\xi)|$ and note that

$$\exp\left[\log|F_{\varphi,c}| + i\mathscr{H}\log|F_{\varphi,c}|\right] = F_{\varphi,c}. \tag{7.73}$$

Therefore

$$M\widehat{\varphi}(\xi)\left[\exp\left[\log|F_{\varphi,c}|(\xi) + i\mathscr{H}\log|F_{\varphi,c}|(\xi)\right]-1\right] = e^{2\pi i c\xi}\widehat{f}(\xi), \tag{7.74}$$

which suffices to reconstruct a translate of f. In $d>1$ dimensions, the 1D Hilbert transform would be applied in the ξ_j-variable to $\log|F_{\varphi,c}|$, with the remainder of the computation exactly as in the 1D-case.

In practice, the determination of $|M\widehat{\varphi}(\xi)|$ would be done by direct measurement. If $\varphi(x)$ is even, then $\widehat{\varphi}(\xi)$ is a real-valued function. The determination of $M\widehat{\varphi}(\xi)$ from the measured magnitudes is then reduced to a much easier "sign retrieval" problem (see Leshem et al. 2018).

Remark 7.13 In Section 13.4 we discuss issues connected to the implementation of the HHT method on sampled data; it gives very good reconstructions regardless of how smooth the image is and does not require an accurate a priori assessment of the support of the object. From our experiments we see that a variety of functions, $\varphi(\boldsymbol{x})$, work very well in practice, even though they do not satisfy all of the hypotheses laid out above. This approach, which has something in common with the heavy ion replacement method used in X-ray crystallography, does not rely on auxiliary information in the sense that most other approaches to phase retrieval do. In fact, this method works well in the 1D-case, for which the standard phase retrieval problem does not even have a unique solution.

Remark 7.14 An older method called *off-axis* or *Fourier transform holography*, considers objects modeled by $f(\boldsymbol{x}) + M\delta(\boldsymbol{x} - \boldsymbol{c})$ (see Maretzke and Hohage 2016; Jacobsen 2019, §10.2). If f is supported in $D_r(\boldsymbol{0})$, then it is assumed $|\boldsymbol{c}| > 3r$; that is the δ-like object should be placed quite far from the support of the object of interest. With these assumptions it is very straightforward to reconstruct a real-valued object. Hohage and Maretzke also consider the case of complex-valued objects, which presents no additional difficulty for the Hilbert transform method herein. For the approach outlined in this section to work well, the δ-like object should lie outside of, but rather near to, the object of interest.

If f is real valued, then the image space autocorrelation function of $F_{M,\boldsymbol{c}} = f(\boldsymbol{x}) + M\delta(\boldsymbol{x} - \boldsymbol{c})$ can be computed directly from the Fourier magnitude measurements; it is

$$F_{M,\boldsymbol{c}} \star F_{M,\boldsymbol{c}}(\boldsymbol{a}) = [f \star f(\boldsymbol{x}) + M^2\delta(\boldsymbol{x})] + Mf(\boldsymbol{x}+\boldsymbol{c}) + Mf(\boldsymbol{c}-\boldsymbol{x}). \quad (7.75)$$

Under the hypotheses $\operatorname{supp} f \subset B_r(\boldsymbol{0})$ and $|\boldsymbol{c}| > 3r$, the terms on the right-hand side of (7.75) are supported in disjoint sets. The term in the brackets is supported in $D_{2r}(\boldsymbol{0})$ and the second and third terms are supported in $D_r(\mp\boldsymbol{c})$, respectively. Thus, we can simply read off f from the formula.

If the δ-function is replaced by a more realistic $\varphi(\boldsymbol{x})$, supported in $D_\rho(\boldsymbol{0})$, then we need to choose $\boldsymbol{c}$ with $3(r+\rho) < |\boldsymbol{c}|$. In this case the second term in (7.75) is replaced by

$$M\int \varphi(\boldsymbol{y}) f(\boldsymbol{x}+\boldsymbol{c}+\boldsymbol{y}) d\boldsymbol{y}, \quad (7.76)$$

which is a slightly smeared out version of f. With an adequate knowledge of φ, this term can be effectively deconvolved, provided that $|\widehat{\varphi}(\boldsymbol{\xi})|$ is large within the effective support of $\widehat{f}(\boldsymbol{\xi})$. This is essentially the condition needed for the success of the Hilbert transform approach.

Building on earlier work of Podorov et al. (2007), a different holographic approach is considered in Guizar-Sicairos and Fienup (2007, 2008), where several specific examples for φ are analyzed. The reconstruction method is essentially algebraic and does not make use of the Hilbert transform. More recently, Barmherzig et al., using a similar algebraic reconstruction approach, have considered the problem of optimizing the external reference for data contaminated by noise (see Barmherzig et al. 2019a, 2019b). The principal experimental differences with the Hilbert transform approach lie in the type of external reference object and its placement. In HERALDO and Barmherzig et al. the external object is a mask placed alongside the object of interest, but at a significant distance.

In another paper, Maretzke and Hohage (2017) consider a similar inverse problem for the weakly diffracting limit in near field (Fresnel) imaging. In this case, the measured amplitude data is again of the form $h \mapsto |1+T(h)|^2$, for a linear operator, T, defined by a unitary Fourier multiplier. In this paper they show that the linearized problem is formally well-posed for compactly supported data, though the condition number grows very rapidly in the "Fresnel parameter."

7.A Appendix: Alternating Projection and Gradient Flows

In this appendix we relate the iterates of AP to the solution of the gradient flow equation

$$\partial_t \boldsymbol{f}_t = -\nabla \rho_{\mathbb{A}_a B_S}(\boldsymbol{f}_t).$$

Here, $\rho_{\mathbb{A}_a B_S}$ is a defining function for $\mathbb{A}_a \cap B_S$. The limit, $\boldsymbol{f}_\infty$, as $t \to \infty$, of the solution to this equation is expected to be a local minimum of the function $\rho_{\mathbb{A}_a B_S}$. While it is not difficult to find explicit defining functions for this intersection, the gradient flow defined by such a function is unlikely to converge to points in $\mathbb{A}_a \cap B_S$, as these functions tend to have many nonzero local minima. As we show below, this is a reflection of the "bad behavior" of AP, which occurs in the "global search phase."

The AP algorithm is defined by choosing a starting point $\boldsymbol{f}^{(0)}$ to be any vector lying on $\mathbb{A}_a$ and then letting

$$\boldsymbol{f}^{(n+1)} = P_{\mathbb{A}_a} \circ P_{B_S}(\boldsymbol{f}^{(n)}). \tag{7.77}$$

For the case of phase retrieval, this algorithm does not stagnate, as per Definition 1.8, but rather seems to converge, very reliably, though not to points in $\mathbb{A}_a \cap B_S$. This is evident in Figure 7.4, which shows *semilog plots* of the error trajectories for 20,000 iterates of 50 random restarts of AP on a single 256×256 image, using a 3-pixel support neighborhood for S. The red curves show the norms of the successive differences $\{\|\boldsymbol{f}^{(n)} - \boldsymbol{f}^{(n+1)}\|\}$, whereas the blue curves show the distance from $\{\boldsymbol{f}^{(n)}\}$ to the nearest point on $\mathbb{A}_a \cap B_{S_3}$. The cyan curves are the residuals, $\{\|P_{\mathbb{A}_a}(\boldsymbol{f}^{(n)}) - P_{B_{S_3}}(\boldsymbol{f}^{(n)})\|\}$, which are,

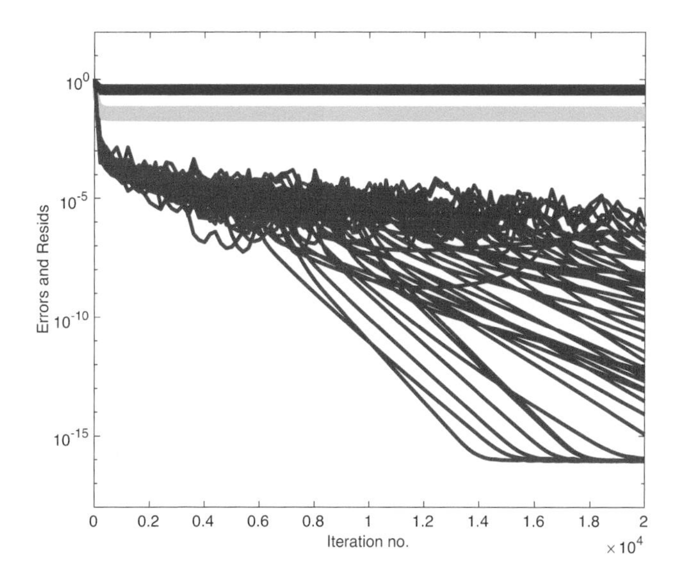

Figure 7.4 Error trajectories for 50 random restarts of the AP algorithm on a 256×256 image. The blue curves are the absolute error, $\|\boldsymbol{f}^{(n)} - \boldsymbol{f}_0\|$, where $\boldsymbol{f}_0$ is the nearest point in $\mathbb{A}_{\boldsymbol{a}} \cap B_{S_3}$. The cyan curves are the residuals, $\|P_{\mathbb{A}_{\boldsymbol{a}}}(\boldsymbol{f}^{(n)}) - P_{B_{S_3}}(\boldsymbol{f}^{(n)})\|$ and the red curves are the distances between successive iterates, $\|\boldsymbol{f}^{(n)} - \boldsymbol{f}^{(n+1)}\|$.

in a practical sense, the true measure of convergence to a point in $\mathbb{A}_{\boldsymbol{a}} \cap B_{S_3}$. After some chaotic behavior, the decrease evident in the red curves clearly demonstrates that the iterates of AP are converging geometrically to points on $\mathbb{A}_{\boldsymbol{a}}$, which according to the blue curves, are quite distant from $\mathbb{A}_{\boldsymbol{a}} \cap B_{S_3}$.

In Figure 7.5 we show the results of running 50,000 iterates of AP on a 256×256 image, using a 1-pixel support constraint. Panel (a) shows the reference image, and panels (b–d) show three images to which AP has converged to 16 digits (that is $\|\boldsymbol{f}^{(n)} - \boldsymbol{f}^{(n+1)}\|_2 < 10^{-16}\|\boldsymbol{f}_{\text{ref}}\|_2$). The distances to the closest trivial associates are shown. These results are typical of those found by running the AP algorithm repeatedly with different random initial conditions.

Allowing A and B to be more or less arbitrary smooth subsets of $\mathbb{R}^N$, we now show that algorithms based on the AP map are closely related to the gradient flow determined by a defining function for $A \cap B$. Such connections were also pointed out in Fienup (1982).

Conceptually, these continuous flows are somewhat easier to understand than the discrete iteration. Suppose that $\boldsymbol{x}^* \in A \setminus B$ is a fixed point for the

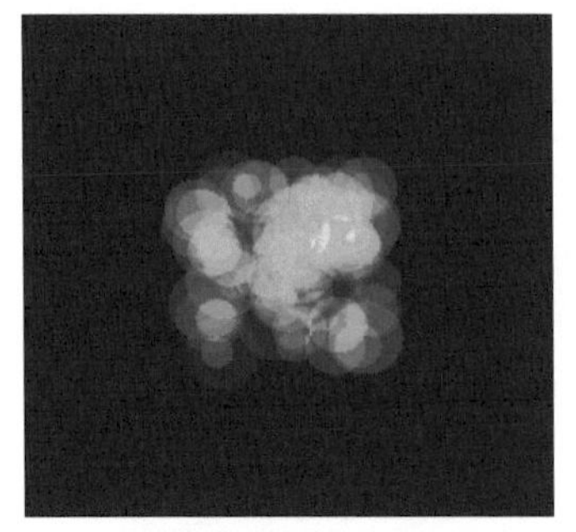

(a) The reference image.

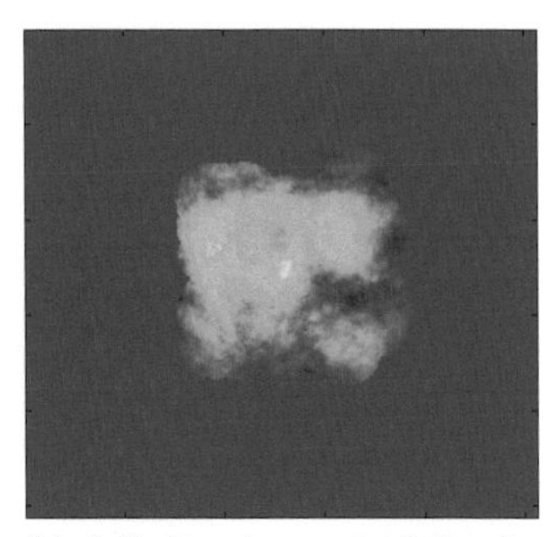

(b) A limiting image at relative distance 0.39 from the nearest trivial associate.

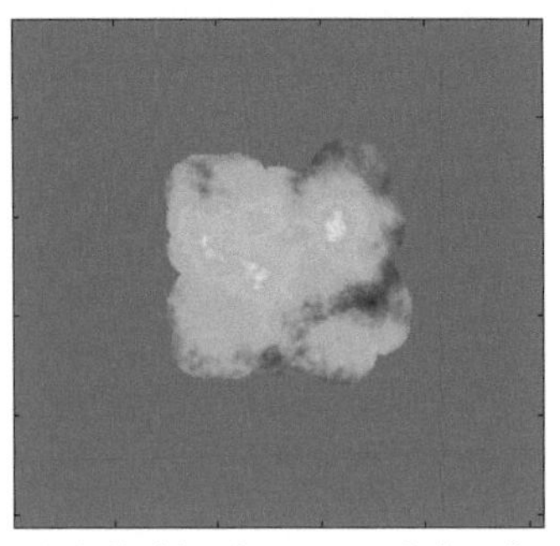

(c) A limiting image at relative distance 0.44 from the nearest trivial associate.

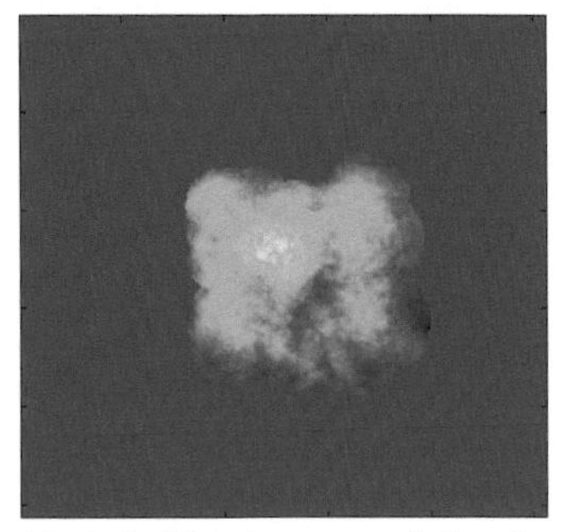

(d) A limiting image at relative distance 0.51 from the nearest trivial associate.

Figure 7.5 Panel (a) shows the target image. Panels (b–d) show the limiting images found after 50,000 iterates of AP starting with different random initial conditions. In all cases the iterates have converged to machine precision, i.e., 10^{-16}.

iteration in (7.77). By definition, the line from $\boldsymbol{x}^*$ to $P_B(\boldsymbol{x}^*)$, is perpendicular to B at $P_B(\boldsymbol{x}^*)$. The fact that $\boldsymbol{x}^* = P_A \circ P_B(\boldsymbol{x}^*)$, means that it is also perpendicular to A at $\boldsymbol{x}^*$. This characterizes the fixed points of this iteration that do not belong to $A \cap B$. For $\boldsymbol{x}^*$ to be a *stable* fixed point, the pair $(\boldsymbol{x}^*, P_B(\boldsymbol{x}^*))$ must be a local minimum of the function

$$d_{AB} : A \times B \to [0, \infty) \text{ defined by } d_{AB}(\boldsymbol{x}, \boldsymbol{y}) = \|\boldsymbol{x} - \boldsymbol{y}\|_2. \tag{7.78}$$

It is not obvious that this function has local minima, other than points in $A \cap B$, but, when $A = \mathbb{A}_{\boldsymbol{a}}$ and $B = B_S$, the behavior of the AP algorithm strongly suggests that there are, in fact, many. This observation is anticipated by remarks in Fienup (1982), Seldin and Fienup (1990), and Elser et al. (2007).

Running the AP algorithm with many different starting points one finds that it reliably converges, to machine precision, to a nonzero minimum of d_{AB}. In a simple experiment we ran the AP map for 8,000 iterates on a 64×64 $k = 0$ image, with a 1-pixel support neighborhood. For each of the 1,000 different random starting points the algorithm converged (or is converging) to a *different* nonzero critical point of $d_{\mathbb{A}_a B_S}$. The minimum distance between these limit points is about 0.316. This should be compared to the diameter of $\mathbb{A}_{\boldsymbol{a}}$, which equals 2.

The iteration step for AP can be interpreted as alternately solving the two gradient flows given by defining functions

$$\varphi_A(\boldsymbol{x}) = \text{dist}(\boldsymbol{x}, A) \text{ and } \varphi_B(\boldsymbol{x}) = \text{dist}(\boldsymbol{x},\ B). \tag{7.79}$$

If $\boldsymbol{x}^{(n)}$ is the current iterate for the AP algorithm, then one step of AP can be interpreted as first solving the differential equation

$$\frac{d\boldsymbol{x}_t}{dt} = -\nabla\varphi_A(\boldsymbol{x}_t), \text{ with } \boldsymbol{x}_0 = \boldsymbol{x}^{(n)}, \tag{7.80}$$

to time $= \text{dist}(\boldsymbol{x}^{(n)}, A)$, to obtain $\widetilde{\boldsymbol{x}}^{(n)}$, and then solving

$$\frac{d\widetilde{\boldsymbol{x}}_t}{dt} = -\nabla\varphi_B(\widetilde{\boldsymbol{x}}_t), \text{with } \widetilde{\boldsymbol{x}}_0 = \widetilde{\boldsymbol{x}}^{(n)}, \tag{7.81}$$

to time $= \text{dist}(\widetilde{\boldsymbol{x}}^{(n)}, B)$ to obtain $\boldsymbol{x}^{(n+1)}$. The Kato–Trotter product formula, which we recall below, then establishes the connection between AP and the gradient flow determined by the defining function for $A \cap B$, given by $\varphi_A(\boldsymbol{x}) + \varphi_B(\boldsymbol{x})$.

Let V be vector a field defined on $\mathbb{R}^N$ and $\exp(tV)$ denote the time t flow on $\mathbb{R}^N$ defined by V. If X and Y are two vector fields so that the commutator $[X,\ Y] \neq 0$, then it is not true that $\exp(tX) \circ \exp(tY) = \exp(t(X+Y))$. However, the Kato–Trotter product formula (Taylor 1996) shows that, under reasonable hypotheses, we have

$$\exp(t(X+Y)) = \lim_{k\to\infty} \left[\exp(\frac{t}{k}X) \circ \exp(\frac{t}{k}Y)\right]^k. \tag{7.82}$$

This formula with $X = -\nabla\varphi_A$, $Y = -\nabla\varphi_B$, and the fact that, for high iterates, the step sizes in equations (7.80) and (7.81) become very small, suggest that the behavior of the solution to

$$\frac{d\boldsymbol{x}_t}{dt} = -\nabla[\varphi_A(\boldsymbol{x}_t) + \varphi_B(\boldsymbol{x}_t)], \tag{7.83}$$

for large times, should behave very much like the high iterates of AP.

Conceptually, it is a little easier to work with the defining function

$$\varphi_{AB}(\boldsymbol{x}) = \sqrt{\varphi_A^2(\boldsymbol{x}) + \varphi_B^2(\boldsymbol{x})}. \tag{7.84}$$

In the experiments below we show that the trajectories of the gradient flows defined by $\varphi_A(\boldsymbol{x}) + \varphi_B(\boldsymbol{x})$ and $\varphi_{AB}(\boldsymbol{x})$ are very similar. As the absolute minima of φ_{AB} are points in $A \cap B$, one might try to find such points by solving the gradient flow equation

$$\frac{d\boldsymbol{x}_t}{dt} = -\nabla\varphi_{AB}(\boldsymbol{x}_t). \tag{7.85}$$

Unfortunately, this equation may have a strongly attracting set, which can be quite distant from the absolute minima of φ_{AB}. A simple calculation shows that

$$\nabla\varphi_{AB} = \frac{\varphi_A \nabla\varphi_A + \varphi_B \nabla\varphi_B \cdot}{\varphi_{AB}}. \tag{7.86}$$

The set of points

$$\mathscr{E}_{AB} = \{\boldsymbol{x} : \varphi_A(\boldsymbol{x}) = \varphi_B(\boldsymbol{x})\} \tag{7.87}$$

is called the *equidistant set*. Regardless of the dimensions of A and B, it is generically a nonempty union of hypersurfaces. This set may have a subset, which we call the *strict* equidistant set

$$\mathscr{E}_{AB}^{\text{str}} = \{\boldsymbol{x} \in \mathscr{E}_{AB} : \nabla\varphi_A(\boldsymbol{x}) = -\nabla\varphi_B(\boldsymbol{x})\}. \tag{7.88}$$

These are points $\boldsymbol{x} \in \mathscr{E}_{AB}$ so that $\boldsymbol{x}$, $P_A(\boldsymbol{x})$ and $P_B(\boldsymbol{x})$ lie along a single line. Since $\|\nabla\varphi_A(\boldsymbol{x})\| = \|\nabla\varphi_B(\boldsymbol{x})\| = 1$ at all points it is clear that, for $\boldsymbol{x} \in \mathscr{E}_{AB}^{\text{str}}$ we have that $\nabla\varphi_{AB}(\boldsymbol{x}) = 0$.

While it is difficult to describe $\mathscr{E}_{AB}^{\text{str}}$ explicitly, or even show that it is nonempty, it is quite clear in numerical experiments (at least for phase retrieval) that the set of points where $\|\nabla\varphi_{AB}\| << 1$ is quickly found by the flow in (7.85) and is very stable under perturbations of the data. Because

$$\frac{d\varphi_{AB}(\boldsymbol{x}_t)}{dt} = -\|\nabla\varphi_{AB}(\boldsymbol{x}_t)\|^2, \tag{7.89}$$

once $\|\nabla\varphi_{AB}(\boldsymbol{x}_t)\|$ becomes small, further progress in reducing φ_{AB} becomes very slow. In fact, the trajectories of (7.85) display exactly this behavior, and it is quite similar to the behavior of the AP algorithm after many iterates. Both $\nabla\varphi_{AB}$ and $\nabla[\varphi_A(\boldsymbol{x}) + \varphi_B(\boldsymbol{x})]$ vanish along $\mathscr{E}_{AB}^{\text{str}}$. The principal difference between these vector fields is that $\nabla\varphi_{AB}$ does not generally vanish off of $\mathscr{E}_{AB}^{\text{str}}$, whereas $\nabla[\varphi_A(\boldsymbol{x}) + \varphi_B(\boldsymbol{x})]$ also vanishes along the lines joining $P_A(\boldsymbol{x})$ to $P_B(\boldsymbol{x})$ for points $\boldsymbol{x} \in \mathscr{E}_{AB}^{\text{str}}$. The flow in (7.85) is strongly attracted to points in $\mathscr{E}^{\text{str}}_{AB}$, corresponding to local minima of d_{AB}, whereas that defined in (7.83) is attracted to this somewhat larger set.

Returning to the case of phase retrieval, if $\boldsymbol{f} \in \mathscr{E}^{\text{str}}_{\mathbb{A}_{\boldsymbol{a}} B_S}$, then $\boldsymbol{f}_{\mathbb{A}_{\boldsymbol{a}}} = P_{\mathbb{A}_{\boldsymbol{a}}}(\boldsymbol{f})$ is a fixed point for the AP algorithm defined by $P_{\mathbb{A}_{\boldsymbol{a}}} \circ P_{B_S}$. While we lack a theoretical proof that that $\mathscr{E}^{\text{str}}_{\mathbb{A}_{\boldsymbol{a}} B_S} \neq \emptyset$, numerical experiments are consistent with this hypothesis. The iterates appear to be converging, to machine precision, to points that are quite distant from $\mathbb{A}_{\boldsymbol{a}} \cap B_S$. In Figures 7.4 these correspond to the red trajectories converging linearly (on a $\log_{10}$-scale) to $-\infty$, while the actual distance to $\mathbb{A}_{\boldsymbol{a}} \cap B_S$ (the blue curves) is essentially a positive constant. For such limit points, $\boldsymbol{f}^*$, the midpoints, $\frac{1}{2}(P_{\mathbb{A}_{\boldsymbol{a}}}(\boldsymbol{f}^*) + P_{B_S}(\boldsymbol{f}^*))$, belong to $\mathscr{E}^{\text{str}}_{\mathbb{A}_{\boldsymbol{a}} B_S}$.

Example 7.15 In these examples we use an algorithm that solves the equations in (7.85) and (7.83) using a second-order Runge–Kutta method. We use 60,000 time steps, with the time step selected so that the solution to the gradient flow equation remains sufficiently accurate. Here, $A = \mathbb{A}_{\boldsymbol{a}}$ is the magnitude torus, defined by the object shown in Figure 7.6(a), and $B = B_{S_1}$, the 1-pixel neighborhood of the true support.

Figure 7.6(b) shows the values, in blue, of $\varphi_{\mathbb{A}_{\boldsymbol{a}} B_{S_1}}(\boldsymbol{f}_t)$, the defining function, in red, of $\|\nabla \varphi_{\mathbb{A}_{\boldsymbol{a}} B_{S_1}}(\boldsymbol{f}_t)\|$, the norm of the gradient, and, in green, the ratios, $\frac{\text{dist}(\boldsymbol{f}, \mathbb{A}_{\boldsymbol{a}})}{\text{dist}(\boldsymbol{f}, B_{S_1})}$, along trajectories of the gradient flow starting from 25 randomly selected initial points on $\mathbb{A}_{\boldsymbol{a}}$. The ratios $\frac{\text{dist}(\boldsymbol{f}, \mathbb{A}_{\boldsymbol{a}})}{\text{dist}(\boldsymbol{f}, B_{S_1})}$ rapidly approach 1 (this curve lies along the top edge of the graph). The "residual," which is measured by $\|\nabla \varphi_{\mathbb{A}_{\boldsymbol{a}} B_{S_1}}(\boldsymbol{f}_t)\|$, can be seen converging geometrically to 0. This strongly indicates that the flow is converging to a stable stationary point, which must, therefore, be a point on $\mathscr{E}^{\text{str}}_{AB}$. Figure 7.6(d) shows the same data as in Figure 7.6(b), but for the flow defined by $-\nabla[\varphi_{\mathbb{A}_{\boldsymbol{a}}}(\boldsymbol{f}) + \varphi_{B_{S_1}}(\boldsymbol{f})]$. While this flow is somewhat less stable (the jitter is reflected in the larger line-widths in the plots); it still converges rapidly in most cases. The other difference between these plots and those in Figure 7.6(b) is that while the ratios $\frac{\text{dist}(\boldsymbol{f}, \mathbb{A}_{\boldsymbol{a}})}{\text{dist}(\boldsymbol{f}, B_{S_1})}$ appear to converge, albeit with some jitter, they do not converge to 1.

In Figure 7.6(c) we show the results of using AP with the same initial data. In this plot the blue curves are the distance to the closest point in $\mathbb{A}_{\boldsymbol{a}} \cap B_{S_1}$, the cyan curves are plots of the "residual": $\|P_{\mathbb{A}_{\boldsymbol{a}}}(\boldsymbol{f}^{(n)}) - P_{B_{S_1}}(\boldsymbol{f}^{(n)})\|$, and the red curves are the distances between successive iterates of the algorithm. The similarities between these trajectories and those in Figures 7.6(b, d) are very striking. It is quite clear that the iterates of the AP are in fact converging geometrically, with $\|\boldsymbol{f}^{(n)} - \boldsymbol{f}^{(n+1)}\|$ decreasing at a fixed rate, and that the limit points are not in the set $\mathbb{A}_{\boldsymbol{a}} \cap B_{S_1}$.

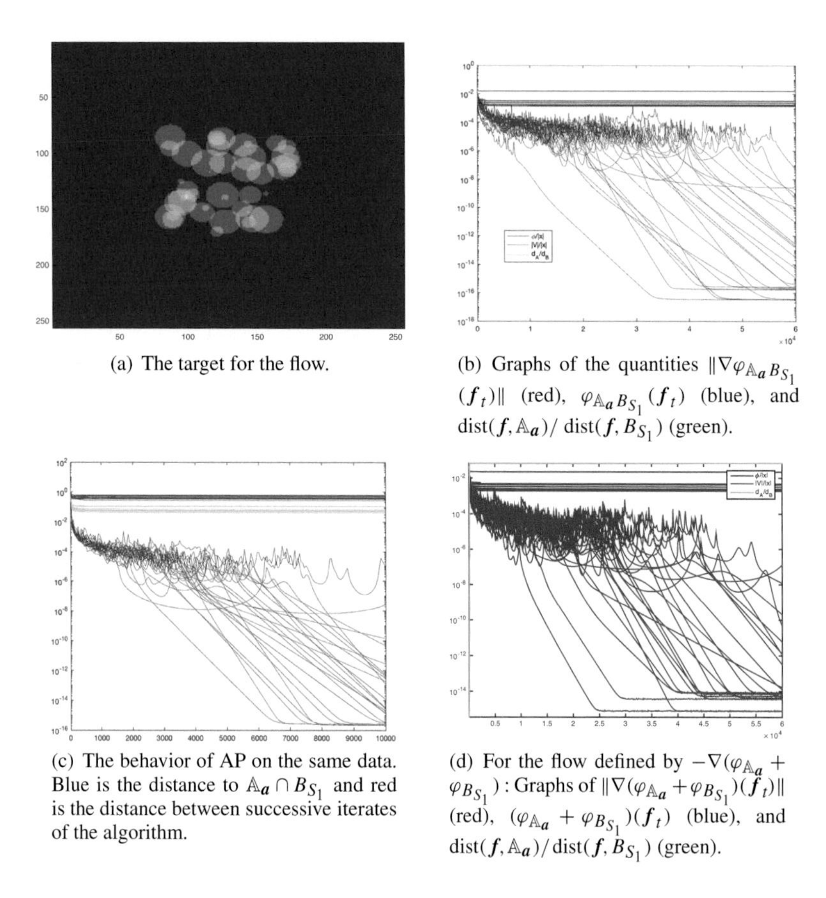

(a) The target for the flow.

(b) Graphs of the quantities $\|\nabla\varphi_{\mathbb{A}_a B_{S_1}}(f_t)\|$ (red), $\varphi_{\mathbb{A}_a B_{S_1}}(f_t)$ (blue), and $\operatorname{dist}(f, \mathbb{A}_a)/\operatorname{dist}(f, B_{S_1})$ (green).

(c) The behavior of AP on the same data. Blue is the distance to $\mathbb{A}_a \cap B_{S_1}$ and red is the distance between successive iterates of the algorithm.

(d) For the flow defined by $-\nabla(\varphi_{\mathbb{A}_a} + \varphi_{B_{S_1}})$: Graphs of $\|\nabla(\varphi_{\mathbb{A}_a} + \varphi_{B_{S_1}})(f_t)\|$ (red), $(\varphi_{\mathbb{A}_a} + \varphi_{B_{S_1}})(f_t)$ (blue), and $\operatorname{dist}(f, \mathbb{A}_a)/\operatorname{dist}(f, B_{S_1})$ (green).

Figure 7.6 Figures (b) and (d) show the error plots for 60,000 time steps of gradient flows defined by (7.85) and (7.83), respectively. Panel (c) shows analogous plots for 10,000 iterates of AP. Each of these plots is computed using the magnitude DFT data from the image in (a) with 25 identical, random initial choices for the phases.

The AP algorithm is no longer used alone in phase retrieval applications. It is a common practice, however, to recover the unmeasured phase information by switching between an HIO-based algorithm, and AP. As noted above, AP is also useful for probing the landscape defined by the function $d_{\mathbb{A}_a B}$. It detects the presence of nonzero local minima, whose prevalence can be considered a reasonable proxy for the complexity of the overall landscape. This is considered further in Example 8.13.

8

The Discrete, Classical, Phase Retrieval Problem

We turn now to a detailed investigation of the behavior of hybrid iterative algorithms in the context of the discrete, classical, phase retrieval problem. For simplicity, we largely restrict our attention to real images, though much of what is said holds equally well for complex images. Recall that the measured data consists of the magnitudes of the discrete Fourier transform (DFT) of an image, $\boldsymbol{f}$, indexed by $J \subset \mathbb{Z}^d$:

$$|\widehat{f}_{\boldsymbol{k}}| = a_{\boldsymbol{k}} \quad \text{for } \boldsymbol{k} \in J. \tag{8.1}$$

Recall also that $\mathbb{A}_{\boldsymbol{a}}$ denotes the real torus in $\mathbb{R}^J$ consisting of all real images that satisfy (8.1). For complex images, $\mathbb{A}_{\boldsymbol{a}}$ denotes the real torus in $\mathbb{C}^J$ consisting of all complex images that satisfy (8.1). The torus itself remains real, as it is defined by equations that involve the modulus. For real images, the symmetry properties of the Fourier transform imply that $\dim \mathbb{A}_{\boldsymbol{a}}$ is approximately $|J|/2$. For complex images, the dimension is approximately $|J|$, which is again half of $\dim \mathbb{C}^J$. In the following sections, we consider three different auxiliary constraints: a support constraint, a nonnegativity constraint, and a bound on the ℓ_1-norm for a nonnegative real image.

The algorithms we consider are constructed from the nearest point maps. In the DFT domain the nearest point map onto the magnitude torus, $\mathbb{A}_{\boldsymbol{a}}$, is defined whenever all coordinates of $\widehat{\boldsymbol{f}}$ are nonzero by

$$\widehat{P}_{\mathbb{A}_a}(\widehat{\boldsymbol{f}})_{\boldsymbol{k}} = a_{\boldsymbol{k}} \frac{\widehat{f}_{\boldsymbol{k}}}{|\widehat{f}_{\boldsymbol{k}}|}, \tag{8.2}$$

and in the image domain by

$$P_{\mathbb{A}_a}(\boldsymbol{f}) = \mathscr{F}^{-1}[\widehat{P}_{\mathbb{A}_a}(\widehat{\boldsymbol{f}})]. \tag{8.3}$$

If $\widehat{f}_{\boldsymbol{k}} = 0$ for some $\boldsymbol{k}$, for which $a_{\boldsymbol{k}} \neq 0$, then we make an arbitrary choice for the phase of $\widehat{P}_{\mathbb{A}_a}(\widehat{\boldsymbol{f}})_{\boldsymbol{k}}$; the map $P_{\mathbb{A}_a}$ cannot be extended continuously to the locus

$$\{\boldsymbol{f} : \text{for some } \boldsymbol{k},\ \widehat{f}_{\boldsymbol{k}} = 0,\ \text{but } a_{\boldsymbol{k}} \neq 0\}.$$

The nearest point map for the support constraint defined by a subset $S \subset J$ is the linear projection given by

$$P_{B_s}(\boldsymbol{f})_{\boldsymbol{j}} = \begin{cases} f_{\boldsymbol{j}} \text{ if } \boldsymbol{j} \in S \\ 0 \text{ if } \boldsymbol{j} \notin S. \end{cases} \tag{8.4}$$

The nearest point map for the nonnegativity constraint is given by

$$P_{B_+}(\boldsymbol{f})_{\boldsymbol{j}} = \begin{cases} f_{\boldsymbol{j}} \text{ if } f_{\boldsymbol{j}} \geq 0 \\ 0 \text{ if } f_{\boldsymbol{j}} < 0. \end{cases} \tag{8.5}$$

This constraint can be extended to complex images with

$$B_+^{\mathbb{C}} = \{\boldsymbol{f} \in \mathbb{C}^J : \operatorname{Re}(f_{\boldsymbol{j}}) \geq 0 \text{ and } \operatorname{Im}(f_{\boldsymbol{j}}) \geq 0 \text{ for all } \boldsymbol{j} \in J\}. \tag{8.6}$$

The closest point map is then

$$P_{B_+^{\mathbb{C}}}(\boldsymbol{f})_{\boldsymbol{j}} = P_{B_+}(\operatorname{Re}(\boldsymbol{f}))_{\boldsymbol{j}} + i P_{B_+}(\operatorname{Im}(\boldsymbol{f}))_{\boldsymbol{j}}. \tag{8.7}$$

A discussion of how to implement the nearest point map onto the boundary of an ℓ_1-ball, $P_{B^1_{r_1}}$, is given in Appendix 9.A.

In this chapter we analyze the algorithms defined by the maps $D_{\mathbb{A}_{\boldsymbol{a}}B}$ and $D_{B\mathbb{A}_{\boldsymbol{a}}}$, with B the subset of $\mathbb{R}^J$ (or $\mathbb{C}^J$) defined by one of these auxiliary conditions. As noted earlier, the iterates of such an algorithm often do not converge, and when they do converge, they do so to a point on the center manifold quite distant from the point in $\mathbb{A}_{\boldsymbol{a}} \cap B$ that defines it. This gives little reason to expect that the linearization of the hybrid iterative map at the true intersection point, $\boldsymbol{f}_0$, will have much predictive value as to the behavior of the iterates in real experiments. Nonetheless, we have found that the angles between $T_{\boldsymbol{f}_0}\mathbb{A}_{\boldsymbol{a}}$ and $T_{\boldsymbol{f}_0}B$ do have a predictable, quantitative effect on the behavior of these iterates.

To understand the phenomena that arise in this setting, we begin by analyzing the hybrid iterative maps in simple model geometries which, nonetheless, capture essential features present in the phase retrieval problem. We then study the linearizations of hybrid iterative maps in the phase retrieval problem, using support as the auxiliary constraint, at limit points of the iteration along the center manifold. These results are less complete than for the simpler models but do provide plausible explanations for the observed behavior in these cases.

8.1 Hybrid Iterative Maps in Model Problems

In this section we consider the behavior of hybrid iterative maps for three simple model geometries:

(i) We consider a circle and a line in $\mathbb{R}^3$ that meet at a point, either transversely or non-transversely. See Figure 8.1(a).
(ii) We consider a sphere and an affine subspace, both of high dimension and codimension, which meet nontransversely.
(iii) We consider a circle and a nonlinear curve in $\mathbb{R}^3$ that do not intersect, but nearly do, or intersect at several points either transversally, or nontransversally. See Figure 8.1(b).

The behavior of the maps in these cases illustrate phenomena observed in experiments using hybrid iterative maps in the context of phase retrieval. For the first two cases, we can provide complete analytic explanations for the observed behaviors; the third case is used to illustrate phenomena connected to having multiple intersections but is not analyzed in detail.

8.1.1 A Circle and a Line

Perhaps the simplest model for a high-dimensional torus meeting a high-dimensional linear subspace is that of a circle in $\mathbb{R}^3$ meeting a line. In these experiments we study the effects of varying the angle, θ, between the circle and the line. For the set A we use the unit circle in the x_1x_2-plane:

$$A = \{(x_1, x_2, x_3) : x_1^2 + x_2^2 = 1 \text{ and } x_3 = 0\}; \tag{8.8}$$

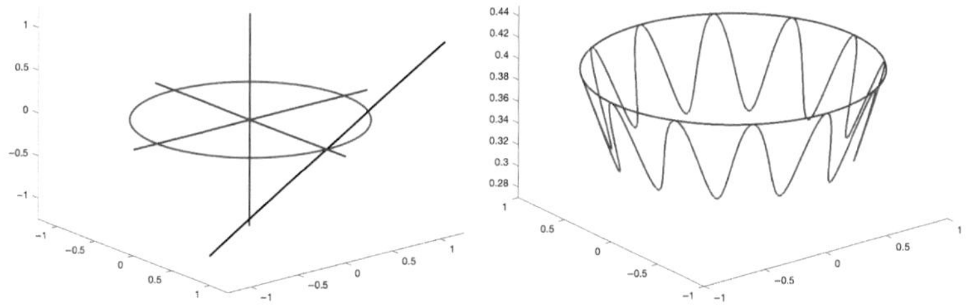

(a) A red circle and a black line meeting transversely.

(b) A blue circle and a red curve meeting nontransversely at several points.

Figure 8.1 Low-dimensional examples considered below.

for the set B we use the line

$$B_\theta = \{(1, t\cos\theta, t\sin\theta) : t \in \mathbb{R}\} \text{ for } \theta \in [-\frac{\pi}{2}, \frac{\pi}{2}). \tag{8.9}$$

There is a unique intersection, $A \cap B_\theta = (1,0,0)$, which is transversal unless $\theta = 0$. The angle θ is the angle between B_θ and the tangent line to the circle at $(1,\ 0,\ 0) : \{(1,\ t,\ 0) : t \in \mathbb{R}\}$. In Figure 8.1(a) $\theta = 0.2\pi$.

We focus on D_{AB_θ} though the behavior of $D_{B_\theta A}$ is quite similar. The map is given by

$$D_{AB_\theta}(x_1,\ x_2,\ x_3) = \begin{pmatrix} x_1 - 1 + \frac{2-x_1}{\eta} \\ \sin^2\theta x_2 - \frac{\sin 2\theta}{2} x_3 + \frac{\cos 2\theta x_2 + \sin 2\theta x_3}{\eta} \\ -\frac{\sin 2\theta}{2} x_2 + \cos^2\theta x_3 \end{pmatrix}, \tag{8.10}$$

where

$$\eta = \left[(2 - x_1)^2 + (\cos 2\theta x_2 + \sin 2\theta x_3)^2\right]^{\frac{1}{2}}. \tag{8.11}$$

In Figure 8.2(a–d) we show the results of iterating this map until the reconstruction error is below 10^{-10} for angles $\theta \in \{10^{-1}\pi,\ 10{-}2\pi,\ 10^{-3}\pi,\ 0\}$. In the left column we show the iterates $(*)$; the plots in the right column show the errors, in blue, and distances between successive iterates in red. When the angle is $10^{-1}\pi$ it requires 140 iterates to get 10 digits of accuracy, when the angle is $10^{-2}\pi$ it requires 4,000 iterates, and when it is $10^{-3}\pi$ it requires about 400,000 iterates. When the angle is 0 it requires only 40 iterates! This example shows that a hybrid iterative algorithm can, under certain circumstances, overcome the difficulties produced by a nontransversal intersection. This is because it converges to a point on the center manifold that is distant from $A \cap B_\theta$. The scalloping error curve in the $\theta = 10^{-1}\pi$ case indicates that the trajectory spirals around the center manifold. This is often observed in phase retrieval problems.

In the analysis below we see that, when $\theta = 0$, the behavior of the iterates is not predicted by the linearization of the map at $A \cap B_0$. In this case the center manifold is the x_1x_3-plane, so only the x_2-coordinate needs to converge to 0, which occurs very rapidly. If $\theta \neq 0$, then the center manifold is the x_1-axis, so both x_2 and x_3 must converge to 0 in order for the iterates to converge.

If θ is small, but nonzero, then, as might be expected from the $\theta = 0$ behavior, the x_2-coordinate becomes small very quickly, but there is a remaining

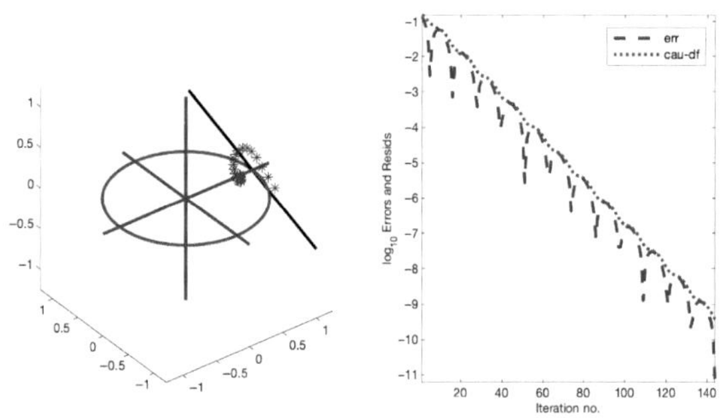

(a) For $\theta = 10^{-1}\pi$, it requires 140 iterates to get an error of 10^{-10}.

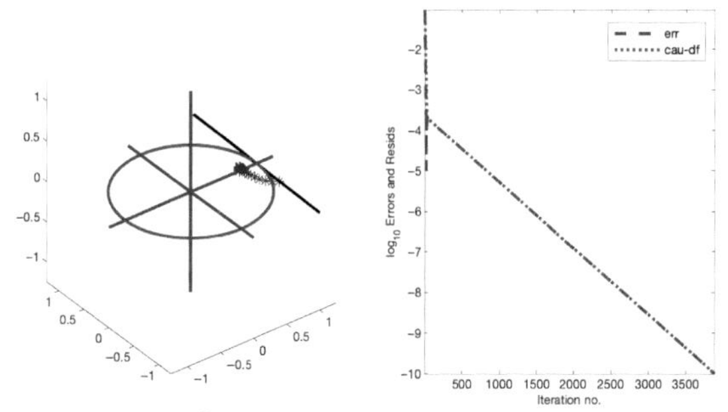

(b) For $\theta = 10^{-2}\pi$, it requires 4,000 iterates to get an error of 10^{-10}.

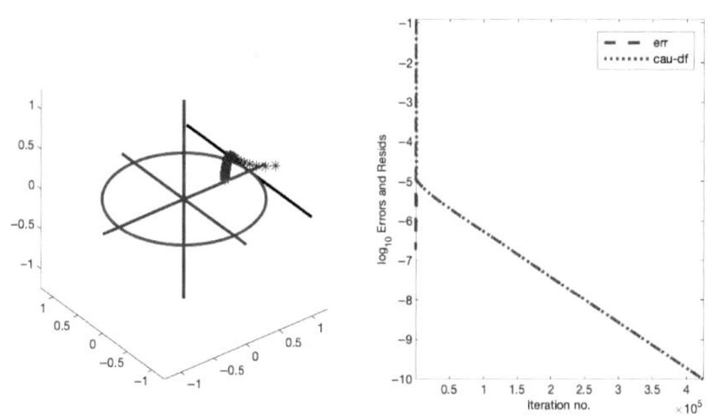

(c) For $\theta = 10^{-3}\pi$, it requires 400,000 iterates to get an error of 10^{-10}.

Figure 8.2 The left column shows the trajectories ($*$) of the iterates of the map D_{AB_θ}. The right column shows the $\log_{10}$-errors, in blue, and the $\log_{10}$-norms of the successive differences, $\{\|\boldsymbol{x}^{(n+1)} - \boldsymbol{x}^{(n)}\|_2\}$, in red. The maps are iterated until the reconstruction error is 10^{-10}.

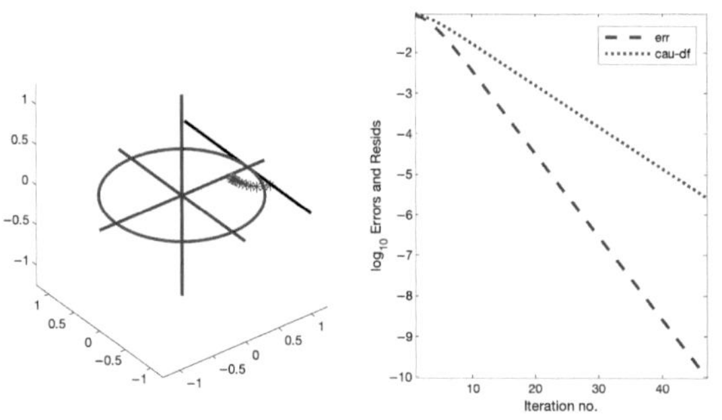

(d) For $\theta = 0$, it requires 40 iterates to get an error of 10^{-10}.

Figure 8.2 (*cont.*)

direction, a bit skew with the x_1x_3-plane, where the error goes to 0 very slowly. That is, the first few iterates behave as if $\theta = 0$, quickly approaching a point near to the x_1x_3-plane, but with a large x_2-coordinate. At this point the reconstruction error is $O(\theta)$. Then the iterates behave more or less as predicted by the linearization at the intersection point. The reason that the reconstruction error tends to 0 so slowly is explained in the analysis below.

The linearization of this map at the intersection point is

$$\begin{pmatrix} x_1 \\ x_2 \\ x_3 \end{pmatrix} \mapsto \begin{pmatrix} 1 & 0 & 0 \\ 0 & \cos^2\theta & \cos\theta\sin\theta \\ 0 & -\cos\theta\sin\theta & \cos\theta^2 \end{pmatrix} \cdot \begin{pmatrix} x_1 \\ x_2 \\ x_3 \end{pmatrix}. \tag{8.12}$$

If $\theta = 0$, then, as expected, this map is the identity, otherwise the components (x_2, x_3) shrink by factor $\cos\theta$ with each iterate.

As we see below, under the nonlinear map, the x_1-coordinate is forced to take a value below 1, (so that $\eta > 1$) and it then converges geometrically. If $\theta \neq 0$, then the (x_2, x_3)-coordinates converge to 0; convergence in one direction is geometric, determined essentially by $1/\eta < 1$, whereas convergence in the other direction is determined by θ. If θ is close to 0, then this occurs at essentially the rate, $\cos^n\theta$, predicted by the linearization at $(1, 0, 0)$.

To start the iteration we choose a random starting point $\boldsymbol{x}^{(0)}$ with nonzero $x_2^{(0)}$ and $x_3^{(0)}$ coordinates. We denote the iterates under D_{AB_θ} by

$$\boldsymbol{x}^{(n)} = \left(x_1^{(n)}, x_2^{(n)}, x_3^{(n)}\right),$$

and set

$$\mu_n^2 = (\cos 2\theta x_2^{(n)} + \sin 2\theta x_3^{(n)})^2$$
$$\eta_n = \sqrt{(2 - x_1^{(n)})^2 + \mu_n^2}. \tag{8.13}$$

As $2 - x_1^{(n)} < \eta_n$, it follows easily that the sequence $\{x_1^{(n)}\}$ is monotonically decreasing and, after finitely many iterates, $x_1^{(n)} < 1$. Hence, η_n is eventually greater than 1. If the intersection is nontransversal, then D_{AB_0} leaves x_3 fixed, and

$$D_{AB_0} : (x_1, x_2) \mapsto \left(x_1 - 1 + \frac{2 - x_1}{\eta}, \frac{x_2}{\eta} \right) \tag{8.14}$$

converges geometrically to a point $(x_1^{(\infty)}, 0)$; the point $(x_1^{(\infty)}, 0, x_3)$ is on the center manifold.

For $\theta \neq 0$, we see that

$$x_1^{(n)} - x_1^{(n+1)} = \frac{\mu_n^2}{\eta_n(2 - x_1^{(n)} + \eta_n)}. \tag{8.15}$$

Below we show that $\mu_n < 1$, and in fact, tends to 0. Once $x_1^{(n)} < 1$ and $\mu_n < 1$ the sequence $\{x_1^{(n)}\}$ is geometrically convergent. In practice this happens after a few iterates. To simplify these calculations, we represent

$$(x_2, x_3) = a\boldsymbol{n}_\theta + b\boldsymbol{t}_\theta, \tag{8.16}$$

where $\boldsymbol{n}_\theta = (-\sin\theta, \cos\theta)$; $\boldsymbol{t}_\theta = (\cos\theta, \sin\theta)$. In terms of this orthonormal basis we have that

$$D_{AB_\theta} : \begin{pmatrix} a \\ b \end{pmatrix} \mapsto \begin{pmatrix} 1 - \frac{\sin^2\theta}{\eta} & \frac{-\sin\theta\cos\theta}{\eta} \\ \frac{\sin\theta\cos\theta}{\eta} & \frac{\cos^2\theta}{\eta} \end{pmatrix} \cdot \begin{pmatrix} a \\ b \end{pmatrix}. \tag{8.17}$$

Though the coefficient η changes with each iterate, as $< x_1^{(n)} >$ is monotonically decreasing, we can assume that it is always greater than a fixed constant $\eta^- > 1$. As a practical matter, the sequence $\{\eta_n\}$ is monotonically increasing, and is bounded above by a positive constant η^+, see Figure 8.3(b). To analyze the behavior of iterates under D_{AB_θ} we diagonalize the matrix in (8.17). The eigenvalues are given by

$$\lambda_\pm(\theta, \eta) = \frac{1}{2}\left[1 + \frac{\cos 2\theta}{\eta} \pm \sqrt{\left(1 - \frac{1}{\eta}\right)^2 - \frac{\sin^2 2\theta}{\eta^2}}\right]. \tag{8.18}$$

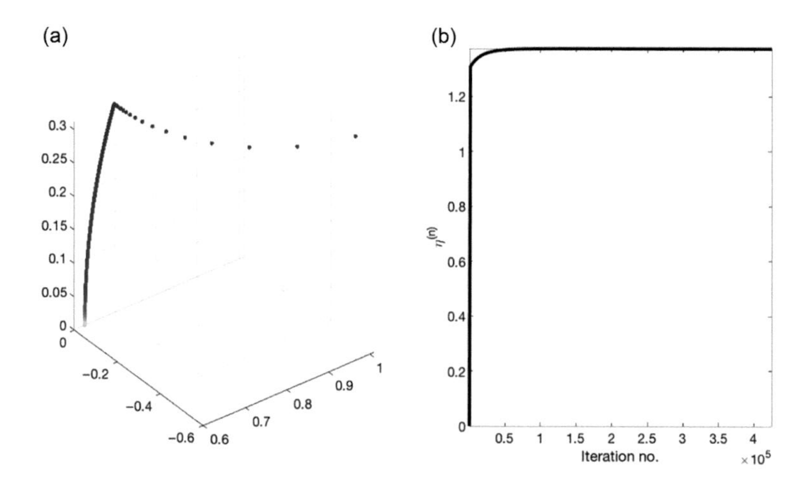

Figure 8.3 Part (a) is a trajectory of the map D_{AB_θ}, with $\theta = 10^{-3}\pi$, ordered from dark blue to cyan. Part (b) shows the values of $\eta^{(n)} = \sqrt{(2-x_1^{(n)})^2 + (\cos 2\theta x_2^{(n)} + \sin 2\theta x_3^{(n)})^2}$.

Assuming that $\eta - 1 > |\sin 2\theta|$, this is approximately

$$\begin{aligned}\lambda_+(\theta,\eta) &\approx 1 - \frac{\sin^2\theta}{\eta} - \frac{\sin^2 2\theta}{4(\eta-1)} \\ \lambda_-(\theta,\eta) &\approx \frac{\cos^2\theta}{\eta} + \frac{\sin^2 2\theta}{2\eta(\eta-1)}.\end{aligned} \tag{8.19}$$

If $v_\pm$ are the corresponding eigenvectors, then we see that the component in the v_+-direction decays like $(1-\theta^2/(\eta-1))^n$, which is very slow if θ is close to 0. The v_--direction decays faster than $1/\eta^k$, which is quite fast. If $\eta - 1 < |\sin 2\theta|$, then the eigenvalues are a conjugate pair of complex numbers with

$$|\lambda_\pm(\theta,\eta)| = \frac{\cos\theta}{\sqrt{\eta}}, \tag{8.20}$$

which predicts a decay rate of $(1-\theta)^n$. This produces a spiraling trajectory as seen in the left-most plots of Figure 8.2. We do not consider this case further.

The true values $\{\eta_n\}$ are bounded above by a constant, η^+, and below by a constant $\eta^- > 1$. Our computation of the eigenvalues shows that each iterate of the map shrinks the (x_2, x_3)-components by at least $\min\{1 - \frac{\theta^2}{\eta^+-1}, \frac{1}{\eta^-}, 1-\theta\}$, verifying our assertion that $\{\mu_n\}$ tends to 0.

For the case that $\eta - 1 > |\sin 2\theta|$, we compute the eigenvector in the λ_+-direction

$$\boldsymbol{v}_+ = \begin{pmatrix} \sin 2\theta \\ \eta - 1 - \sqrt{(\eta-1)^2 - \sin^2 2\theta} \end{pmatrix}. \tag{8.21}$$

For very small θ this gives

$$\boldsymbol{v}_+ \approx \sin 2\theta \begin{pmatrix} 1 \\ \frac{\sin 2\theta}{2(\eta-1)} \end{pmatrix}. \tag{8.22}$$

Recalling that we are using the basis defined by $\boldsymbol{n}_\theta, \boldsymbol{t}_\theta$, we see that the direction of slow decay is $\boldsymbol{n}_\theta + \frac{\sin 2\theta}{2(\eta-1)}\boldsymbol{t}_\theta$. After n iterates this becomes $\boldsymbol{u}^{(n)} = \lambda_+^n(\boldsymbol{n}_\theta + \frac{\sin 2\theta}{2(\eta-1)}\boldsymbol{t}_\theta)$, whose projection to B is the point

$$\pi_B(\boldsymbol{u}^{(n)}) = \left(1, \frac{\lambda_+^n \sin 2\theta}{2(\eta-1)}\boldsymbol{t}_\theta\right). \tag{8.23}$$

To leading order in θ, the distance of this point to the unit circle is given by

$$\mathrm{dist}(\pi_B(\boldsymbol{u}^{(n)}), S^1) \approx C\frac{\lambda_+^n \sin 2\theta}{2(\eta-1)} \approx C\frac{\theta}{\eta-1}\left(1 - \frac{\theta^2}{\eta}\right)^n. \tag{8.24}$$

Even though $\{\eta_n\}$ is not constant, this analysis gives a very accurate description of the behavior of the iterates of D_{AB_θ}, when θ is very small. The error, after n iterates, is bounded above by a constant times $\theta/(\eta^- - 1)(1-\theta^2/\eta^+)^n$. This shows that if $\theta = 10^{-\alpha}$ radians, then each additional digit of accuracy requires about $O(10^{2\alpha})$ additional iterates of the algorithm, which is in good agreement with the plots in Figure 8.2.

To summarize: when θ is small, but nonzero, the iterates quickly approach a point $\boldsymbol{x}^{(n_0)}$ such that $\mathrm{dist}(\boldsymbol{r}^{(n_0)}, S^1) \propto |\theta|$. After that the iterates, $\{\boldsymbol{x}^{(n_0+n)}\}$ follow a trajectory along which $\mathrm{dist}(\boldsymbol{r}^{(n_0+n)}, S^1) \propto |\theta|(1-\theta^2)^n$. Behavior qualitatively similar to this is quite often observed in phase retrieval problems when the iterates converge, and there are many directions in which $T_{f_0}\mathbb{A}_{\boldsymbol{a}}$ makes a small angle with B_S.

8.1.2 A Sphere and an Affine Space

The next problem we consider takes place in the ambient space $\mathbb{R}^{n_1+n_2+n_3}$; here, $\boldsymbol{x} = (\boldsymbol{x}_1, \boldsymbol{x}_2, \boldsymbol{x}_3)$ with $\boldsymbol{x}_j \in \mathbb{R}^{n_j} : j = 1,2,3$. For the numerical examples we take $n_1 \approx n_3 \approx 2{,}000$, and $n_2 \approx 100$. In this example A is a sphere in $\mathbb{R}^{n_1+n_2} \times \{\boldsymbol{0}\}$:

$$A = \{(\boldsymbol{x}_1, \boldsymbol{x}_2, \boldsymbol{0}) : \|\boldsymbol{x}_1\|_2^2 + \|\boldsymbol{x}_2\|_2^2 = 1\}. \tag{8.25}$$

B is an affine subspace that meets A at a single point, $\boldsymbol{i} = (\boldsymbol{0},\ \boldsymbol{e}_1,\ \boldsymbol{0})$, where $\boldsymbol{e}_j$ is the vector in $\mathbb{R}^{n_2}$ with j^{th} coordinate 1, and the rest 0. We choose B so that the intersection is nontransversal with

$$T_{\boldsymbol{i}} A \cap T_{\boldsymbol{i}} B = \text{span}\{(\boldsymbol{0}, \boldsymbol{e}_j, \boldsymbol{0}) : j = 2, \ldots, n_2\}. \tag{8.26}$$

The affine space B is the translate by $\boldsymbol{i}$ of a linear subspace defined as the span of the vectors

$$(\boldsymbol{0},\ \boldsymbol{e}_j,\ \boldsymbol{0}) \quad \text{for } j = 2, \ldots, n_2 \tag{8.27}$$

along with the orthonormal vectors

$$\boldsymbol{u}_l = (\boldsymbol{u}_l^1,\ \boldsymbol{0},\ \boldsymbol{u}_l^3) \text{ with } l = 1, \ldots, n_3, \tag{8.28}$$

here, $\boldsymbol{u}_l^1 \in \mathbb{R}^{n_1}$, $\boldsymbol{0} \in \mathbb{R}^{n_2}$, and $\boldsymbol{u}_l^3 \in \mathbb{R}^{n_3}$. We assume that the vectors $\{\boldsymbol{u}_1^3, \ldots, \boldsymbol{u}_{n_3}^3\}$ are a basis for $\mathbb{R}^{n_3}$. This is to ensure that (8.26) holds.

As stated earlier, $A \cap B = \boldsymbol{i}$ and this intersection is nontransversal if $n_2 > 1$. A moment's consideration shows that the center manifold is one-dimensional and equals

$$C = \{(\boldsymbol{0}, t\boldsymbol{e}_1,\ \boldsymbol{0}) : t \in \mathbb{R}\}. \tag{8.29}$$

Letting

$$\boldsymbol{x}_2 = (x_{21}, \boldsymbol{x}_2'), \tag{8.30}$$

$x_{21} \in \mathbb{R}$ is a coordinate along the center manifold. In order for the iterates of the hybrid iterative map to converge, the $\boldsymbol{x}_1$, $\boldsymbol{x}_2'$, and $\boldsymbol{x}_3$ components must all go to 0, and the x_{21}-coordinate must converge.

Let U_1 be the $n_1 \times n_3$matrix with columns $\{\boldsymbol{u}_1^{1t}, \ldots, \boldsymbol{u}_{n_3}^{1t}\}$ and U_3 be the $n_3 \times n_3$-matrix with columns $\{\boldsymbol{u}_1^{3t}, \ldots, \boldsymbol{u}_{n_3}^{3t}\}$. Let P_U denote the orthogonal projection of $\mathbb{R}^{n_1+n_3}$ onto the subspace spanned by the vectors $\{\widetilde{\boldsymbol{u}}_1, \ldots, \widetilde{\boldsymbol{u}}_{n_3}\}$, where $\widetilde{\boldsymbol{u}}_j = (\boldsymbol{u}_j^1, \boldsymbol{u}_j^3)$. P_U is given by the block matrix

$$P_U = \begin{pmatrix} U_1 U_1^t & U_1 U_3^t \\ U_3 U_1^t & U_3 U_3^t \end{pmatrix}. \tag{8.31}$$

We let

$$(\boldsymbol{x}_1^1,\ \boldsymbol{x}_3^1) = P_U(\boldsymbol{x}_1,\ \boldsymbol{x}_3) \text{ and } (\boldsymbol{x}_1^0,\ \boldsymbol{x}_3^0) = (\text{Id} - P_U)(\boldsymbol{x}_1,\ \boldsymbol{x}_3), \tag{8.32}$$

be the orthogonal decomposition of $\mathbb{R}^{n_1+n_3}$ with respect to this subspace. With this notation, we see that the hybrid iterative map is given by

$$D_{AB}\begin{pmatrix} \boldsymbol{x}_1 \\ x_{21} \\ \boldsymbol{x}'_2 \\ \boldsymbol{x}_3 \end{pmatrix} = \begin{pmatrix} \left(1-\frac{1}{\eta}\right)\boldsymbol{x}_1^0 + \frac{1}{\eta}\boldsymbol{x}_1^1 \\ x_{21} - 1 + \frac{2-x_{21}}{\eta} \\ \frac{\boldsymbol{x}'_2}{\eta} \\ \boldsymbol{x}_3^0 \end{pmatrix}, \tag{8.33}$$

where

$$\eta = \sqrt{(2-x_{21})^2 + \mu^2} \text{ and } \mu^2 = \|\boldsymbol{x}_1^1 - \boldsymbol{x}_1^0\|_2^2 + \|\boldsymbol{x}'_2\|_2^2. \tag{8.34}$$

As in the previous case, we can easily show that $\{x_{21}^{(n)}\}$ is a monotonically decreasing sequence, which is less than 1 after a finite number of iterates. Thereafter, $\eta_n = \sqrt{(2-x_{21}^{(n)})^2 + \mu_n^2}$ is greater than a constant, greater than 1. The sequence $\{\eta_n\}$ in this case behaves very much like to sequence $\{\eta_n\}$ in the previous section. From this discussion and (8.33) it is clear that the $\boldsymbol{x}'_2$-components, which parameterize the projection into $T_{\tilde{i}}A \cap B$, converge geometrically to 0. The difference of successive x_{21}-components is given by

$$x_{21}^{(n)} - x_{21}^{(n+1)} = \frac{\mu_n^2}{\eta_n\left[\eta_n + (2-x_{21}^{(n)})\right]} \tag{8.35}$$

Below, we see that D_{AB} is contracting in the $(\boldsymbol{x}_1, \boldsymbol{x}_3)$-directions. From (8.35), it then follows that the sequence $\{x_{21}^{(n)}\}$ must be bounded and is, therefore, convergent.

What remains is to understand the behavior of the $(\boldsymbol{x}_1, \boldsymbol{x}_3)$-components. As in the previous example we see that it is not the failure of transversality, per se, that leads to stagnation, or slow convergence. To understand the behavior of the remaining components, we extract the top and bottom blocks of D_{AB} to get

$$A_\eta\begin{pmatrix} \boldsymbol{x}_1 \\ \boldsymbol{x}_3 \end{pmatrix} = \begin{pmatrix} \left(1-\frac{1}{\eta}\right)\boldsymbol{x}_1^0 + \frac{1}{\eta}\boldsymbol{x}_1^1 \\ \boldsymbol{x}_3^0 \end{pmatrix}. \tag{8.36}$$

A computation shows that

$$\left\| A_\eta\begin{pmatrix} \boldsymbol{x}_1 \\ \boldsymbol{x}_3 \end{pmatrix}\right\|_2^2 = \left(1-\frac{1}{\eta}\right)^2 \|\boldsymbol{x}_1^0\|_2^2 + \|\boldsymbol{x}_3^0\|_2^2 + \frac{2}{\eta}\left(1-\frac{1}{\eta}\right)\left\langle \boldsymbol{x}_1^0, \boldsymbol{x}_1^1\right\rangle + \frac{1}{\eta^2}\|\boldsymbol{x}_1^1\|_2^2. \tag{8.37}$$

Applying the Cauchy–Schwarz and arithmetic-geometric mean inequalities we obtain that

$$\left\| A_\eta \begin{pmatrix} \boldsymbol{x}_1 \\ \boldsymbol{x}_3 \end{pmatrix} \right\|_2^2 \leq \left(1 - \frac{1}{\eta^2}\right) \|\boldsymbol{x}_1^0\|_2^2 + \|\boldsymbol{x}_3^0\|_2^2 + \frac{1}{\eta}\|\boldsymbol{x}_1^1\|_2^2. \tag{8.38}$$

The $(\boldsymbol{x}_1^0,\ \boldsymbol{x}_3^0)$-components satisfy the equation $U_1^t \boldsymbol{x}_1^0 + U_3 \boldsymbol{x}_3^0 = 0$, and therefore, $\boldsymbol{x}_3^0 = -[U_3^t]^{-1} U_1^t \boldsymbol{x}_1^0$, where we recall that U_3 is an invertible matrix. This implies that there is a constant M so that $\|\boldsymbol{x}_3^0\|_2 < M\|\boldsymbol{x}_1^0\|_2$, which, in turn, shows that if C satisfies

$$1 - \frac{1}{\eta^2(1+M)} < C < 1, \tag{8.39}$$

then

$$\left(1 - \frac{1}{\eta^2}\right) \|\boldsymbol{x}_1^0\|_2^2 + \|\boldsymbol{x}_3^0\|_2^2 \leq C\left(\|\boldsymbol{x}_1^0\|_2^2 + \|\boldsymbol{x}_3^0\|_2^2\right), \tag{8.40}$$

demonstrating that A_η is a contraction, with

$$\left\| A_\eta \begin{pmatrix} \boldsymbol{x}_1 \\ \boldsymbol{x}_3 \end{pmatrix} \right\|_2^2 \leq \max\left\{C, \frac{1}{\eta}\right\} \left(\|\boldsymbol{x}_1\|_2^2 + \|\boldsymbol{x}_3\|_2^2\right). \tag{8.41}$$

From this estimate and (8.35) it follows that $\{\eta_n\}$ is bounded from above and, therefore, the $(\boldsymbol{x}_1,\ \boldsymbol{x}_3)$-components do, in fact, converge to 0. Loosely speaking, the rate at which this occurs is regulated by the value of M, which is, in turn, mainly determined by $\|[U_3^t]^{-1}\|$. It is complicated to analyze the details of the asymptotics of this iteration in general, but for an interesting subclass of examples, this can be done quite explicitly.

For these examples we let $n_1 = n_3$ and set

$$U_1 = \sqrt{\mathrm{Id} - B} \text{ and } U_3 = \sqrt{B}, \tag{8.42}$$

where B is a positive definite, self-adjoint matrix with norm less than 1. With these choices $(U_1,\ U_3)^t$ has orthonormal columns. Let $\{(\boldsymbol{v}_j, \sigma_j) : j = 1, \ldots, n_3\}$ be the eigenvectors and eigenvalues of the matrix B :

$$B\boldsymbol{v}_j = \sigma_j \boldsymbol{v}_j. \tag{8.43}$$

The analogue of the "H"-matrix, introduced in (7.3) is $\sqrt{\mathrm{Id} - B}$, from which it follows that the principal-angles $\{\theta_j\}$ between $T_i A$ and $T_i B$ satisfy

$$\cos\theta_j = \sqrt{1 - \sigma_j} \leftrightarrow \sigma_j = \sin^2\theta_j. \tag{8.44}$$

For each j, the subspace $\{(\alpha \boldsymbol{v}_j, \beta \boldsymbol{v}_j) : (\alpha,\beta) \in \mathbb{R}^2\}$ is invariant under the action of A_η, and mapped into itself by P_U :

$$A_\eta \begin{pmatrix} \alpha \boldsymbol{v}_j \\ \beta \boldsymbol{v}_j \end{pmatrix} = \begin{pmatrix} \frac{1}{\eta} + \left(1 - \frac{2}{\eta}\right)\sigma_j & \left(\frac{2}{\eta} - 1\right)\sqrt{\sigma_j(1-\sigma_j)} \\ -\sqrt{\sigma_j(1-\sigma_j)} & (1-\sigma_j) \end{pmatrix} \begin{pmatrix} \alpha \boldsymbol{v}_j \\ \beta \boldsymbol{v}_j \end{pmatrix} \tag{8.45}$$

and

$$P_U \begin{pmatrix} \alpha \boldsymbol{v}_j \\ \beta \boldsymbol{v}_j \end{pmatrix} = \begin{pmatrix} [(1-\sigma_j)\alpha + \sqrt{\sigma_j(1-\sigma_j)}\beta]\boldsymbol{v}_j \\ [\sqrt{\sigma_j(1-\sigma_j)}\alpha + \sigma_j\beta]\boldsymbol{v}_j \end{pmatrix}. \tag{8.46}$$

As above, η_n actually changes with each iterate, and is eventually larger than 1; it is bounded and converges as $n \to \infty$. As shown in Figure 8.3, in the asymptotic regime, the $\{\eta_n\}$ are very nearly constant. As our principal interest is in the asymptotic behavior in the case where $\sigma_j \ll 1$, we proceed as in Section 8.1.1, taking for η the $\lim_{n\to\infty}\eta_n$, and analyzing the eigenvalues and eigenvectors of A_η on the subspaces $V_j = \{(\alpha \boldsymbol{v}_j, \beta \boldsymbol{v}_j) : \alpha,\beta \in \mathbb{C}\}$. As we shall see, the results of this slightly simplified analysis agree very well with the results of numerical experiments.

A calculation shows that the eigenvalues of A_η on V_j are given by

$$\lambda_j^{\pm} = \frac{1}{2}\left[1 + \frac{1}{\eta} - \frac{2\sigma_j}{\eta} \pm \left(1 - \frac{1}{\eta}\right)\sqrt{1 - \frac{4\sigma_j(1-\sigma_j)}{(\eta-1)^2}}\right]. \tag{8.47}$$

For small σ_j with $4\sigma_j(1-\sigma_j) < (\eta-1)^2$, we have two real eigenvalues given by

$$\lambda_j^{\pm} = \begin{cases} 1 - \frac{\sigma_j}{\eta} - \frac{\sigma_j(1-\sigma_j)}{\eta(\eta-1)} + o(\sigma_j^2) & + \\ \frac{1-\sigma_j}{\eta} + \frac{\sigma_j(1-\sigma_j)}{\eta(\eta-1)} + o(\sigma_j^2) & - \end{cases} \tag{8.48}$$

The distance along the center manifold from the exact intersection point is bounded below by $\eta - 1$. As in the previous case, there is a direction, $\boldsymbol{v}_j^+ = (\alpha^+ \boldsymbol{v}_j, \beta^+ \boldsymbol{v}_j)$, with

$$\begin{aligned} \begin{pmatrix} \alpha^+ \\ \beta^+ \end{pmatrix} &= \begin{pmatrix} \left(1 - \frac{2}{\eta}\right)\sqrt{\sigma_j(1-\sigma_j)} \\ \left(1 - \frac{1}{\eta}\right)\left[\sigma_j - \frac{1}{2} - \frac{1}{2}\sqrt{1 - \frac{4\sigma_j(1-\sigma_j)}{(\eta-1)^2}}\,\right] \end{pmatrix} \\ &= \frac{(\eta-1)(1-\sigma_j)}{\eta}\left[\begin{pmatrix} \left(\frac{\eta-2}{\eta-1}\right)\sqrt{\sigma_j} \\ -1 \end{pmatrix} + O(\sigma_j)\right], \end{aligned} \tag{8.49}$$

where the contraction rate is nearly 1 and essentially determined by σ_j as well as a direction, $\boldsymbol{v}_j^-$, where the rate is essentially $1/\eta < 1$. Even far from the

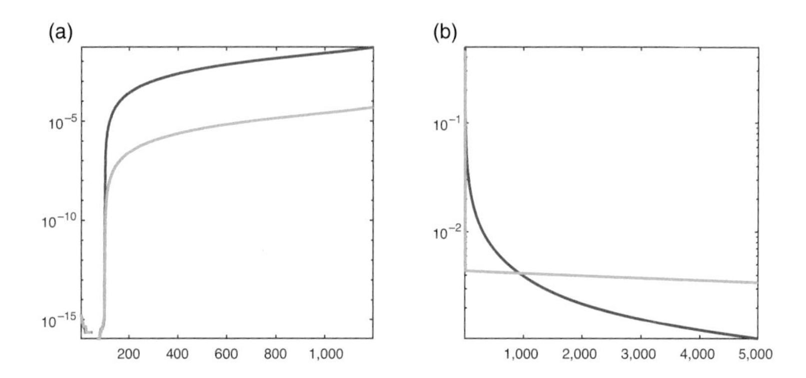

Figure 8.4 Part (a) shows the $\{\sin^2\theta_j\}$, where $\{\theta_j\}$ are the angles between $T_{\boldsymbol{i}}A$ and $T_{\boldsymbol{i}}B$. In both panels, the $\{\theta_j\}$ are closer to 0 in the green curves and larger in the magenta curves. Part (b) shows the reconstruction errors for 5,000 iterates of D_{AB}. The smaller angles produce the rapid initial decay in the green curve, which is followed by a much slower decay.

intersection point, where η is significantly greater than 1, the overall rate of convergence is determined by the angles between the tangent spaces to A and B at the intersection point. The direction $\boldsymbol{v}_j^+$, which lies in $T_{\boldsymbol{i}}A + T_{\boldsymbol{i}}B$, is the difference of two directions where these subspaces almost coincide. Somewhat paradoxically, the resultant direction $\boldsymbol{v}_j^+$ is almost orthogonal to $T_{\boldsymbol{i}}B$. Indeed, $\|P_U\boldsymbol{v}_j^+\|_2 \propto \|\boldsymbol{v}_j^+\|_2\sqrt{\sigma_j}$. Once we are in the asymptotic regime, the norm $\|P_U A_\eta^n \boldsymbol{v}_j^+\|_2 \propto (1-\sigma_j/\eta)^n\sqrt{\sigma_j}$, which is a good estimate for the true error, i.e., the distance between the current reconstruction $\boldsymbol{r}^{(n)} = P_B(\boldsymbol{x}^{(n)})$ and the intersection point $\boldsymbol{i} = A\cap B$.

We close this section with a numerical experiment that illustrates the analysis above. Figure 8.4 shows the results of 5,000 iterates of D_{AB} with two different choices of affine space, as defined in (8.42). In these experiments $n_1 = n_3 = 1{,}099$, and $n_2 = 100$. Part (a) shows $\{\sin^2\theta_j\}$, on a semilog scale, and part (b) shows the errors $\{\|\boldsymbol{r}^{(n)} - \boldsymbol{i}\|_2\}$. There is a 99-dimensional subspace where $T_{\boldsymbol{i}}A$ and $T_{\boldsymbol{i}}B$ intersect, producing the initial part of part (a), showing angles $\sin^2\theta_j = O(10^{-15})$. The components in these directions converge to $O(10^{-16})$ in about 50 iterates. As predicted by the analysis above, the errors are initially smaller when the $\{\theta_j\}$ are closer to 0, (green curves), but the subsequent decay in the error occurs faster when these angles are larger (magenta curves).

Remark 8.1 If we were to replace D_{AB} with D_{BA} in this and the previous section, then the x_1-coordinate in the previous section and the x_{21}-coordinate in this section would always converge to a value greater than 1. It would be

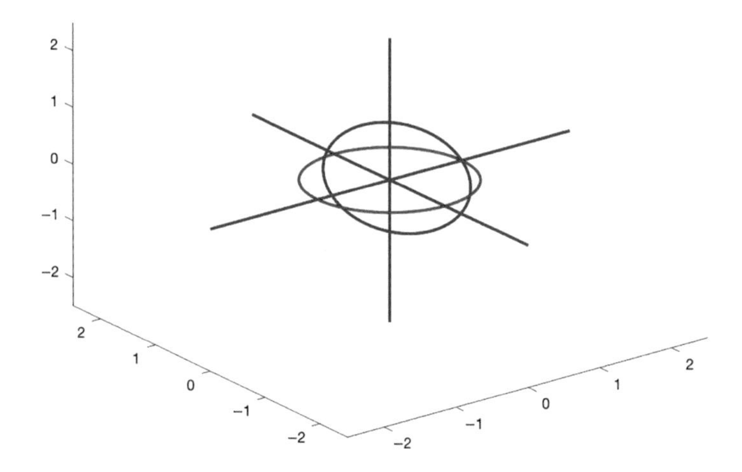

Figure 8.5 A plot of two circles of radii 1 (red) and 0.9 (blue) meeting at the point (1, 0, 0) on the x_1-axis. Their tangent lines make an angle $\theta = \pi/4$ at the intersection point.

tempting to conclude that this is a general phenomenon, but it depends on the fact that B is a linear subspace and A is convex. One can examine this question by considering a pair of circles in $\mathbb{R}^3$, one of radius 1 in the x_1x_2-plane (the set A) and one of radius r (the set B), which meet at the point (1, 0, 0) on the x_1-axis, collinear with their centers. At the intersection point their tangent lines make an angle θ. An example is shown in Figure 8.5. If $\theta = 0$, then, for $r < 1$, the iterates of D_{AB} converge to a point with $x_1 > 1$ and, if $r > 1$, to a point with $x_1 < 1$. However, when θ is not 0 (or very close to 0), then the sign $x_1 - 1$ is not determined by that of $r - 1$ but depends on the initial data in a complicated way. If r is very small or very large, then the sign of $x_1 - 1$ is again determined by that of $r - 1$, as one would expect from the limiting cases: As $r \to \infty$, the circle B tends to a straight line passing through $(1, 0, 0)$. As $r \to 0$, we can rescale so that B tends to a circle of radius 1 and A tends to a straight line.

8.1.3 Nonlinear Curves in Three Dimensions

The final set of low-dimensional examples illustrates additional phenomena that arise in the iteration of a hybrid iterative map when the distance, d_{AB}, between the two sets has critical points other than global minima and global maxima. For these experiments we let A be a circle in $\mathbb{R}^3$ and B be a nonlinear curve, given parametrically by

$$(\alpha + \beta \cos(pt))(\cos t,\ \sin t,\ \gamma) \tag{8.50}$$

An example is shown in Figure 8.1(b).

By adjusting the parameters $(\alpha,\ \beta)$, we can arrange for A and B

(i) to just barely miss intersecting;
(ii) to have multiple nontransversal intersections; and
(iii) to have multiple transversal intersections.

Because of the periodicity of B, the distance function between A and B has many nontrivial critical points. As detailed in Section 7.2.1, any critical point of d_{AB} defines a generalized center manifold for D_{AB}. In these examples we see that even center manifolds defined by a saddle point can be attracting.

Figure 8.6 shows the results of iterating D_{AB}, with B defined using $p = 12$ and by four different choices of values for $(\alpha,\ \beta,\ \gamma)$

$$\{(0.9,\ 0.099,\ 0.4), (0.9,\ 0.1,\ 0.4), (0.9,\ 0.101,\ 0.4),\ (0.9,\ 0.102,\ 0.4)\}.$$

(i) In the first case, the curves just fail to intersect. A careful examination shows that the iterates are entrained to a center manifold defined by a local, nonzero minimum of the distance function. They are spiraling out to infinity in a fixed neighborhood of this center manifold.
(ii) In the second case, there are 12 nontransversal intersections. The iterates converge to a point on the (two-dimensional) center manifold defined by one such intersection point. This point is quite distant from $A \cap B$; as predicted by the analysis in the previous two sections, this convergence occurs quite rapidly.
(iii) In the third case, there are 12 pairs of very nearby transversal intersections. The iterates have become entrained to the center manifold defined by a saddle point of the distance function and are slowly moving out towards infinity along this center manifold. The red curve in the right-hand plots for this example shows that the distance between successive iterates is very nearly constant, which agrees well with the prediction provided by (7.43) in the linear case.
(iv) In the fourth case, there are 12 pairs of not-so-nearby transversal intersections. The iterates are converging to a point on the center manifold defined by one of these intersection points. The scalloping error curve indicates that, in this case, the iterates are slowly spiraling inward to the center manifold.

These last examples illustrate much of the range of behavior that occurs when using hybrid iterative maps to find the intersections of nonlinear sets.

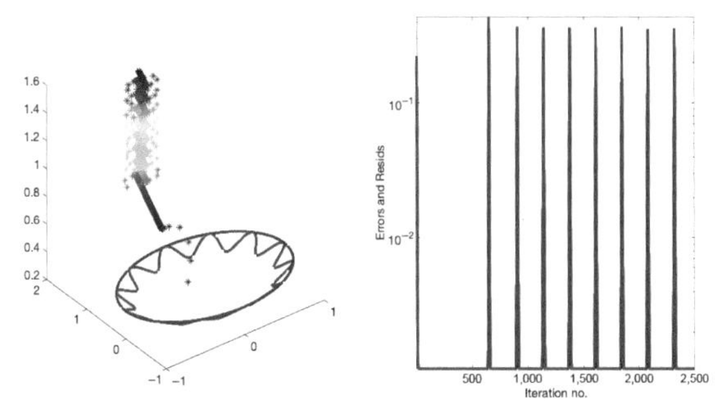

(a) In this example ($\alpha = 0.9$, $\beta = 0.099$, and $\gamma = 0.4$) the blue and red curves in the left panel do not intersect. The iterates are attracted to a center manifold defined by a local minimum of d_{AB} and are tending to infinity.

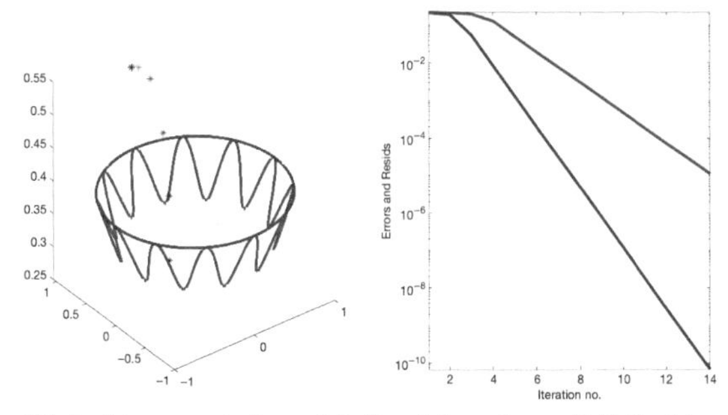

(b) In this example ($\alpha = 0.9, \beta = 0.1$, and $\gamma = 0.4$) the blue and red curves in the left panel have only nontransversal intersections. The iterates have found the center manifold defined by such an intersection and have converged rapidly to it.

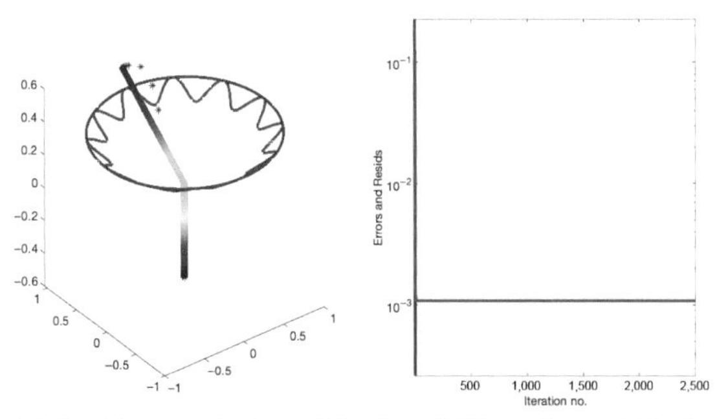

(c) In this example ($\alpha = 0.9$, $\beta = 0.101$, and $\gamma = 0.4$) the blue and red curves in the left panel have very nearby pairs of transversal intersections. The iterates have found a generalized center manifold defined by a saddle point of d_{AB}.

Figure 8.6 The plots on the left show the iterates $*$ for hybrid map, D_{AB}, with the indicated parameter values used to define B. The earlier iterates are in shades of blue, and later iterates in reds. The plots on the right show the errors, in blue, and the difference between successive iterates, in red.

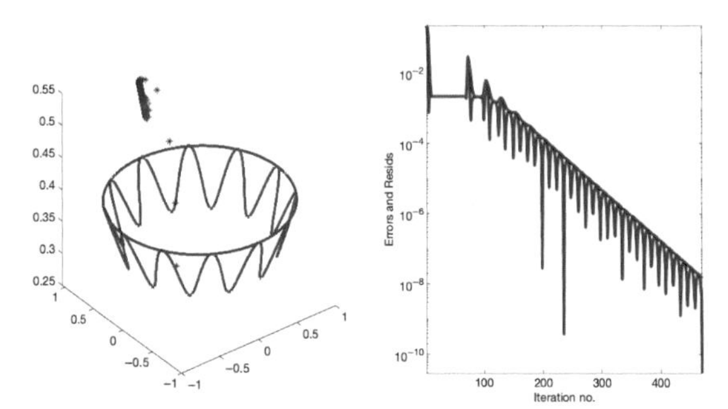

(d) In this example ($\alpha = 0.9$, $\beta = 0.102$, and $\gamma = 0.4$) there are pairs of transversal intersections, with more separation than in (c). The iterates have found one and are converging to its center manifold along a scalloping trajectory.

Figure 8.6 (*cont.*)

A notable consequence of the first two examples is that an algorithm for finding $\boldsymbol{x}_0 = A \cap B$ based on a hybrid iterative map can *sometimes* overcome the difficulties posed by a nontransversal intersection. In these examples the landscape, as defined by the function d_{AB}, is very simple, and there is a unique intersection point, which makes these problems quite different from phase retrieval. The basic reason that iterating D_{AB} or D_{BA} can readily find $A \cap B$, even when the intersection is nontransversal, is that the iterates converge to points on a center manifold that are quite distant from $A \cap B$ itself. The effects at the limit point of the local geometry of A and B near to $\boldsymbol{x}_0$ are therefore some sort of "action-at-a-distance." In these two examples, the convergence in the nontransversal directions is governed by a single parameter, which automatically assumes a value that assures convergence. In the phase retrieval problem, there are many parameters involved in the behavior of iterates in the nontransversal directions; experimentally the iterates do not always find values that lead to convergence.

It appears to be a robust result that small, nonzero angles between $T_{\boldsymbol{x}_0}A$ and $T_{\boldsymbol{x}_0}B$ produce very slow convergence. In the context of phase retrieval, we see below that, most of the time, a nontransversal intersection between $\mathbb{A}_{\boldsymbol{a}}$ and B does prevent the iterates of a hybrid iterative map from converging. Small angles significantly exacerbate the problem. Moreover, as indicated by the third example, a complicated d_{AB}-landscape can have a significant effect on the behavior of the iterates.

The landscape defined by the function, $d_{\mathbb{A}_{\boldsymbol{a}}B_S}$ is often extremely complicated, with a vast array of critical points of different sorts. This leads to large numbers of center manifolds and generalized center manifolds that may be weakly attracting. When the nearby intersections are nontransversal, this seems to cause the iterates of a hybrid iterative map to wander somewhat chaotically in a small neighborhood of several such attracting basins, or less chaotically, but nonconvergently, in a single attracting basin. This is, in either case, a very stable phenomenon. See Chapter 10.

8.2 Linearization of Hybrid Iterative Maps Along the Center Manifold

We now consider algorithms based on hybrid iterative maps for the phase retrieval problem for real images with auxiliary data defined by a support condition, that is $B = B_S$. We first identify the center manifolds for these maps and then analyze additional conditions that are required for a point on the center manifold to define a fixed point of the map. We then compute the linearization of $D_{\mathbb{A}_a B_S}$ at an empirically located attracting fixed point. These maps are difficult to analyze, as the map $P_{\mathbb{A}_a}$ takes a simple form in the DFT domain, whereas the map P_{B_S} takes a simple form in the image domain, and *not* vice versa. Beyond these computational difficulties, the linearization at an attracting fixed point often turns out to be a nonnormal map, which is *not a contraction*. Its eigenvalues are complex numbers of modulus less than 1, but the eigenbasis is very far from orthogonal. In light of this difficulty, we have performed a variety of numerical experiments to understand the properties of the linearization near points to which the iterates converge. A similar analysis can be done for $D_{B_S\mathbb{A}_a}$; we leave this to the interested reader.

Let $f_0 \in \mathbb{A}_a \cap B_S$ then center manifold $\mathscr{C}^{f_0}_{\mathbb{A}_a B_S}$ is locally given by

$$\mathscr{C}^{f_0}_{\mathbb{A}_a B_S} = R_B(N_{f_0}\mathbb{A}_a) \cap [f_0 + B_S^{\perp}], \tag{8.51}$$

which is easily seen to equal the affine subspace $f_0 + N^0_{f_0}\mathbb{A}_a \cap B_S^{\perp}$. The underlying linear subspace is, of course, just $(T^0_{f_0}\mathbb{A}_a + B_S)^{\perp}$, and therefore,

$$\dim \mathscr{C}^{f_0}_{\mathbb{A}_a B_S} = |J| - [\dim T^0_{f_0}\mathbb{A}_a + \dim B_S] + \dim T^0_{f_0}\mathbb{A}_a \cap B_S. \tag{8.52}$$

The dimension of the center manifold increases with $\dim T^0_{f_0}\mathbb{A}_a \cap B_S$. For a reasonable estimate S of S_{f_0}, and an adequate degree of oversampling for the magnitude data to determine the object, we have, for a d-dimensional image, that $\dim B_S \approx 2^{-d}|J|$, so

$$\dim \mathscr{C}^{f_0}_{\mathbb{A}_a B_S} \approx |J| \frac{2^{d-1}-1}{2^d}, \tag{8.53}$$

which is typically quite large.

As noted above, a necessary condition for $\boldsymbol{f}$ to be a fixed point (near to $\boldsymbol{f}_0$) of $D_{\mathbb{A}_a B_S}$ is for $\boldsymbol{f} \in \mathscr{C}^{f_0}_{\mathbb{A}_a B_S}$. This means that $\boldsymbol{f} = \boldsymbol{f}_0 + \boldsymbol{n}$, where $\boldsymbol{n} \in N^0_{f_0} \mathbb{A}_{\boldsymbol{a}} \cap B_S^{\perp}$. A moment's thought shows that

$$D_{\mathbb{A}_a B_S}(\boldsymbol{f}_0 + \boldsymbol{n}) = \boldsymbol{n} + P_{\mathbb{A}_a}(\boldsymbol{f}_0 - \boldsymbol{n}). \tag{8.54}$$

We need to check the conditions on $\boldsymbol{n}$ so that $P_{\mathbb{A}_a}(\boldsymbol{f}_0 - \boldsymbol{n}) = \boldsymbol{f}_0$; this is nontrivial because $\mathbb{A}_{\boldsymbol{a}}$ is nonlinear. In the DFT representation, $\widehat{\mathbb{A}}_{\boldsymbol{a}}$ is the product of circles in complex coordinate lines

$$\widehat{\mathbb{A}}_{\boldsymbol{a}} = \prod_{\boldsymbol{k} \in J} \{a_{\boldsymbol{k}} e^{i\theta_{\boldsymbol{k}}} : \theta_{\boldsymbol{k}} \in \mathbb{R}\}, \tag{8.55}$$

which satisfy $e^{i\theta_{\boldsymbol{k}}} = e^{-i\theta_{\boldsymbol{k}'}}$, where $\boldsymbol{k}'$ is the index conjugate to $\boldsymbol{k}$, see (2.17). If we let $\widehat{\boldsymbol{f}}_0$ and $\widehat{\boldsymbol{n}}$ be the DFT representatives of $\boldsymbol{f}_0$ and $\boldsymbol{n}$, respectively, then equation (2.21) shows that, for each $\boldsymbol{k} \in J$, there is a $\rho_{\boldsymbol{k}} \in \mathbb{R}$ so that

$$\widehat{n}_{\boldsymbol{k}} = \rho_{\boldsymbol{k}} \widehat{f}_{0\boldsymbol{k}}. \tag{8.56}$$

The condition $P_{\mathbb{A}_a}(\boldsymbol{f}_0 - \boldsymbol{n}) = \boldsymbol{f}_0$ is, therefore, equivalent to the requirement

$$\rho_{\boldsymbol{k}} < 1 \qquad \text{for all } \boldsymbol{k} \in J. \tag{8.57}$$

For, if $\rho_{\boldsymbol{k}} > 1$, then the point on the circle of radius $a_{\boldsymbol{k}} = |\widehat{f}_{0\boldsymbol{k}}|$ closest to $(1-\rho_{\boldsymbol{k}})\widehat{f}_{0\boldsymbol{k}}$ is $-\widehat{f}_{0\boldsymbol{k}}$. If $\rho_{\boldsymbol{k}} = 1$, for any index $\boldsymbol{k}$, then the value of $P_{\mathbb{A}_a}(\boldsymbol{f}_0 - \boldsymbol{n})$ is not uniquely defined. This proves the following proposition:

Proposition 8.2 *Let $\mathbb{A}_{\boldsymbol{a}}$ be a magnitude torus, with $\boldsymbol{f}_0 \in \mathbb{A}_{\boldsymbol{a}} \cap B_s$. A point $\boldsymbol{f}_0 + \boldsymbol{n} \in \mathscr{C}^{f_0}_{\mathbb{A}_a B_S}$ is a fixed point of the map $D_{\mathbb{A}_a B_S}$ if and only if* (8.57) *holds where $\{\rho_{\boldsymbol{k}}\}$ are defined in* (8.56).

It is not hard to see that the set of fixed points is an open set, which is contained in an orthant in a high-dimensional Euclidean space.

We can now compute the linearization of $D_{\mathbb{A}_a B_S}$ around such a fixed point. The only part of this mapping that is not already linear is $P_{\mathbb{A}_a}$. This calculation is best done in the DFT representation. We need to compute the leading order part of

$$P_{\mathbb{A}_a}(\widehat{\boldsymbol{f}}_0 - \widehat{\boldsymbol{n}} + \delta_1 \widehat{\boldsymbol{\tau}} + \delta_2 \widehat{\boldsymbol{\nu}}) - (\widehat{\boldsymbol{f}}_0 - \widehat{\boldsymbol{n}}) = dP_{\mathbb{A}_a}(\widehat{\boldsymbol{f}}_0 - \widehat{\boldsymbol{n}})[\delta_1 \widehat{\boldsymbol{\tau}} + \delta_2 \widehat{\boldsymbol{\nu}}] + o(|\delta_1| + |\delta_2|), \tag{8.58}$$

where $\widehat{\boldsymbol{\tau}} \in T^0_{\widehat{f}_0}\widehat{\mathbb{A}}_{\boldsymbol{a}}$ and $\widehat{\boldsymbol{v}} \in N^0_{\widehat{f}_0}\widehat{\mathbb{A}}_{\boldsymbol{a}}$ are unit vectors and $|\delta_1| + |\delta_2| \ll 1$. The analysis in Section 2.1 shows that for each $\boldsymbol{k}$, there exist $\alpha_{\boldsymbol{k}}, \beta_{\boldsymbol{k}} \in \mathbb{R}$ so that

$$\widehat{\tau}_{\boldsymbol{k}} = i\alpha_{\boldsymbol{k}}\widehat{f}_{\boldsymbol{k}} \text{ and } \widehat{v}_{\boldsymbol{k}} = \beta_{\boldsymbol{k}}\widehat{f}_{\boldsymbol{k}}. \tag{8.59}$$

The computation of $P_{\mathbb{A}_a}(\widehat{\boldsymbol{f}}_0 - \widehat{\boldsymbol{n}} + \delta_1\widehat{\boldsymbol{\tau}} + \delta_2\widehat{\boldsymbol{v}})$ can be done a coordinate at a time. We see that the $\boldsymbol{k}$th coordinate is given by

$$\begin{aligned} P_{\widehat{\mathbb{A}}_a}(\widehat{\boldsymbol{f}}_0 - \widehat{\boldsymbol{n}} + \delta_1\widehat{\boldsymbol{\tau}} + \delta_2\widehat{\boldsymbol{v}})_{\boldsymbol{k}} &= \frac{(1-\rho_{\boldsymbol{k}} + \delta_2\beta_{\boldsymbol{k}} + i\delta_1\alpha_{\boldsymbol{k}})\widehat{f}_{\boldsymbol{k}}}{\sqrt{(1-\rho_{\boldsymbol{k}} + \delta_2\beta_{\boldsymbol{k}})^2 + \delta_1^2\alpha_{\boldsymbol{k}}^2}} \\ &= \widehat{f}_{\boldsymbol{k}} + \frac{i\delta_1\alpha_{\boldsymbol{k}}\widehat{f}_{\boldsymbol{k}}}{1-\rho_{\boldsymbol{k}}} + O(\delta_1^2 + \delta_2^2). \end{aligned} \tag{8.60}$$

Hence, at a fixed point, the first order term is given by

$$dP_{\widehat{\mathbb{A}}_a}(\widehat{\boldsymbol{f}}_0 - \widehat{\boldsymbol{n}})[\delta_1\widehat{\boldsymbol{\tau}} + \delta_2\widehat{\boldsymbol{v}}]_{\boldsymbol{k}} = \frac{|\widehat{f}_{\boldsymbol{k}}|}{|\widehat{f}_{\boldsymbol{k}} - \widehat{n}_{\boldsymbol{k}}|}\delta_1\widehat{\tau}_{\boldsymbol{k}} = \frac{1}{|1-\rho_{\boldsymbol{k}}|}\delta_1\widehat{\tau}_{\boldsymbol{k}}. \tag{8.61}$$

Observe that, to first order, the $\delta_2\widehat{\boldsymbol{v}}$ terms makes no contribution, and that $dP_{\widehat{\mathbb{A}}_a}$ takes values in the fiber of the tangent space, $T_{\widehat{f}_0}\widehat{\mathbb{A}}_{\boldsymbol{a}}$. The linearization of $P_{\mathbb{A}_a}$ is obtained by conjugating $dP_{\widehat{\mathbb{A}}_a}$ with the DFT. We denote this map by $dP_{\mathbb{A}_a}(\boldsymbol{f}_0 - \boldsymbol{n})[\delta_1\boldsymbol{\tau} + \delta_2\boldsymbol{v}]$. When $\boldsymbol{n} = 0$ the map $dP_{\mathbb{A}_a}(\boldsymbol{f}_0)$ reduces to the orthogonal projection onto $T^0_{f_0}\mathbb{A}_{\boldsymbol{a}}$, which is denoted by $\pi_{\mathbb{A}_a f_0}$. In all cases, $dP_{\mathbb{A}_a}(\boldsymbol{f}_0 - \boldsymbol{n})$ takes values in $T^0_{f_0}\mathbb{A}_{\boldsymbol{a}}$.

Let $\boldsymbol{f}_0 + \boldsymbol{n} \in \mathscr{C}^{f_0}_{\mathbb{A}_a B_S}$ be a fixed point as above and $\boldsymbol{v} \in T^0_{f_0}\mathbb{A}_{\boldsymbol{a}} + B_S$, with $\|\boldsymbol{v}\|_2 \ll 1$. The computation above shows that

$$\begin{aligned} &D_{\mathbb{A}_a B_S}(\boldsymbol{f}_0 + \boldsymbol{n} + \boldsymbol{v}) \\ &\quad = \boldsymbol{f}_0 + \boldsymbol{n} + (\mathrm{Id} - P_{B_s})[\boldsymbol{v}] + dP_{\mathbb{A}_a}(\boldsymbol{f}_0 - \boldsymbol{n})[R_{B_S}(\boldsymbol{v})] + O(\|\boldsymbol{v}\|_2^2). \end{aligned} \tag{8.62}$$

If the intersection is nontransversal, then we let $F = T^0_{f_0}\mathbb{A}_{\boldsymbol{a}} \cap B_S$, as suggested by the analysis in Section 7.2.2. We then define

$$A_0 = T^0_{f_0}\mathbb{A}_{\boldsymbol{a}} \cap F^{\perp} \text{ and } B_0 = B_S \cap F^{\perp}. \tag{8.63}$$

With the splitting $T^0_{f_0}\mathbb{A}_{\boldsymbol{a}} + B_S = A_0 + B_0 + F$, we write $\boldsymbol{v} = \boldsymbol{v}_A + \boldsymbol{v}_B + \boldsymbol{v}_F$. Note that $\boldsymbol{v}_F$ is orthogonal to $\boldsymbol{v}_A + \boldsymbol{v}_B$, but $\boldsymbol{v}_A$ and $\boldsymbol{v}_B$ are generally not orthogonal to one another. We can reexpress $dD_{\mathbb{A}_a B_S}$ as

$$\begin{aligned} dD_{\mathbb{A}_a B_S}(\boldsymbol{f}_0 + \boldsymbol{n})[\boldsymbol{v}] = (\mathrm{Id} - P_{B_s})[\boldsymbol{v}_A] + dP_{\mathbb{A}_a}(\boldsymbol{f}_0 - \boldsymbol{n})[\pi_{\mathbb{A}_a f_0}(\boldsymbol{v}_B + R_{B_S}(\boldsymbol{v}_A))] \\ + dP_{\mathbb{A}_a}(\boldsymbol{f}_0 - \boldsymbol{n})[\boldsymbol{v}_F]. \end{aligned} \tag{8.64}$$

When $\boldsymbol{n} = \boldsymbol{0}$ this map reduces to

$$dD_{\mathbb{A}_a B_S}(\boldsymbol{f}_0)[\boldsymbol{v}] = \pi_{\mathbb{A}_a f_0}(\boldsymbol{v}_B + 2P_{B_s}(\boldsymbol{v}_A)) - P_{B_s}(\boldsymbol{v}_A) + \boldsymbol{v}_F. \tag{8.65}$$

In the F-directions it is the identity. From the results in Section 7.2.2 it follows that the map $\boldsymbol{v}_A+\boldsymbol{v}_B \mapsto \pi_{\mathbb{A}_a f_0}(\boldsymbol{v}_B+2P_{B_s}(\boldsymbol{v}_A))-P_{B_s}(\boldsymbol{v}_A)$ is a contraction with contraction rates$\{\cos^2\theta_j\}$ where $\{\theta_j\}$ are the angles between the pair of subspaces A_0, B_0, as defined in (2.8). See also Remark 7.7.

In numerical examples below, we consider the linearized map

$$\boldsymbol{v} \mapsto dD_{\mathbb{A}_a B_S}(\boldsymbol{f}_0+\boldsymbol{n})[\boldsymbol{v}],$$

where the iterates of $D_{\mathbb{A}_a B_S}$ have converged to the point $\boldsymbol{f}_0+\boldsymbol{n} \in \mathscr{C}^{\boldsymbol{f}_0}_{\mathbb{A}_a B_S}$. The linearized map, in these examples, is *not a contraction*; it is a nonnormal map with complex eigenvalues of modulus less than 1. This explains part of the difficulty one encounters trying to analyze the linearization and finding simple, explicit conditions for a point $\boldsymbol{f}_0+\boldsymbol{n} \in \mathscr{C}^{\boldsymbol{f}_0}_{\mathbb{A}_a B_S}$ to be an attracting fixed point.

The linearized action at $\boldsymbol{f}_0+\boldsymbol{n}$, in the F-directions, is given by

$$\boldsymbol{v}_F \mapsto dP_{\mathbb{A}_a}(\boldsymbol{f}_0-\boldsymbol{n})[\boldsymbol{v}_F].$$

A sufficient condition for this map to be contracting is for $\rho_{\boldsymbol{k}}<0$, see (8.56), for all $\boldsymbol{k} \in J$. This is, however, not a necessary condition as F is a very low-dimensional subspace. Recall that in order for $\boldsymbol{f}_0+\boldsymbol{n}$ to be a fixed point we needed to require that $\rho_{\boldsymbol{k}}<1$. While points satisfying this condition certainly exist, it is by no means clear that there exist points $\boldsymbol{n} \in N^0_{\boldsymbol{f}_0}\mathbb{A}_{\boldsymbol{a}} \cap B_S^{\perp}$ with $\rho_{\boldsymbol{k}}<0$ for all $\boldsymbol{k}$. Note that these are conditions on the DFT of $\boldsymbol{f}_0$ and $\boldsymbol{n}$.

While we have not found simple conditions that characterize the stable fixed points, it is suggestive to rewrite the linearization in the following form

$$\begin{aligned} dD_{\mathbb{A}_a B_S}(\boldsymbol{f}_0+\boldsymbol{n})[\boldsymbol{v}] &= dP_{\mathbb{A}_a}(\boldsymbol{f}_0-\boldsymbol{n})[\pi_{\mathbb{A}_a f_0}(\boldsymbol{v}_B+2P_{B_S}(\boldsymbol{v}_A))-P_{B_s}(\boldsymbol{v}_A)] \\ &\quad +\big(dP_{\mathbb{A}_a}(\boldsymbol{f}_0)-dP_{\mathbb{A}_a}(\boldsymbol{f}_0-\boldsymbol{n})\big)[\pi_{\mathbb{A}_a f_0}\circ(\mathrm{Id}-P_{B_S})\boldsymbol{v}_A] \\ &\quad +dP_{\mathbb{A}_a}(\boldsymbol{f}_0-\boldsymbol{n})[\boldsymbol{v}_F]-[(\mathrm{Id}-\pi_{\mathbb{A}_a f_0})\circ P_{B_S}](\boldsymbol{v}_A). \end{aligned} \tag{8.66}$$

The first three terms lie in $T^0_{\boldsymbol{f}_0}\mathbb{A}_{\boldsymbol{a}}$, whereas the final term lies in $N^0_{\boldsymbol{f}_0}\mathbb{A}_{\boldsymbol{a}}$. As noted above, the map $\boldsymbol{v}_A+\boldsymbol{v}_B \mapsto \pi_{\mathbb{A}_a f_0}(\boldsymbol{v}_B+2P_{B_S}(\boldsymbol{v}_A))-P_{B_s}(\boldsymbol{v}_A)$ is a contraction, with contraction rates $\{\cos^2\theta_j\}$. The map $\boldsymbol{v}_A \mapsto -[(\mathrm{Id}-\pi_{\mathbb{A}_a f_0})\circ P_{B_S}](\boldsymbol{v}_A)$ is even more strongly contracting, with contraction rates $\{\sin^2\theta_j\cos^2\theta_j\}$. Finally, the mapping $\boldsymbol{v}_A \mapsto \pi_{\mathbb{A}_a f_0}\circ(\mathrm{Id}-P_{B_S})\boldsymbol{v}_A$ is only weakly contracting. For $\boldsymbol{v}_A \in T^0_{\boldsymbol{f}_0}\mathbb{A}_{\boldsymbol{a}} \cap B_S^{\perp}$ this map is an isometry. For 2D images this subspace has dimension about $|J|/4$. In order for the term

$$\big(dP_{\mathbb{A}_a}(\boldsymbol{f}_0)-dP_{\mathbb{A}_a}(\boldsymbol{f}_0-\boldsymbol{n})\big)[\pi_{\mathbb{A}_a f_0}\circ(\mathrm{Id}-P_{B_S})\boldsymbol{v}_A]$$

to be a contraction it is sufficient that $\rho_{\boldsymbol{k}}<\frac{1}{2}$, for all $\boldsymbol{k}$. These observations suggest that:

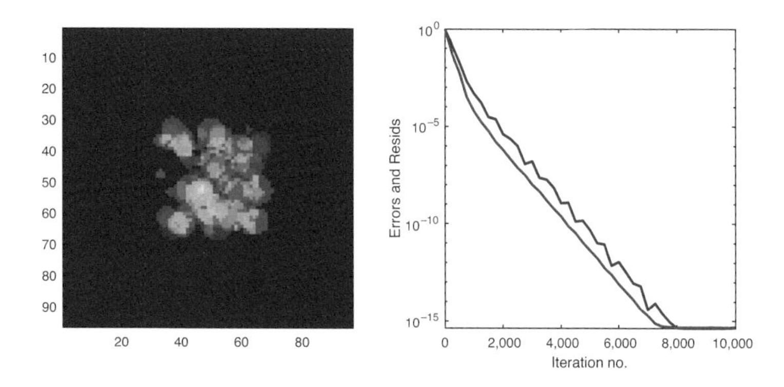

Figure 8.7 The reconstructed image and error trajectories for a 96×96, piecewise constant image. The blue curve is the error, and the red curve is the residual.

(i) The angles, $\{\theta_j\}$, between A_0 and B_0 will continue to affect the behavior of iterates near to the center manifold, even at points distant from $\boldsymbol{f}_0$ itself.

(ii) The existence of a nontransversal intersection may obstruct the iterates from converging, as there may not exist points, $\boldsymbol{f}_0 + \boldsymbol{n} \in \mathscr{C}^{f_0}_{\mathbb{A}_a B_S}$ where $\boldsymbol{v}_F \mapsto dP_{\mathbb{A}_a}(\boldsymbol{f}_0 - \boldsymbol{n})[\boldsymbol{v}_F]$ is contracting.

(iii) In order for $\boldsymbol{f}_0 + \boldsymbol{n}$ to be a stable fixed point, it seems quite likely that $\rho_{\boldsymbol{k}} < \frac{1}{2}$ must hold for a preponderance of $\boldsymbol{k} \in J$.

Below, it is shown that all of these predictions are borne out in a numerical example. We compute the linearization at an attracting fixed point defined by a piecewise constant 96×96 image, $\boldsymbol{f}_0$ using the 1-pixel neighborhood of S_{f_0} as the support constraint.

Example 8.3 In Figure 8.7 we show the reconstruction of the piecewise constant image, $\boldsymbol{f}_0$, and the error trajectories for 10,000 iterates, $\{\boldsymbol{f}^{(n)}\}$, of $D_{\mathbb{A}_a B_{S_1}}$. The reconstructions are $\boldsymbol{r}^{(n)} = P_{B_{S_1}}(\boldsymbol{f}^{(n)})$. The sequence $\{\boldsymbol{f}^{(n)}\}$ has largely converged to a point $\boldsymbol{f}_0^{[1,1]} + \boldsymbol{n}$, where $\boldsymbol{f}_0^{[1,1]}$, is a trivial associate for which $T_{\boldsymbol{f}_0^{[1,1]}} \mathbb{A}_{\boldsymbol{a}} \cap B_{S_1} = \boldsymbol{f}_0$, i.e., the intersection is transversal. The vector $\boldsymbol{n} \in N^0_{\boldsymbol{f}_0^{[1,1]}} \mathbb{A}_{\boldsymbol{a}} \cap B_{S_1}^{\perp}$. The blue curve in Figure 8.7 shows the true errors $\{\|\boldsymbol{r}^{(n)} - \boldsymbol{f}_0^{[1,1]}\|_2\}$ and the red curve shows the residual

$$\|P_{\mathbb{A}_a} \circ R_{B_{S_1}}(\boldsymbol{f}^{(n)}) - P_{B_{S_1}}(\boldsymbol{f}^{(n)})\|_2 = \|\boldsymbol{f}^{(n)} - \boldsymbol{f}^{(n+1)}\|_2. \tag{8.67}$$

In this example $dP_{\mathbb{A}_a}(\boldsymbol{f}_0^{[1,1]} - \boldsymbol{n})$ is defined by ratios $\left\{\rho_{\boldsymbol{k}} = \frac{\widehat{n}_{\boldsymbol{k}}}{\widehat{f}_{0\boldsymbol{k}}}\right\}$. This set has 9,216 elements of which all but 92 are smaller than 1/2; about half of these

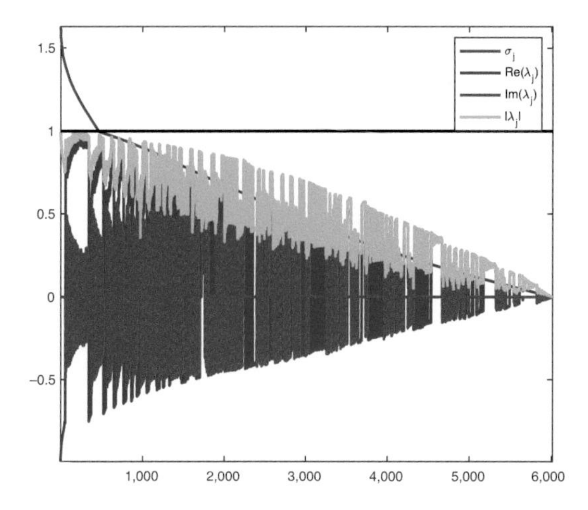

Figure 8.8 The spectrum/singular values of the linearization, $dD_{\mathbb{A}_a B_{S_1}}(\boldsymbol{f}_0^{[1,1]} + \boldsymbol{n})$, of $D_{\mathbb{A}_a B_{S_1}}$ at an attracting fixed point. The real and imaginary parts of the eigenvalues of $dD_{\mathbb{A}_a B_{S_1}}(\boldsymbol{f}_0^{[1,1]} + \boldsymbol{n})$ are shown in blue and magenta, respectively, with their modulus in lime green; the singular values are shown in red.

ratios are less than 0. Evidently having $\rho_{\boldsymbol{k}} < 0$ for all $\boldsymbol{k}$ is not a necessary condition for a fixed point to be attracting. Figure 8.8 shows the spectral data connected to $dD_{\mathbb{A}_a B_{S_1}}(\boldsymbol{f}_0^{[1,1]} - \boldsymbol{n})$. The red curve shows the singular values $\{\sigma_j\}$ of the linearization, whose maximum is 1.631. There are 466 singular values greater than 1. The blue and magenta curves are the real and imaginary parts of the eigenvalues $\{\lambda_j\}$, whose maximum modulus is 0.9961. The moduli of the eigenvalues are shown in lime green. The scalloping in the blue curve in Figure 8.7 is a reflection of the fact that the linearized map has complex eigenvalues and therefore the iterates spiral in toward the eventual fixed point. This example illustrates the inherent difficulty in analyzing the linearization of a hybrid iterative map, even at an attracting fixed point.

8.3 Further Numerical Examples

In this section, we continue our experimental investigation of the behavior of hybrid iterative maps. We use real images defined by (1.46), which are similar to those shown in Figure 1.5. In all cases, these algorithms are initialized by choosing an image $\boldsymbol{f}^{(0)} \in \mathbb{A}_{\boldsymbol{a}}$ with randomly selected phases, which usually are required to satisfy the symmetry requirements for a real image. The detailed behavior of the iterates depends on this choice. The results of the experiments in this section would not be too different if we used complex-valued images, though, generally speaking, it is somewhat more difficult to reconstruct complex-valued images.

Most practical approaches to image reconstruction involve running the same iterative scheme with many such initial guesses for the phases. Our first experiment illustrates the (well-known) sensitivity of a hybrid iterative map to the initial data, and its continuing sensitivity even after it has found an attracting basin. In the experiments in this section, we show typical cases of "good" behavior for the algorithms under study. The statistical properties of these algorithms as a function of the starting point are considered in Chapter 12.

Theorem 3.3 shows that near a nontransversal intersection, $\boldsymbol{f}_0$ the problem of finding points in $\mathbb{A}_{\boldsymbol{a}} \cap B_S$ is itself ill-conditioned. Suppose that $\boldsymbol{w}$ is a unit vector belonging to $T^0_{\boldsymbol{f}_0}\mathbb{A}_{\boldsymbol{a}} \cap T_{\boldsymbol{f}_0} B_S$. Because this direction is tangent to both submanifolds it is clear that, $\boldsymbol{f}_0 + t\boldsymbol{w} \in B_S$ and for small t,

$$\begin{aligned} \& \operatorname{dist}(\boldsymbol{f}_0 + t\boldsymbol{w}, \mathbb{A}_{\boldsymbol{a}}) \propto t^2, \\ \operatorname{dist}(\boldsymbol{f}_0 + t\boldsymbol{w}, \mathbb{A}_{\boldsymbol{a}} \cap B) \approx t. \end{aligned} \tag{8.68}$$

Hence, if our numerical computations are accurate to ϵ_{mach}, then the set of vectors

$$\{\boldsymbol{f}_0 + t\boldsymbol{w} : |t| \le \sqrt{\epsilon_{\text{mach}}} \text{ and } \boldsymbol{w} \in T^0_{\boldsymbol{f}_0}\mathbb{A}_{\boldsymbol{a}} \cap T_{\boldsymbol{f}_0} B_S \text{ with } \|\boldsymbol{w}\|_2 = 1\} \tag{8.69}$$

all lie in $\mathbb{A}_{\boldsymbol{a}} \cap B_S$ to accuracy ϵ_{mach}. That is, these are images, contained in B_S, which satisfy

$$\frac{\|\mathcal{M}(\boldsymbol{f}_0 + t\boldsymbol{w}) - \mathcal{M}(\boldsymbol{f}_0)\|_2}{\|\boldsymbol{f}_0\|_2} \le Ct^2 \le C\epsilon_{\text{mach}}; \tag{8.70}$$

hence, to precision ϵ_{mach} they are solutions to the phase retrieval problem. As noted earlier in Remark 2.12, given $0 < p$, with $S = S_{\boldsymbol{f}_0, 2p}$, one can find sequences of perturbations, $\{\delta \boldsymbol{f}_n\} \subset B_S$ for which $\|\delta \boldsymbol{f}_n\|_2 \to 0$ that satisfy an estimate of the form

$$\frac{\|\mathcal{M}(\boldsymbol{f}_0 + \delta \boldsymbol{f}_n) - \mathcal{M}(\boldsymbol{f}_0)\|_2}{\|\boldsymbol{f}_0\|_2} \le C\|\delta \boldsymbol{f}_n\|_2^{2p+2}. \tag{8.71}$$

Showing that, with a loose enough estimate on $S_{\boldsymbol{f}_0}$, one can find find points $\boldsymbol{f}_0 + \delta \boldsymbol{f} \in B_S$ so that $\|\delta \boldsymbol{f}\|_2 \propto \epsilon_{\text{mach}}^{\frac{1}{2p+2}}$, but

$$\frac{\|\mathcal{M}(\boldsymbol{f}_0 + \delta \boldsymbol{f}) - \mathcal{M}(\boldsymbol{f}_0)\|_2}{\|\boldsymbol{f}_0\|_2} \le C\epsilon_{\text{mach}}. \tag{8.72}$$

This analysis also suggests that, with an accurate enough support constraint, we might expect to find algorithms capable of finding points that are within $\sqrt{\epsilon_{\text{mach}}}$ of true intersection points. In practice, this proves difficult to do. These algorithms usually stagnate at distances of 10^{-1}–10^{-2} from true intersection points. Part of the difficulty is that, beyond the failure of transversality, the maps $D_{\mathbb{A}_a B_S}$ and $D_{B_S \mathbb{A}_a}$ often have large-dimensional subspaces on which they contract very weakly. Experimentally, algorithms based on these maps stagnate as they approach attracting basins. Greater smoothness and/or a looser support constraint result in more pronounced stagnation. In the experiments below, we see that, once the weakly contracting subspace reaches an appreciable size, the distances between successive iterates do not decrease after a certain point. Nonetheless, these iterates do remain in a fixed sized neighborhood of the center manifold or center manifolds of a few nearby exact intersection points (Figures 8.12 and 8.13). The iterates remain in a subset of small diameter, even though the successive step sizes remain comparatively large. This phenomenon is stable over millions of iterates (Section 10.2).

From the numerical experiments in this section (see Figures 8.12 and 8.13) it is clear that the reconstructions obtained from hybrid map iterations have a very strong tendency to approach an intersection point in $\mathbb{A}_{\boldsymbol{a}} \cap B_S$ along trajectories that satisfy an estimate like (8.68). That is, if $\{\boldsymbol{f}^{(n)}\}$ is a sequence of iterates, then, typically

$$\operatorname{dist}(P_{\mathbb{A}_a} \circ R_{B_S}(\boldsymbol{f}^{(n)}), P_{B_S}(\boldsymbol{f}^{(n)})) \propto \operatorname{dist}(\boldsymbol{r}^{(n)}, \mathbb{A}_{\boldsymbol{a}} \cap B_S)^2. \tag{8.73}$$

Unlike the simple models in Section 8.1, in the phase retrieval problem, a nontransversal intersection often has a very deleterious effect on the behavior of iterations using a hybrid map. Indeed, once the $\dim T^0_{\boldsymbol{f}_0} \mathbb{A}_{\boldsymbol{a}} \cap B_S$ becomes appreciable, the iterates rarely get close to a center manifold defined by a point in $\mathbb{A}_{\boldsymbol{a}} \cap B_S$. Given the analysis in Section 8.2, this outcome is not surprising. Somewhat more surprising is the existence of images $\boldsymbol{f}_0$ for which $\dim T_{\boldsymbol{f}_0} \mathbb{A}_{\boldsymbol{a}} \cap B_{S_{\boldsymbol{f}_0}} > 0$, and yet a hybrid iterative map *can* reconstruct $\boldsymbol{f}_0$ with nearly machine precision. These examples are constructed as convolutions $\boldsymbol{f}_0 = \boldsymbol{g} * \boldsymbol{h}$, where $\boldsymbol{h}_j = \boldsymbol{h}_{-j}$ and $\boldsymbol{g}$ is supported at a sparse set of isolated pixels. This phenomenon is illustrated in Example 8.13. The nontransversality in this case is of the sort predicted by Proposition 2.15 and the discussion surrounding it: $\boldsymbol{f}_0$ "inherits" the nontransversality from $\boldsymbol{h}$. It may be notable that the construction, used in Remark 2.12, of curves with higher-order contact cannot be applied in this case. While the nontransversality of the intersection $\mathbb{A}_{\boldsymbol{a}} \cap B$ is clearly consequential for the solution of the phase retrieval problem, the full range of possible outcomes and their exact causes are not yet clearly understood.

In experiments we do sometimes encounter sequences of iterates that satisfy

$$\operatorname{dist}(P_{\mathbb{A}_a} \circ R_{B_S}(\boldsymbol{f}^{(n)}), P_{B_S}(\boldsymbol{f}^{(n)})) \propto \operatorname{dist}(\boldsymbol{r}^{(n)}, \mathbb{A}_a \cap B_S)^m, \tag{8.74}$$

for a power $m > 2$. It may be the case that the iterates "find" the curves of higher-order contact described in Remark 2.12. It could also be related to the fact that nonzero critical points of $d_{\mathbb{A}_a B_{S_p}}$ can also define generalized center manifolds that are themselves attracting. Both phenomena are intimately related to the structures of $T^0_{\boldsymbol{f}}\mathbb{A}_a$ and $N^0_{\boldsymbol{f}}\mathbb{A}_a$, as described in Sections 2.1–2.2.

Occasionally it happens that the distances $\operatorname{dist}(P_{\mathbb{A}_a} \circ R_{B_S}(\boldsymbol{f}^{(n)}), P_{B_S}(\boldsymbol{f}^{(n)}))$ and $\operatorname{dist}(\boldsymbol{r}^{(n)}, \mathbb{A}_a \cap B_S)$ remain proportional, though this behavior is unlikely once either the support constraint becomes too loose, or the unknown object becomes too smooth. Upon close examination, one finds that if the sequence $\{\boldsymbol{r}^{(n)}\}$ converges to a point, $\widetilde{\boldsymbol{f}}_0 \in \mathbb{A}_a \cap B_S$, to a very high order of accuracy, then this point is usually a trivial associate of $\boldsymbol{f}_0$ at which

$$\dim T_{\widetilde{\boldsymbol{f}}_0}\mathbb{A}_a \cap B_S = 0, \tag{8.75}$$

precluding the behavior described in (8.68). An example of this is shown in Figure 8.11.

A hybrid iterative map algorithm produces a sequence of iterates $\{\boldsymbol{f}^{(n)}\}$ that ideally converges to a point on the center manifold. From this sequence one constructs a sequence of approximate reconstructions $\{\boldsymbol{r}^{(n)}\}$, see (1.41). In the error plots below, the y-axis is logarithmic, so that a decreasing straight line is indicative of geometric convergence toward 0. The blue curves in these plots show the true errors, $\{\|\boldsymbol{r}^{(n)} - \widetilde{\boldsymbol{f}}\|_2\}$, where $\widetilde{\boldsymbol{f}}$ is the point in $\mathbb{A}_a \cap B$ closest to $\boldsymbol{r}^{(n)}$. In a real experiment this information is not available. The red curves show the *residuals*, which is a reasonable empirical measure of the reconstruction error. For an algorithm based on $D_{\mathbb{A}_a B_{S_p}}$ the residual is defined to be

$$\| P_{\mathbb{A}_a} \circ R_{B_{S_p}}(\boldsymbol{f}^{(n)}) - P_{B_{S_p}}(\boldsymbol{f}^{(n)})\|_2. \tag{8.76}$$

For a hybrid iterative map this also equals the distance between successive iterates $\{\|\boldsymbol{f}^{(n)} - \boldsymbol{f}^{(n+1)}\|_2\}$. If we use $\boldsymbol{r}^{(n)} = P_{B_{S_p}}(\boldsymbol{f}^{(n)})$ as the current reconstruction, then, as $P_{\mathbb{A}_a}(r^{(n)})$ is defined to be the point on $\mathbb{A}_a$ closest to $\boldsymbol{r}^{(n)}$, we have the estimate

$$\begin{aligned}\|\boldsymbol{f}^{(n+1)} - \boldsymbol{f}^{(n)}\|_2 &= \|P_{\mathbb{A}_a} \circ R_{B_{S_p}}(\boldsymbol{f}^{(n)}) - \boldsymbol{r}^{(n)}\|_2 \\ &\geq \|P_{\mathbb{A}_a}(\boldsymbol{r}^{(n)}) - \boldsymbol{r}^{(n)}\|_2 \geq \|\boldsymbol{a} - \mathscr{M}(\boldsymbol{r}^{(n)})\|_2,\end{aligned} \tag{8.77}$$

see (3.9). Hence, the residual provides an upper bound on what would ordinarily be called the data-error for the solving the equation $\mathscr{M}(\boldsymbol{r}) = \boldsymbol{a}$, for an $\boldsymbol{r} \in B_{S_p}$. As we see below in Example 8.9, the residual does not always give a good estimate for the true error.

Remark 8.4 Recall that in numerical examples we use the set $S_f^{10^{-14}}$ as the support of the image $\boldsymbol{f}$. To simplify the notation in these examples, we often write S_f instead of the more accurate $S_f^{10^{-14}}$. When we speak of a p-pixel neighborhood of the support of $\boldsymbol{f}$, it is a p-pixel neighborhood of $S_f^{10^{-14}}$.

Before exploring how these algorithms behave on different types of objects, with different types of support masks, we examine the average distances between the 1-pixel translates of typical images. The outcomes of these experiments provide insight into the distance between different exact solutions of the phase retrieval problem and thereby provide us with natural units to calibrate the results of the numerical experiments that follow.

Example 8.5 In Figure 8.9 we show graphs of the quantity

$$\frac{1}{4}\left[\frac{\|\boldsymbol{f}^{(\boldsymbol{e}_1)} - \boldsymbol{f}\|_2}{\|\boldsymbol{f}\|_2} + \frac{\|\boldsymbol{f}^{(-\boldsymbol{e}_1)} - \boldsymbol{f}\|_2}{\|\boldsymbol{f}\|_2} + \frac{\|\boldsymbol{f}^{(\boldsymbol{e}_2)} - \boldsymbol{f}\|_2}{\|\boldsymbol{f}\|_2} + \frac{\|\boldsymbol{f}^{(-\boldsymbol{e}_2)} - \boldsymbol{f}\|_2}{\|\boldsymbol{f}\|_2}\right], \tag{8.78}$$

averaged over 10 random examples each having a given level of smoothness. The smoother images are either obtained by summing smoother functions, as in (1.46), or by convolving a piecewise constant image with a Gaussian.

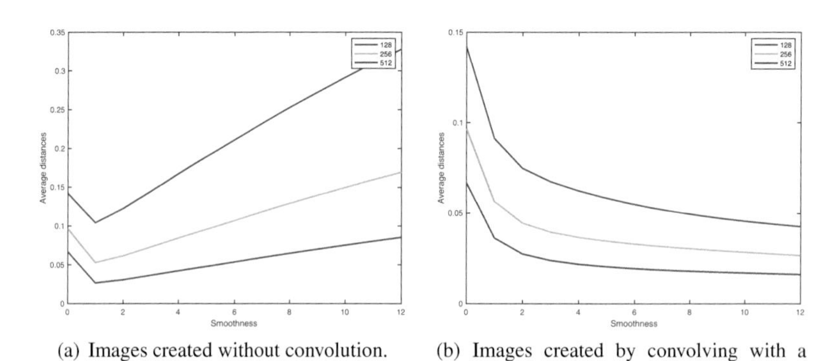

(a) Images created without convolution.

(b) Images created by convolving with a Gaussian.

Figure 8.9 Plots showing the average distances, defined by equation (8.78), to 1-pixel translates of a fixed reference point on a magnitude torus, as a function of the smoothness, k, for images of size 128×128 (red), 256×256 (green), and 512×512 (blue). In part (a), greater smoothing is obtained by summing smoother functions, and by convolution with Gaussians having larger effective support in part (b).

We also consider this quantity as a function of the size of the example, for $N = 128, 256, 512$. For Figure 8.9(a), with images generated using (1.46), but for the transition from $k = 0$ to $k = 1$, this is a monotonely increasing function of smoothness, and a decreasing function of the overall size of the image. If, instead, greater smoothness is obtained by using convolution, then these functions become monotonely *decreasing* functions of smoothness as well, as shown in Figure 8.9(b). This difference in behavior is easily explained: the supports of the functions $\psi_k(t)$, used in (1.46), do not depend on k. Hence, as the smoothness increases, more derivatives of ψ_k must vanish at $t = \pm 1$, forcing these functions to become more peaked around 0 as k increases. Hence, translations produce larger norm differences as these functions get smoother. Convolving with Gaussian attains greater smoothness at the expense of larger support, which leads the difference of translates to have smaller norms as the smoothness increases.

Comparison of this experiment with those below shows that the distances between 1-pixel translates are often comparable to the distances between the successive iterates, $\{\boldsymbol{f}^{(n)}\}$. From this perspective, it is somewhat surprising that these algorithms have attracting basins at all. As remarked above, the iterates remain in more or less fixed sized neighborhoods of the center manifolds of a few points in $\mathbb{A}_{\boldsymbol{a}} \cap B_S$. This behavior is often stable over millions of iterates. See Chapter 10.

We next consider how the behavior of the iterates is affected by small perturbations at several points in the trajectory of a *convergent* sequence.

Example 8.6 For this experiment we use a 256×256 piecewise constant image, $\boldsymbol{f}_0$, and a 1-pixel neighborhood of $S_{\boldsymbol{f}_0}$ to define the support constraint. We examine the effects of perturbations of the data using an algorithm based on iterating the map $D_{\mathbb{A}_{\boldsymbol{a}} B_{S_1}}$. In the first experiment, we choose a collection of 200 random initial conditions of the form $\{\boldsymbol{f}^{(0)} + \boldsymbol{i}_j : j = 1, \ldots, 200\}$. The image $\boldsymbol{f}^{(0)}$ lies on $\mathbb{A}_{\boldsymbol{a}}$ and is of unit norm. The $\{\boldsymbol{i}_j\}$ are random real images with norms less than 1.48×10^{-13}. We let $\boldsymbol{f}_j^{(200)}$ denote the result of 200 iterates of $D_{\mathbb{A}_{\boldsymbol{a}} B_{S_1}}$ starting with $\boldsymbol{f}^{(0)} + \boldsymbol{i}_j$, and set

$$d_{jk} = \|\boldsymbol{f}_j^{(200)} - \boldsymbol{f}_k^{(200)}\|_2. \tag{8.79}$$

These pairwise distances take values in the interval $[0.1, 0.17]$, and are, more or less, normally distributed. All of the iterates have fallen into the same attracting basin defined by the center manifold of the trivial associate $\boldsymbol{f}_0^{[-1,-1]}$, for which the intersection $\mathbb{A}_{\boldsymbol{a}} \cap B_{S_1}$ happens to be transverse. In the initial phase of this iteration the outcome depends very sensitively on the starting point. One might

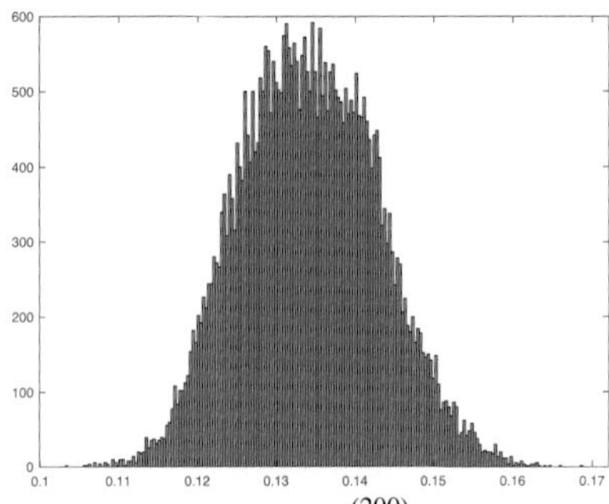

(a) Distances between $\{f_j^{(200)}\}$, with random perturbations of size $O(10^{-13})$ in the initial conditions.

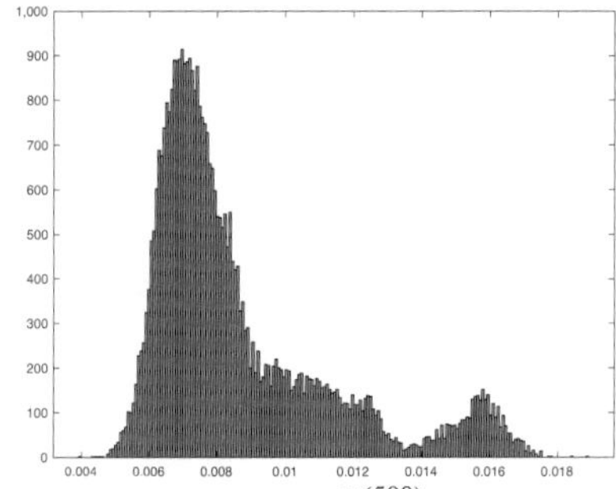

(b) Distances between $\{\widetilde{f}_j^{(500)}\}$ starting with perturbations of size $O(10^{-13})$ to $f_1^{(500)}$.

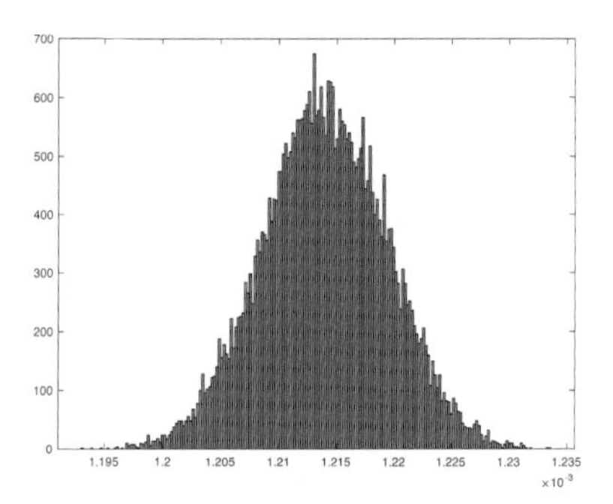

(c) Distances between images obtained after 500 iterates from random perturbations of size $O(10^{-3})$ to $f_1^{(1000)}$.

Figure 8.10 Histograms of the pairwise distances of the results of perturbing a convergent sequence of iterates of the $D_{\mathbb{A}_a B_{S_1}}$ at three different points in the iteration.

even say that the map is chaotic! A histogram of these results is shown in Figure 8.10(a).

We let $f_1^{(\infty)}$ denote the result of running 20,000 iterates of $D_{\mathbb{A}_a S_{B_1}}$ with the initial data $f^{(0)} + i_1$, at which point the iteration has converged, to machine precision. The distance $\|f_1^{(500)} - f_1^{(\infty)}\|_2 \approx 0.0253$. For the second experiment, we use the initial data $\{f_1^{(500)} + i_j : j = 1, \ldots, 200\}$ for 500 additional iterates of $D_{\mathbb{A}_a B_{S_1}}$, and let $\{\widetilde{f}_j^{(500)}\}$ denote the outputs of these iterations. As before, we set

$$\widetilde{d}_{jk} = \|\widetilde{f}_j^{(500)} - \widetilde{f}_k^{(500)}\|_2. \tag{8.80}$$

These distances take values in the interval [0.004, 0.019], indicating that even after 500 iterates the behavior of the algorithm remains quite chaotic. This is likely due to a profusion of attracting fixed points on the center manifold $\mathscr{C}_{\mathbb{A}_a B_{S_1}}^{f_0^{[-1,-1]}}$, and of the weakness of the attraction to any particular one. A histogram of these results is shown in Figure 8.10(b).

We finally redo the experiment with initial data $\{\boldsymbol{f}_1^{(1000)} + \boldsymbol{i}_j : j = 1\ldots,200\}$. The difference $\|\boldsymbol{f}_1^{(1000)} - \boldsymbol{f}_1^{(\infty)}\|_2 \approx 0.0033$. At this point we use perturbations $\boldsymbol{i}_j$ of size 1.48×10^{-3}. The pairwise distances after 500 additional iterates are in the interval $[1.19,\ 1.23] \times 10^{-3}$, and they do not appear to be converging to the same point on the center manifold. Indeed, using a subset of this data for 10,000 iterates leads to essentially the same pairwise distances. The fact that these pairwise distances are not decreasing indicates that there is an open set of attracting fixed points on $\mathscr{C}_{\mathbb{A}_a B_{S_1}}^{f_0^{[-1,-1]}}$, close to $\boldsymbol{f}_1^{(\infty)}$, and that $D_{\mathbb{A}_a B_{S_1}}$ acts more or less like the identity in directions parallel to the center manifold. This map reduces to the identity on a neighborhood of $\boldsymbol{f}_1^{(\infty)}$ in the center manifold, $\mathscr{C}_{\mathbb{A}_a B_{S_1}}^{f_0^{[-1,-1]}}$, but it is by no means obvious that the nonlinear map $D_{\mathbb{A}_a B_{S_1}}$ will behave similarly at a nontrivial distance from the center manifold. A histogram of these results is shown in Figure 8.10(c).

The next example clearly illustrates the effect that the nontransversality of the intersection between $\mathbb{A}_{\boldsymbol{a}}$ and B_S can have on the behavior of an algorithm based on $D_{\mathbb{A}_a B_S}$.

Example 8.7 Our theory predicts that the conditioning of the problem of finding a point $f \in \mathbb{A}_{\boldsymbol{a}} \cap B_S$ depends on the angles between $T_f \mathbb{A}_{\boldsymbol{a}}$ and B_S (see (8.68) and the surrounding discussion). If the target image is piecewise constant, then the theory in Section 2.1 and numerical examples in Example 2.18 suggest that *only* the exact intersection between $T_f \mathbb{A}_{\boldsymbol{a}}$ and B_S produces very small angles. Moreover, the $\dim T_f \mathbb{A}_{\boldsymbol{a}} \cap B_S$ equals the cardinality of the set $\mathscr{T}_{f,S}^{\text{sym}}$. This, in turn, should govern the conditioning of the problem near to this intersection point.

For illustration, we examine the trajectories of the hybrid iteration based on the map $D_{\mathbb{A}_a B_{S_2}}$ for a single 256×256, piecewise constant image with three different random starting points. We let f_0 denote the reference image centered in the 2-pixel neighborhood S_2. Figure 8.11 shows the errors and residuals over 50,000 iterates for 3 different starting points. In Figure 8.11(a) the iterates are in the attracting basin defined by $f_0^{(2,-2)}$. At this translate we have that $|\mathscr{T}_{f_0^{(2,-2)},S_2}^{\text{sym}}| = 0$ and so $\dim T_{f_0^{(2,-2)}} \mathbb{A}_{\boldsymbol{a}} \cap B_{S_2} = 0$. The iterates are clearly converging geometrically to the center manifold defined by this

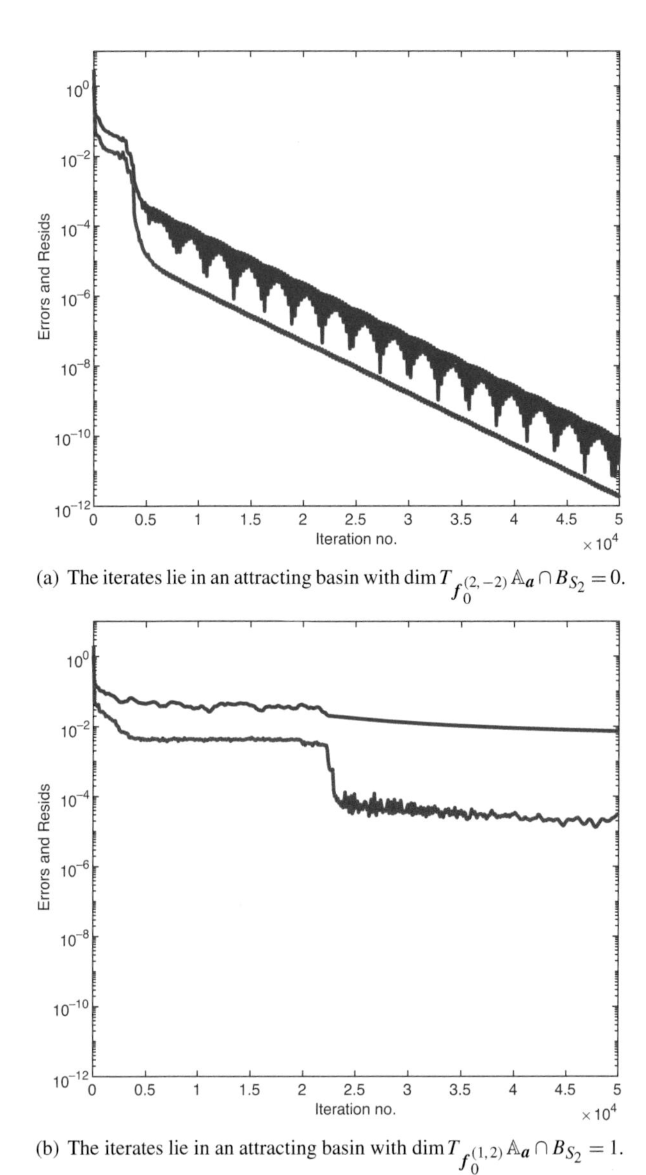

(a) The iterates lie in an attracting basin with $\dim T_{f_0^{(2,-2)}} \mathbb{A}_{\boldsymbol{a}} \cap B_{S_2} = 0$.

(b) The iterates lie in an attracting basin with $\dim T_{f_0^{(1,2)}} \mathbb{A}_{\boldsymbol{a}} \cap B_{S_2} = 1$.

Figure 8.11 An illustration of how the convergence properties of the hybrid iterative map $D_{\mathbb{A}_{\boldsymbol{a}} B_{S_2}}$ depend on the dimension of the $\dim T_{f_0^{(v)}} \mathbb{A}_{\boldsymbol{a}} \cap B_{S_2}$. Errors shown in blue, residuals in red. In these plots the x-axis ranges from 0 to 5×10^4 iterates, and the y-axis ranges from 10^{-12} to 10^1.

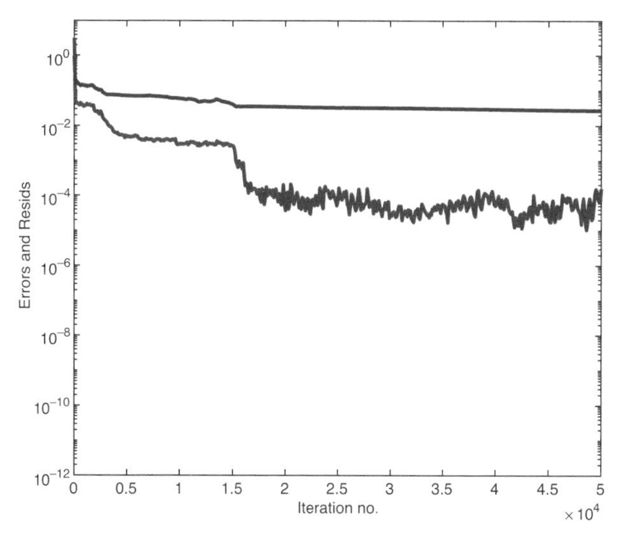

(c) The iterates lie in an attracting basin with $\dim T_{f_0^{(0,0)}} \mathbb{A}_{\boldsymbol{a}} \cap B_{S_2} = 12$.

Figure 8.11 (*cont.*)

exact intersection point. The asymptotic error is close to machine precision, 7×10^{-12}. Note that the error and the residual remain essentially proportional throughout the iterations. In Figure 8.11(b) the iterates are in the attracting basin defined by $f_0^{(1,2)}$. At this translate, we have that $|\mathscr{T}^{\text{sym}}_{f_0^{(1,2)}, S_2}| = 1$ and so

$$\dim T_{f_0^{(1,2)}} \mathbb{A}_{\boldsymbol{a}} \cap B_S = 1. \tag{8.81}$$

The iterates may have stagnated, with the distance to the nearest point in $\mathbb{A}_{\boldsymbol{a}} \cap B_{S_2}$ either essentially constant, or very slowly decreasing. The trajectory appears to lie along directions in $T_{f_0^{(1,2)}} \mathbb{A}_{\boldsymbol{a}} \cap B_{S_2}$, i.e., the trajectory eventually satisfies (8.73). The asymptotic error is 7.3×10^{-3}. Finally in Figure 8.11(c) the iterates are in the attracting basin defined by $f_0^{(0,0)}$. At this translate,

$$\dim T_{f_0^{(0,0)}} \mathbb{A}_{\boldsymbol{a}} \cap B_S = 12. \tag{8.82}$$

These iterates have clearly stagnated and, as before, the trajectory appears to lie along a common tangent direction. The asymptotic error is 2.7×10^{-2}, almost an order of magnitude worse than the previous case. The residuals are more variable, but of roughly the same size as the previous case.

In Examples 8.9–8.10, we explore the effects of different levels of smoothness and different support neighborhood sizes on the performance of the

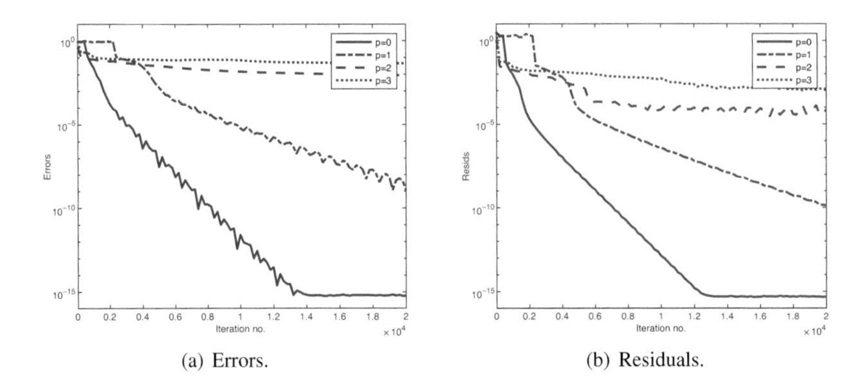

(a) Errors.

(b) Residuals.

Figure 8.12 The dependence of the geometry near intersection points in $\mathbb{A}_{\boldsymbol{a}} \cap B_{S_p}$ on the size of support neighborhood is revealed by the behavior of 20,000 iterates of $D_{\mathbb{A}_a B_{S_p}}$, on a $k = 0$ image using p-pixel neighborhoods, with $p = 0, 1, 2, 3$. In these plots the x-axis ranges from 0 to 2×10^4 iterates and the y-axis ranges from 10^{-15} to 10^0.

algorithms based on the maps $D_{\mathbb{A}_a B_{S_p}}$. As expected, this performance degrades rapidly as the object becomes smoother, or the size of the support neighborhood grows. In Examples 8.9–8.10, for each experiment, we select the initial conditions that produced the best result from 20 trials. These results are often much better than the typical outcome.

Remark 8.8 Fienup (1987) investigated the effects of the smoothness of the unknown object on the performance of hybrid input output (HIO) reconstruction techniques. This paper contains an early demonstration that an accurate support constraint could be used to recover the phase of the Fourier transform of a complex-valued function.

Example 8.9 In these examples, shown in Figure 8.12, we study the dependence of the errors and residuals on the size of the support neighborhood for algorithms based on $D_{\mathbb{A}_a B_{S_p}}$. We use a piecewise constant object ($k = 0, N = 256$), with $p = 0, 1, 2, 3$. Using 20,000 iterates of the hybrid map $D_{\mathbb{A}_a B_{S_p}}$, we obtain trajectories $\{\boldsymbol{f}^{(n)}\}$ heading toward points on a center manifold based at a point in $\boldsymbol{f} \in \mathbb{A}_{\boldsymbol{a}} \cap B_{S_p}$. As noted above, for an algorithm based on a hybrid iterative map, the residual gives an estimate for the

$$\max\{\text{dist}(\boldsymbol{r}^{(n)}, \mathbb{A}_{\boldsymbol{a}}), \text{dist}(\boldsymbol{r}^{(n)}, B_{S_p})\}.$$

If $p = 0, 1$, then the ratio of the error to the residual is essentially a positive constant, and the iterates are converging geometrically. In these cases, we see evidence that the trajectory is spiraling toward the center manifold.

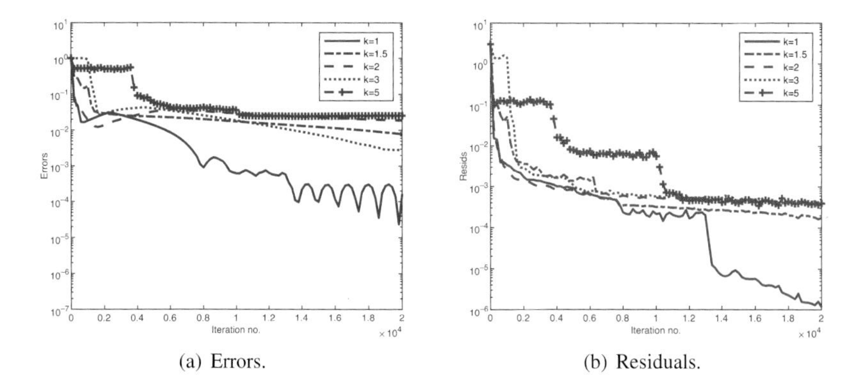

(a) Errors. (b) Residuals.

Figure 8.13 The dependence of the geometry near intersection points in $\mathbb{A}_{\boldsymbol{a}} \cap B_{S_1}$ on the smoothness of the image is revealed by the behavior of iterates of $D_{\mathbb{A}_{\boldsymbol{a}} B_{S_1}}$. We use images with $k \in \{1,\ 1.5,\ 2, 3,\ 5\}$, and a 1-pixel support neighborhood. In these plots the x-axis ranges from 0 to 2×10^4 iterates and the y-axis ranges from 10^{-7} to 10^1.

It is quite apparent in the trajectory corresponding to $p = 2, 3$ that, after about the 6,000th iterate, the residuals, $\|\boldsymbol{f}^{(n)} - \boldsymbol{f}^{(n+1)}\|_2$, are very close to $\text{dist}(\boldsymbol{r}^{(n)}, \mathbb{A}_{\boldsymbol{a}} \cap B_{S_p})^2$. For $p = 2$, the error seems to still be slowly decreasing at the end of the experiment.

Example 8.10 These examples, shown in Figure 8.13, are similar to the previous ones except that we vary the smoothness of the object, while always using the 1-pixel support neighborhood. We use objects with $k \in \{1,\ 1.5, 2, 3, 5\}$. In these experiments we use 20,000 iterates of the map $D_{\mathbb{A}_{\boldsymbol{a}} B_{S_1}}$. We have selected the initial conditions producing the best outcomes from 20 trials. The plots in Figure 8.13(a) show the errors and those in Figure 8.13(b) show the residuals. For $k = 1, 1.5,\ 2$ the iterates appear to still be converging at the end of the experiment, albeit at decreasing rates as the smoothness increases. The $k = 3$ case seems to be stagnating around the 18,000th iterate, and, in the $k = 5$ case, the iterates seem to have stagnated at around the 10,000th iterate. The ratio, at the end of the experiment, of the square of the error to the residual is 1.66.

In Example 8.11, we compare the behavior of hybrid iterative maps using p-pixel neighborhoods of the smallest rectangle containing the set $S_{\boldsymbol{f}}^{10^{-14}}$, versus using p-pixel neighborhoods of the $S_{\boldsymbol{f}}^{10^{-14}}$.

Example 8.11 Thus far we have examined the behavior of the algorithms based on hybrid iterative maps where the support constraint is specified as a p-

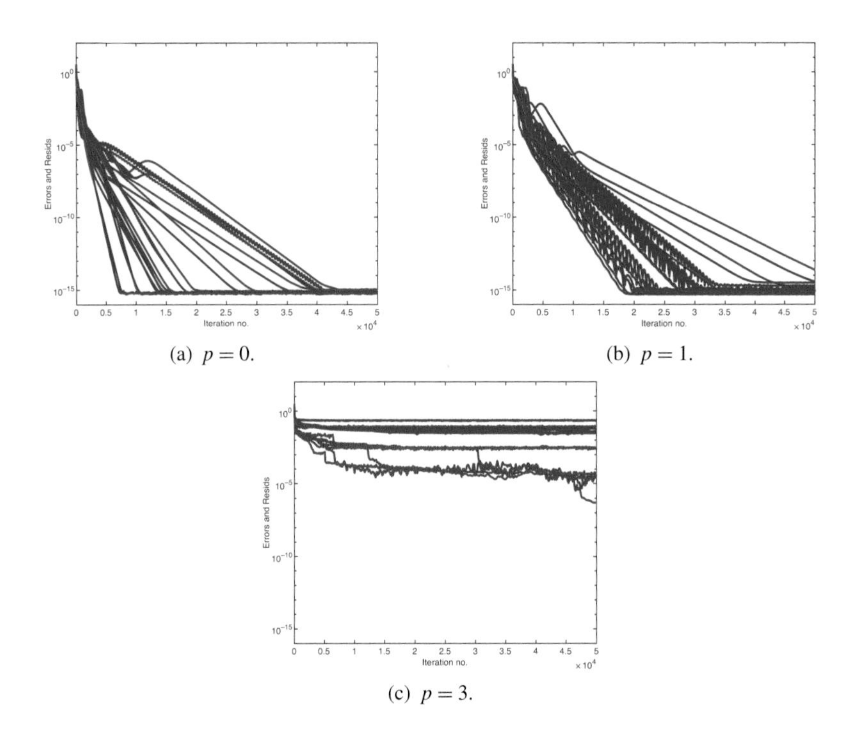

(a) $p = 0$.

(b) $p = 1$.

(c) $p = 3$.

Figure 8.14 Plots show the errors (blue) and residuals (red) for 50,000 iterates of $D_{\mathbb{A}_a S_p}$ with 10 random starting points. We use p-pixel neighborhoods of the actual support, with $p = 0, 1, 3$. In these plots the x-axis ranges from 0 to 5×10^4 iterates and the y-axis ranges from 10^{-15} to 10^1.

pixel neighborhood, S_p, of $S_{f_0}^{10^{-14}}$. In practice such precise information may be difficult to obtain, whereas, for nonnegative images, the autocorrelation image itself determines the smallest rectangle, R_0, that contains $S_{f_0}^{10^{-14}}$. In Example 2.18 we have computed the dimensions of the intersections $T_{f_0}\mathbb{A}_{\boldsymbol{a}} \cap B_{S_p}$ and $T_{f_0}\mathbb{A}_{\boldsymbol{a}} \cap B_{R_p}$ for various values of p and levels of smoothness. The dimensions of the exact intersections are the same, but the number of directions in which these subspaces meet at small angles is much larger for a rectangular support constraint. One might therefore expect that an algorithm based on $D_{\mathbb{A}_a R_p}$ will work less well than one based on $D_{\mathbb{A}_a S_p}$.

In this example we compare the results of 50,000 iterates with 10 initial conditions for both these algorithms with a piecewise constant object ($k = 0$) and $p = 0, 1, 3$. The results for S_p are shown in Figure 8.14 and those for R_p are shown in Figure 8.15. The rectangular constraint certainly contains less information, and the algorithms using S_p give consistently better results, as expected from our analysis in Section 8.2.

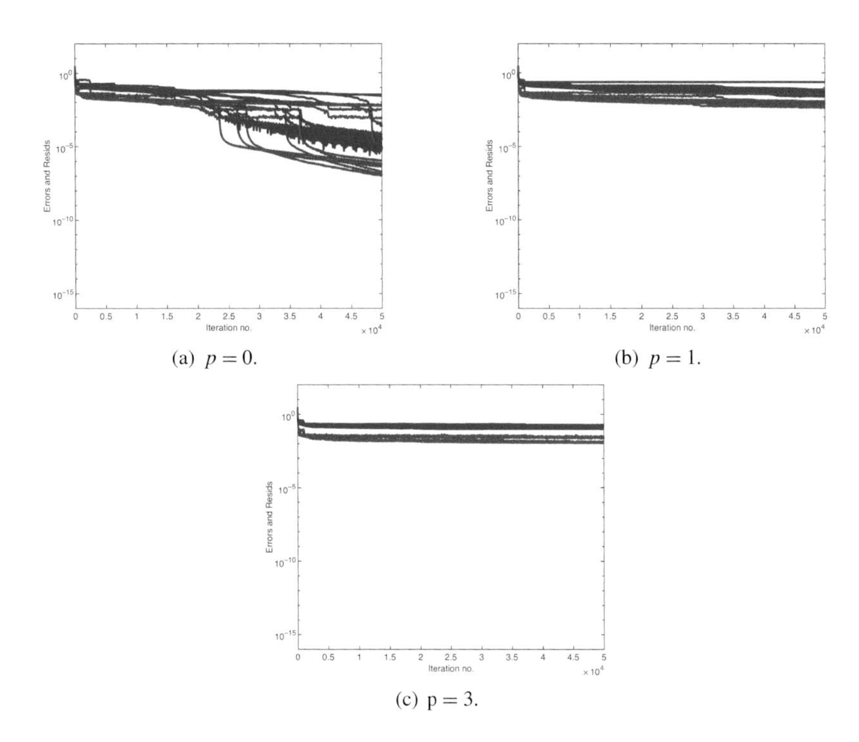

(a) $p = 0$. (b) $p = 1$.

(c) p $= 3$.

Figure 8.15 Plots show the errors (blue) and residuals (red) for 50,000 iterates of $D_{\mathbb{A}_a R_p}$ with 10 random starting points. We use p-pixel neighborhoods of the smallest rectangle containing the support of the object, with $p = 0, 1, 3$. In these plots the x-axis ranges from 0 to 5×10^4 iterates and the y-axis ranges from 10^{-15} to 10^1.

The next example illustrates what is meant by the statement that the trajectory of an iterative algorithm is "approaching the center manifold along common tangent directions." As a proxy for this we look for trials where the errors and residuals satisfy a relation like that in (8.73). Looking at the error and residual trajectories in plots like Figure 8.11(b, c), Figure 8.14(c), and Figure 8.15(b, c) this statement seems qualitatively correct, but it is a difficult statement to assess quantitatively. There are several reasons for this:

(i) Many trials result in trajectories that fall into attracting basins defined by trivial associates, $\boldsymbol{f}^{(v)}$, that do not satisfy the support constraint. That is, $\boldsymbol{f}^{(v)} \notin \mathbb{A}_{\boldsymbol{a}} \cap B_S$, so there is no intersection, let alone a nontransversal intersection.
(ii) As shown in Example 8.3, the linearization of $D_{\mathbb{A}_a B_S}$ at an attracting fixed point can fail to be contraction, leading to nonmonotone error and/or residual trajectories, i.e., scalloping.

(iii) Among trajectories that do find a nontransversal intersection, the error often remains rather large. If, asymptotically, the error is $O(10^{-1})$ and the residual is $O(10^{-2})$, then the residuals are proportional to the square of the errors, but they are also proportional to the errors, just with a worse constant.

In the next example, with these caveats in mind, we try to give some quantitative justification for the statement that many trajectories approach along common tangent directions, or at least satisfy a relationship like that in (8.73).

Example 8.12 Noting the caveats above, to address this question, we have run 250 trials of 10,000 iterates each of $D_{\mathbb{A}_a B_{S_p}}$ on a $k=0$ and a $k=1, 256 \times 256$ image. We use p-pixel support constraints, for $p=1,2,3,$ and let $\{\boldsymbol{f}_{kq}^{(n)}\}$ denote the iterates of the qth trial for the image of smoothness k, and $\{\boldsymbol{r}_{kq}^{(n)} = P_{B_{S_p}}(\boldsymbol{f}_{kq}^{(n)})\}$ the approximate reconstructions.

From these trials we extract the subset, Q_{kp}, that find an attracting basin defined by a nontransversal intersection, $\boldsymbol{f}_{kq} \in \mathbb{A}_{\boldsymbol{a}} \cap B_{S_p}$. For these trials we plot histograms of the averages

$$\overline{\frac{\log_{10} \| P_{\mathbb{A}_a}(\boldsymbol{r}_{kq}^{(n)}) - \boldsymbol{r}_{kq}^{(n)} \|}{\log_{10}(\|\boldsymbol{r}_{kq}^{(n)} - \boldsymbol{f}_{kq}\|)}}. \tag{8.83}$$

The $\overline{}$ indicates that for each $q \in Q_{kp}$, we average the $\log_{10}$ of the errors and residuals over the last 4,000 iterates. This reduces the effects of scalloping and jitter in these quantities. In these experiments we use the true residuals, $\{\| P_{\mathbb{A}_a}(\boldsymbol{r}_{kq}^{(n)}) - \boldsymbol{r}_{kq}^{(n)} \|\}$, rather than the proxy, $\{\boldsymbol{f}_{kq}^{(n+1)} - \boldsymbol{f}_{kq}^{(n)} \|\}$, used in most of our plots, as this gives a more reliable result. The results of these experiments are shown in Figures 8.16–8.17; the statistical data relating to these experiments is in Table 8.1. Except for the case $k=0, p=1$ these experiments provide strong evidence that the relationship between the errors and residuals

Table 8.1. *Statistics underlying the trials for which* $\overline{\frac{\log_{10} \| P_{\mathbb{A}_a}(\boldsymbol{r}_{1q}^{(n)}) - \boldsymbol{r}_{1q}^{(n)} \|}{\log_{10}(\|\boldsymbol{r}_{1q}^{(n)} - \boldsymbol{f}_{1q}\|)}}$ *is shown in Figures 8.16–8.17.*

$k=0$	$p=1$	$p=2$	$p=3$	$k=1$	$p=1$	$p=2$	$p=3$
no. trials/250	26	104	138		46	70	47
mean	1.41	2.1	2.14		2.03	1.92	1.84

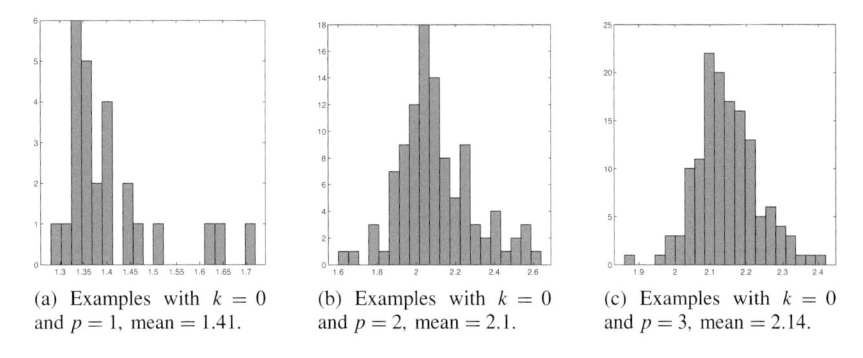

(a) Examples with $k = 0$ and $p = 1$, mean $= 1.41$.

(b) Examples with $k = 0$ and $p = 2$, mean $= 2.1$.

(c) Examples with $k = 0$ and $p = 3$, mean $= 2.14$.

Figure 8.16 Plots showing $\frac{\log_{10} \| P_{\mathbb{A}_a}(\boldsymbol{r}_{0q}^{(n)}) - \boldsymbol{r}_{0q}^{(n)} \|}{\log_{10}(\| \boldsymbol{r}_{0q}^{(n)} - \boldsymbol{f}_{0q} \|)}$ for trials that lie in attracting basins defined by nontransversal intersections.

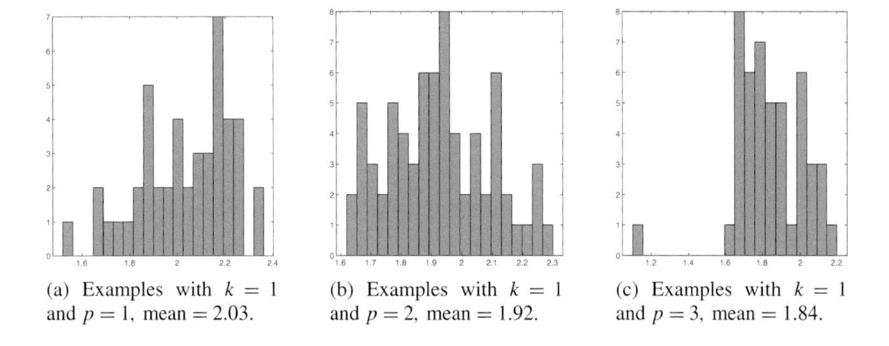

(a) Examples with $k = 1$ and $p = 1$, mean $= 2.03$.

(b) Examples with $k = 1$ and $p = 2$, mean $= 1.92$.

(c) Examples with $k = 1$ and $p = 3$, mean $= 1.84$.

Figure 8.17 Plots showing $\frac{\log_{10} \| P_{\mathbb{A}_a}(\boldsymbol{r}_{1q}^{(n)}) - \boldsymbol{r}_{1q}^{(n)} \|}{\log_{10}(\| \boldsymbol{r}_{1q}^{(n)} - \boldsymbol{f}_{1q} \|)}$ for trials that lie in attracting basins defined by nontransversal intersections.

in (8.73) holds for trials that find an attracting basin defined by a nontransversal intersection. In the case $k = 0, p = 1$ many of the trials converge more or less geometrically, which explains the values closer to 1.

The final example in the section shows that an algorithm based on a hybrid iterative map is sometimes able to defeat nontransversality, beyond what one would expect, given that the inverse map is Hölder-$\frac{1}{2}$ in some directions.

Example 8.13 In this example we look at piecewise constant ($k = 0$) 256×256-images of the form $\boldsymbol{f}_q = \boldsymbol{\varphi}_q * \boldsymbol{h}$, where $\boldsymbol{\varphi}_q$ is supported on a set of q well–separated pixels. For $\boldsymbol{h}$ we use samples of the characteristic function of a disk centered at $(0,0)$. The important feature of $\boldsymbol{h}$ is that it is inversion symmetric: $\boldsymbol{h}_j = \boldsymbol{h}_{-j}$, for all $\boldsymbol{j} \in J$. It follows from Proposition 2.15 that

$$\dim T_{\boldsymbol{f}_q} \mathbb{A}_{\boldsymbol{a}_{f_q}} \cap B_{S_{f_q}} \approx \frac{|S_{\boldsymbol{h}}|}{2} = 262. \tag{8.84}$$

That is, even if we use the exact support of $\boldsymbol{f}_q$ as the support constraint, the intersections, $\mathbb{A}_{\boldsymbol{a}_{f_q}} \cap B_{S_{f_q}}$, are far from transversal. Note that in this example the dimension of $T_{\boldsymbol{f}_q} \mathbb{A}_{\boldsymbol{a}_{f_q}} \approx 16,000$. For all of the experiments in this example we do use the exact support of $\boldsymbol{f}_q$, as the support constraint, so that any failure to converge can be unambiguously attributed to the nontransversality of the intersection described above.

The behavior of algorithms for recovering the phase of $|\widehat{\boldsymbol{f}}_q|$ has a strong dependence on the number of disks, which indicates that the effect of non-transversality may be modulated by the complexity of the landscape defined by $d_{\mathbb{A}_{\boldsymbol{a}_{f_q}} B_{S_{f_q}}}$. Recall that the iterates of alternating projection (AP) converge to local minima of $d_{\mathbb{A}_{\boldsymbol{a}_{f_q}} B_{S_{f_q}}}$. To probe this landscape, we examine the behavior of AP on this type of image. As before, we use the prevalence of *nonzero* local minima as a proxy for the complexity of this landscape.

To see how the complexity of the $d_{\mathbb{A}_{\boldsymbol{a}_{f_q}} B_{S_{f_q}}}$-landscape depends on q we run 10,000 iterates of AP with 20 random initial conditions, for $q = 4, 16$. The results of these experiments are shown in Figure 8.18. When $q = 4$, all 20 trials converge to the unique point in $\mathbb{A}_{\boldsymbol{a}_{f_4}} \cap B_{S_{f_4}}$. Because the iterates are converging to the intersection point, they cannot "avoid" the consequences of the nontransversal intersection. As the residuals are very close to the squares of the errors, it is very likely that the trajectories of these iterates lie along directions in the intersection $T_{\boldsymbol{f}_4} \mathbb{A}_{\boldsymbol{a}_{f_4}} \cap B_{S_{f_4}}$. The convergence is very slow, and it appears that the asymptotic error will be larger than $\sqrt{\epsilon_{\text{mach}}}$, see Figure 8.20(b). The 20 error plots are almost identical.

When $q = 16$, all of the trials converge geometrically to nonzero local minima of $d_{\mathbb{A}_{\boldsymbol{a}_{f_4}} B_{S_{f_4}}}$. Note, in addition, that each error plot in Figure 8.18(b) is different from the others. These are strong indications that the landscape has become considerably more complicated. This phenomenon is example dependent; different choices of disk radii and locations produce different results. In Figure 8.18 the cyan curves are the residuals, the blue curves show the true error, and the red curves are the norms of the successive differences.

We next consider the behavior of the algorithm based on iterating hybrid iterative maps, $D_{\mathbb{A}_{\boldsymbol{a}_{f_q}} B_{S_{f_q}}}$. If they exist, then the attracting fixed points of a hybrid iterative map lie on the center manifold, typically at a considerable distance from the intersection point. Our analysis in Section 8.2 of the

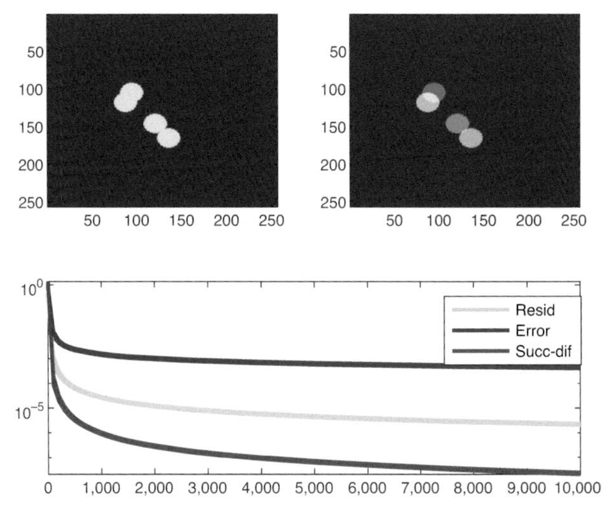

(a) 20 examples of 10,000 iterates of AP on an example composed of 4 equal sized disks. The upper left panel is the support constraint S_{f_4} and the upper right is the image. The lower panel shows plots of the true error, in blue, the residuals, in cyan, and the norms of the successive differences, in red. These curves indicate that all trials are converging along common tangent directions to the true intersection point.

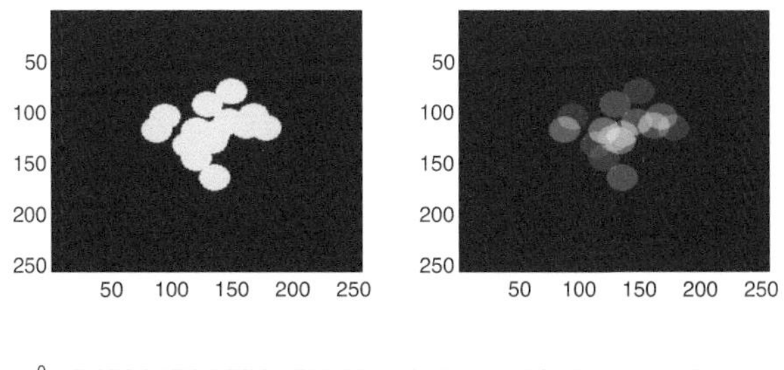

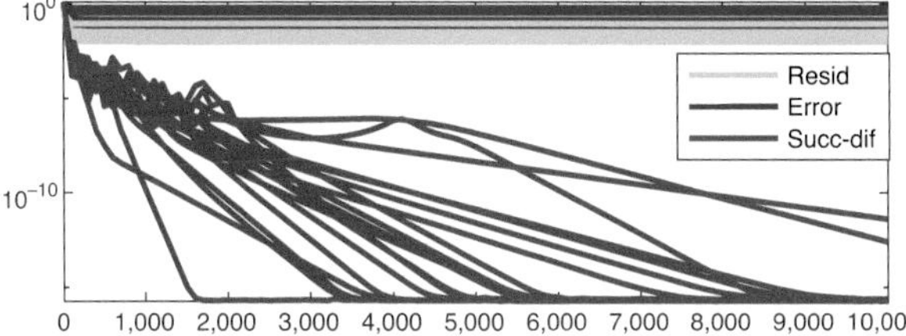

(b) 20 examples of 10,000 iterates of AP on an example composed of 16 equal sized disks. The upper left panel is the support constraint $S_{f_{16}}$ and the upper right is the image. The lower panel shows plots of the true error, in blue, the residuals, in cyan, and the norms of the successive differences, in red. These curves indicate that all trials are converging geometrically to nonzero local minima of $d_{\mathbb{A}_{a_{f_{16}}} B_{S_{f_{16}}}}$.

Figure 8.18 The results of experiments using the AP algorithm to probe the complexity of the landscape defined by $d_{\mathbb{A}_{a_{f_q}} B_{S_{f_q}}}$ for examples with $q = 4, 16$.

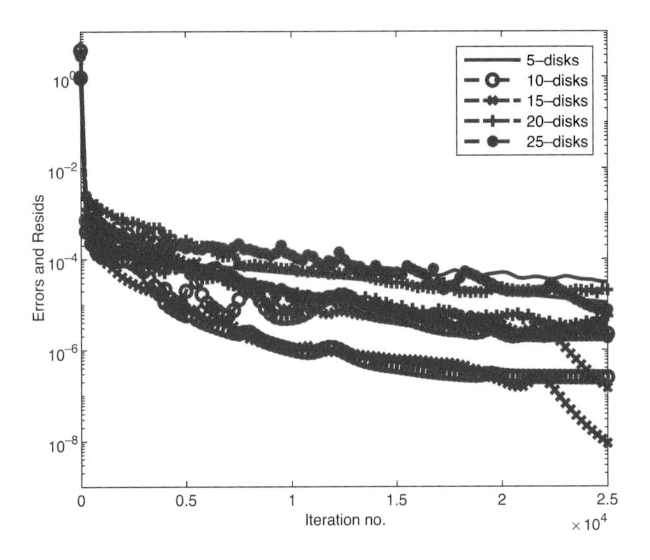

Figure 8.19 2 trials each of 25,000 iterates of hybrid iterative maps on examples comprised of 5, 10, 15, 20, and 25 equal sized disks. The plot legend shows the number of disks used to produce each curve. The plots of the true error are in blue, and the residuals, which is also the norms of the successive differences, in red. These curves indicate that all trials are converging geometrically to points on the center manifold.

linearization of $D_{\mathbb{A}_a B_S}$ at such fixed points, while not definitive, shows that, even when the intersection is nontransversal, it *may* be possible for there to exist points on the center manifold that are attracting. In Figure 8.19 we show the results of running 2 trials of 25,000 iterates each, for $q = 5, 10, 15, 20,$ and 25 disks. In all cases the iterates appear to be slowly converging. Unlike the previous experiments with AP, the errors are about 10 times the residuals in all cases, not the squares. This indicates that the trajectories are not approaching the center manifold along common tangent directions.

For the final experiment we compare 200,000 iterates of the hybrid iterative map on an example composed of 10 disks, to 400,000 iterates of AP on the same example. The results of this experiment are shown in Figure 8.20. It is striking that the iterates of the hybrid map have converged, geometrically, to 16 digits, with the residual very close to 10 times the error throughout the entire experiment. Whereas, with twice as many iterates, the AP reconstruction gets only four digits; its trajectory has the residual very close to the square of error throughout, and the curves are somewhat concave. These are strong indications that the iterates are not converging geometrically.

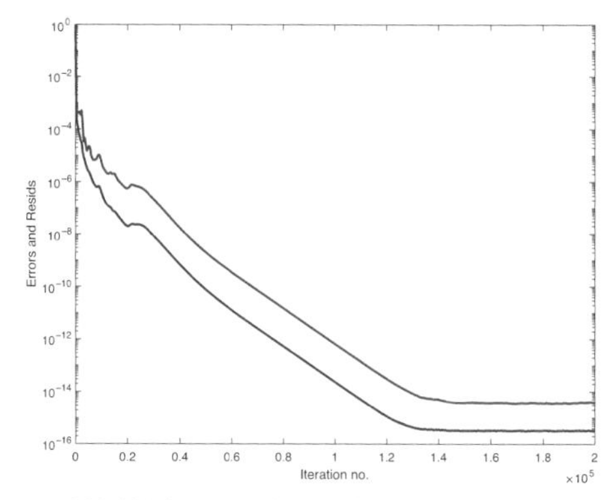

(a) 200,000 iterates of a hybrid iterative map on an example comprised of 10 equal radius disks. The errors are in blue and the residuals are in red. The trajectory shows that the iterates have converged (geometrically) to a point on the center manifold.

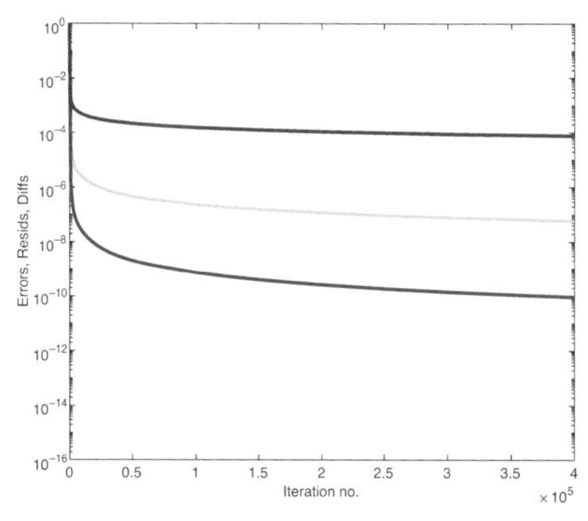

(b) 400,000 iterates of AP on an example comprised of 10 equal radius disks. The errors are in blue, the residuals are in cyan, and the norms of the differences of successive iterates in red. The trajectory shows that the iterates are approaching the intersection point along a trajectory that lies along common tangent directions.

Figure 8.20 An experiment that illustrates the difference in the behaviors of hybrid iterative maps and AP on an example where the intersection $T_{f_{10}}\mathbb{A}_{a_{f_{10}}} \cap B_{S_{f_{10}}}$ is nontransversal due to convolution with an inversion image.

The final experiment is a clear demonstration that a hybrid iterative map is sometimes able to avoid the deleterious effects of a nontransversal intersection and find a machine precision reconstruction. In these examples, the failure of transversality is due to convolution with an inversion symmetric object. This favorable outcome has not been observed when the failure of transversality is due to a loose support constraint.

As yet, we do not have a compelling explanation for these results. For larger q values the hybrid iterative map behaves much like it does on generic piecewise constant examples, with a looser support constraint: it still seems to be geometrically converging, but rather slowly. For small values of q, the Z-transform, see (3.42), of images like the $\boldsymbol{f}_q$ can be factored as a low-degree polynomial times the Z-transform of an inversion symmetric object, $\boldsymbol{F}_q(z) = p_q(z)\boldsymbol{F}_{\text{disk}}(z)$. The DFT of the inversion symmetric object is real valued, so the phase retrieval problem for this factor reduces to a much simpler sign retrieval problem. In this sense the data for these examples is simpler than in the generic case, though it is difficult to explain how the hybrid iterative map, using only the support constraint as auxiliary information, would "know" this. The fact that there exist cases of nontransversal intersections, where the phase retrieval problem is effectively solvable gives hope that there might be a way to harness this more generally.

9

Phase Retrieval with the Nonnegativity Constraint

In this chapter, we consider the behavior of hybrid iterative maps when it is known that the image is of a single sign, that is $f_0 \in \mathbb{A}_a \cap B_+$. Recall that $B_+ = \mathbb{R}_+^J$, where $\mathbb{R}_+ = [0, \infty)$. In addition, we either assume that f_0 has small support, or the support of its autocorrelation image $S_{f_0 \star f_0}$ satisfies the conditions of Theorem 4.5, guaranteeing generic uniqueness up to trivial associates. This requires that many coordinates of f_0 be 0. As described earlier (introduction to Chapter 4), the boundary of B_+ is not a smooth hypersurface, but instead a stratified space. See (4.2). If f lies on the boundary of B_+, then there is a nonempty subset $L \subset J$ so that $f \in \partial B_+^L$, that is,

$$f_l = 0 \text{ if } l \in L \text{ and } f_j > 0 \text{ if } j \in J \setminus L. \tag{9.1}$$

Near to f, the set B_+ is isomorphic to a product of half-closed intervals and open intervals: $[0, 1)^{|L|} \times (-\epsilon, \epsilon)^{|J|-|L|}$, for some $\epsilon > 0$. The number, $|L|$, of coordinates that vanish at f is the *codimension* of the stratum of ∂B_+ to which f belongs.

A nonnegative image with many coordinates equal to 0 belongs to a stratum of ∂B_+ of high codimension. It should be noted that the algorithms we consider below do not make explicit use of an estimate for the support of f_0. Algorithms that use both a support constraint, and nonnegativity are easily constructed, and may well lead to significant improvements with real measured data (see Fienup et al. 1990).

As the maps P_{B_+} and R_{B_+} and the boundary ∂B_+ are not smooth, it is quite difficult to "linearize" hybrid iterative maps based on the nonnegativity condition near to their center manifolds. The analysis for the case $B = B_S$ suggests that the relationship between $T_{f_0}\mathbb{A}_a$ and ∂B_+ near to f_0 will dictate the behavior of these maps close enough to f_0. In Section 4.5 we analyzed the "linearization" of the intersection of $\mathbb{A}_a$ with B_+ near a point like f_0.

This amounted to assessing the intersection of the affine subspace $T_{f_0}\mathbb{A}_{\boldsymbol{a}}$ with the piecewise affine subset ∂B_+. If $r_1 = \|f_0\|_1$, and $B^1_{r_1} = \{f : \|f\|_1 \le r_1\}$, then we showed that

$$T_{f_0}\mathbb{A}_{\boldsymbol{a}} \cap \partial B_+ = T_{f_0}\mathbb{A}_{\boldsymbol{a}} \cap \partial B^1_{r_1} \tag{9.2}$$

is a convex, conic subset of an orthant in some Euclidean space. Other than images obtained by convolving with a centrally symmetric image, these intersections seem to consist of f_0 alone, i.e., the intersections of $\mathbb{A}_{\boldsymbol{a}}$ with B_+ are generically transversal. One might, therefore, expect that an algorithm based on iterating $D_{\mathbb{A}_{\boldsymbol{a}}B_+}$ should work better than one based on $D_{\mathbb{A}_{\boldsymbol{a}}B_S}$, which, empirically, seems correct. At the very least, the algorithm based on nonnegativity converges more reliably than one based on a support constraint.

Remark 9.1 The empirical fact that the nonnegativity constraint leads to a more robust reconstruction algorithm than a support constraint alone is mentioned in Fienup (1987).

If S is an estimate for the support of f_0, then the geometry near to one point in $\mathbb{A}_{\boldsymbol{a}} \cap B_S$ can differ quite a lot from the geometry near another. This is clearly reflected in the differences in the dimensions of $T_{f_0^{(\boldsymbol{v})}}\mathbb{A}_{\boldsymbol{a}} \cap B_S$ as $\boldsymbol{v}$ varies. This, in turn, explains why different runs of algorithms based on $D_{\mathbb{A}_{\boldsymbol{a}}B_S}$ can be expected to produce starkly different results, i.e., convergence to machine precision versus stagnation (see Example 8.7). In this regard, the nonnegativity and ℓ_1 constraints are quite different; the local geometry near every point in $\mathbb{A}_{\boldsymbol{a}} \cap \partial B_+$ or $\mathbb{A}_{\boldsymbol{a}} \cap \partial B^1_{r_1}$, is the same. The fact that different runs sometimes produce rather different results is probably a reflection of the extreme sensitivity to initial conditions and the complicated landscape of the center manifold itself.

9.1 Hybrid Iterative Maps Using Nonnegativity

When using a sign condition to define the B set, the coordinates of the nearest point map are given by $P_{B_+}(f)_j = \max\{f_j, 0\}$. Thus, the reflection is given by

$$R_{B_+}(f) = 2P_{B_+}(f)_j - f_j = |f_j|. \tag{9.3}$$

This image lies entirely in B_+, and as expected, B_+ consists of the fixed points for this reflection. The hybrid iterative maps, utilizing nonnegativity, are given by

$$\begin{aligned} D_{\mathbb{A}_a B_+}(f) &= f + P_{\mathbb{A}_a}(R_{B_+}(f)) - P_{B_+}(f) \text{ and} \\ D_{B_+ \mathbb{A}_a}(f) &= f + P_{B_+}(R_{\mathbb{A}_a}(f)) - P_{\mathbb{A}_a}(f). \end{aligned} \tag{9.4}$$

Either map can be used to find points in $\mathbb{A}_{\boldsymbol{a}} \cap B_+$. It should be emphasized that the map used in Fienup's hybrid input output (HIO) algorithm, see (6.7), is quite different from these maps, and it iterates behave quite differently.

We let $\boldsymbol{f}_0 \in \mathbb{R}^J$ denote a nonnegative image in $\mathbb{A}_{\boldsymbol{a}} \cap B_+$, with small support. As noted, the ∂B_+ is stratified by the number of coordinates that vanish. Suppose that after "vectorizing" $\boldsymbol{f}_0$ and permuting indices, we have $\boldsymbol{f}_0 = (x_1^{**}, \dots, x_m^{**};\ 0, \dots, 0) \in \partial B_+$, with the first m coordinates positive, and the remaining $l = |J| - m$ equal to 0. We use this labeling of the coordinates in the remainder of this section. The set

$$L_{B+} = \{(x_1^{**}, \dots, x_m^{**}; y_1, \dots, y_l) : \ y_j \leq 0 \quad \text{for all } j\}; \tag{9.5}$$

this is an orthant of dimension l. The more entries of $\boldsymbol{f}_0$ that vanish, the larger the dimension of this set becomes.

On the other hand, $\mathbb{A}_{\boldsymbol{a}}$ is a smooth submanifold of real dimension bounded by $|J|/2$. At least near to $\boldsymbol{f}_0$, the set $L_{\mathbb{A}_a}$ is an affine subspace of real dimension at least $|J|/2$. It is given by the normal bundle to $\mathbb{A}_{\boldsymbol{a}}$ at $\boldsymbol{f}_0$, $N_{\boldsymbol{f}_0}\mathbb{A}_{\boldsymbol{a}}$, thought of as an affine subspace of $\mathbb{R}^J$.

As $\boldsymbol{f}_0$ lies on a stratum of ∂B_+ of codimension l, the set L_{B_+} is an orthant of dimension l. Let $< L_{B_+} >$ be the minimal affine subspace of X that contains L_{B_+}, which of course has dimension l. The $\dim L_{\mathbb{A}_a} \cap < L_{B_+} >$ is at least of dimension $|J|/2 - m$ as well. If we oversample the discrete Fourier transform (DFT) sufficiently to assure uniqueness (with the support condition), then $m \leq |J|/2^d$, so that $|J|/2 - m$ is quite large. The analysis in Section 4.5 shows that there is a basis for $N^0_{\boldsymbol{f}_0}\mathbb{A}_{\boldsymbol{a}}$ that consists entirely of vectors with nonnegative entries. If $\boldsymbol{f}_0$ has small support, then there is a large-dimensional subspace of this vector space consisting of vectors of the form $(\mathbf{0}_m; y_1, \dots, y_l)$ with all $y_j \leq 0$. This shows that $N_{\boldsymbol{f}_0}\mathbb{A}_{\boldsymbol{a}} \cap L_{B_+}$, which contains the center manifold, is an orthant in a positive dimensional subspace C.

The reflection $R_{\mathbb{A}_a} = 2P_{\mathbb{A}_a} - \mathrm{Id}$, restricted to $L_{\mathbb{A}_a}$ is a linear, orthogonal map, so that $R^{-1}_{\mathbb{A}_a} = R_{\mathbb{A}_a}$. The affine subspace $L_{\mathbb{A}_a}$ is invariant under $R_{\mathbb{A}_a}$, and therefore,

$$R^{-1}_{\mathbb{A}_a}(L_{B_+}) \cap L_{\mathbb{A}_a} = R_{\mathbb{A}_a}(L_{B_+} \cap L_{\mathbb{A}_a}). \tag{9.6}$$

Near to $\boldsymbol{f}_0$, the set $L_{\mathbb{A}_a} \cap L_{B_+}$ consists of points of the form $\boldsymbol{f}_0 + (0;\ \boldsymbol{v})$, where $(0;\ \boldsymbol{v}) \in N_{\boldsymbol{f}_0}\mathbb{A}_{\boldsymbol{a}}$, and all coordinates of $\boldsymbol{v}$ are nonpositive. Thus,

$$R_{\mathbb{A}_a}(\boldsymbol{f}_0 + (0; \boldsymbol{v})) = \boldsymbol{f}_0 - (0; \boldsymbol{v}) \in B_+. \tag{9.7}$$

The center manifold for $D_{B_+\mathbb{A}_a}$ is

$$\mathscr{C}^{f_0}_{B_+\mathbb{A}_a} = R^{-1}_{\mathbb{A}_a}(L_{B_+}) \cap L_{\mathbb{A}_a} = R_{\mathbb{A}_a}(L_{B_+} \cap L_{\mathbb{A}_a}) \subset B_+. \tag{9.8}$$

On the other hand, if we use $R_{B_+} = 2P_{B_+} - \mathrm{Id}$ then, because the image of R_{B_+} is contained in B_+, we see that $R^{-1}_{B_+}(L_{\mathbb{A}_a}) = R^{-1}_{B_+}(L_{\mathbb{A}_a} \cap B_+)$. For a point f, in the interior of the positive orthant, $R^{-1}_{B_+}(f)$ consists of $2^{|J|}$ points. In particular, it is clear that $R^{-1}_{B_+}(L_{\mathbb{A}_a} \cap B_+)$ contains $L_{\mathbb{A}_a} \cap L_{B_+}$, and possibly other points. The center manifold for $D_{\mathbb{A}_a B_+}$ is

$$\mathscr{C}^{f_0}_{\mathbb{A}_a B_+} = R^{-1}_{B_+}(L_{\mathbb{A}_a}) \cap L_{B_+} \supset L_{\mathbb{A}_a} \cap L_{B_+}. \tag{9.9}$$

From dimensional considerations, it is not obvious why one algorithm might work better than the other. An obvious difference between these two cases is that

$$\mathscr{C}^{f_0}_{B_+\mathbb{A}_a} \subset B_+ \text{ whereas } \mathscr{C}^{f_0}_{\mathbb{A}_a B_+} \subset B_+^c. \tag{9.10}$$

In practice, this sometimes proves to be an important difference, perhaps because having $\mathscr{C}^{f_0}_{\mathbb{A}_a B_+} \subset B_+^c$ implicitly enforces a support constraint, which allows $D_{\mathbb{A}_a B_+}$ to converge more reliably. Such behavior is suggested by numerical experiments.

As we saw in Section 8.2, not every point in the center manifold is a fixed point of $D_{\mathbb{A}_a B_+}$ or $D_{B_+\mathbb{A}_a}$, respectively. A point on the center manifold, $\mathscr{C}^{f_0}_{B_+\mathbb{A}_a}$, takes the form $\boldsymbol{f}_0 + \boldsymbol{n}$, where $\boldsymbol{n} \in N^0_{f_0}\mathbb{A}_a$ has all coordinates nonnegative. A calculation shows that in order for $\boldsymbol{f}_0 + \boldsymbol{n}$ to be a fixed point of $D_{B_+\mathbb{A}_a}$ it is necessary and sufficient that $P_{\mathbb{A}_a}(\boldsymbol{f}_0 + \boldsymbol{n}) = \boldsymbol{f}_0$. A similar calculation shows that to be a fixed point of $D_{\mathbb{A}_a B_+}$ it is necessary and sufficient that $P_{\mathbb{A}_a}(\boldsymbol{f}_0 - \boldsymbol{n}) = \boldsymbol{f}_0$. In this case the coefficients of $\boldsymbol{n}$ are nonpositive. More explicit conditions, in the DFT representation, follow as before from (8.56)–(8.57).

9.2 Numerical Examples

In this section, we investigate the behavior of hybrid iterative maps and the nonnegativity constraint; the initialization is obtained by choosing an image $\boldsymbol{f}^{(0)} \in \mathbb{A}_a$ with randomly selected DFT phases, satisfying the symmetries required of a real image. The results in Section 4.5 show that usually $T_{f_0}\mathbb{A}_a \cap \partial B_+ = \boldsymbol{f}_0$, provided that $\boldsymbol{f}_0$ is not the result of convolution with an inversion symmetric image. This suggests that algorithms based on nonnegativity should work somewhat better; an expectation borne out in the experiments below.

In the generic (nonconvolution) case, these algorithms are much less sensitive to the smoothness of the image than algorithms relying on the support constraint. Though as the images become smoother, the number of directions in which $T_{f_0}\mathbb{A}_{\boldsymbol{a}}$ and ∂B_+ make small angles tends to increase, and this does slow the convergence of these algorithms.

One can base a hybrid iterative algorithm, using nonnegativity as the auxiliary information, on either $D_{\mathbb{A}_{\boldsymbol{a}} B_+}$ or $D_{B_+ \mathbb{A}_{\boldsymbol{a}}}$. In experiments we have conducted the performance of the two algorithms are quite similar, though it seems that an algorithm based on $D_{\mathbb{A}_{\boldsymbol{a}} B_+}$ may converge a little more reliably than one based on $D_{B_+ \mathbb{A}_{\boldsymbol{a}}}$. For economy of space, our subsequent numerical experiments only utilize $D_{\mathbb{A}_{\boldsymbol{a}} B_+}$.

Example 9.2 We now study the behavior of the algorithm $D_{\mathbb{A}_{\boldsymbol{a}} B_+}$ on randomly selected 256×256 images, with varying degrees of smoothness. We first examine less smooth examples, with $k = 0, 0.25,\ 0.5,\ 0.75, 1,$ and then smoother examples, with $k = 1, 1.5, 2,\ 3,\ 5$. For each level of smoothness we use the best result from 20 trials. Figures 9.1(a, b) show that, in all cases, the errors and residuals remain proportional, and the iterates appear to be converging geometrically, with a rate that decreases with increasing smoothness. Figures 9.1(c, d) shows that, for $k = 1, 1.5, 2,\ 3,\ 5,$ the errors and residuals remain comparable, and the convergence is geometric, albeit very slow. Once $k > 1.5$, the rate of convergence is essentially independent of k in this range. The results shown here are typical for generic, nonnegative objects with the given smoothness.

The plots in Figure 9.1(c, d) should be compared with those in Figure 8.13, where we examine the effects of smoothness on the geometry of the intersections using a support constraint, $\mathbb{A}_{\boldsymbol{a}} \cap B_{S_1}$. Both sets of experiments use the same objects. For the experiments using a support constraint, we use a very tight 1-pixel constraint, which is unlikely to be available in a practical experiment. In Figure 9.1(c), the nonnegativity constraint continues to provide about two digits of accuracy through $k = 5$, while, with a 1-pixel support constraint, we obtain slightly less for $k = 2, 5$. In both cases we use the best results obtained from 20 trials, and this allows the $D_{\mathbb{A}_{\boldsymbol{a}} B_{S_1}}$ to find a transversal intersection. Though the experiment does not show it, there is more variability in the outcomes when using a support constraint. This is a reflection of the fact that, for most points in $\mathbb{A}_{\boldsymbol{a}} \cap B_{S_p}$, the intersection is not transversal. With the nonnegativity constraint, the geometry near all the points in $\mathbb{A}_{\boldsymbol{a}} \cap B_+$ is similar and the intersections are all transversal.

The next example explores the effects of nontransversality in the context of the nonnegativity constraint.

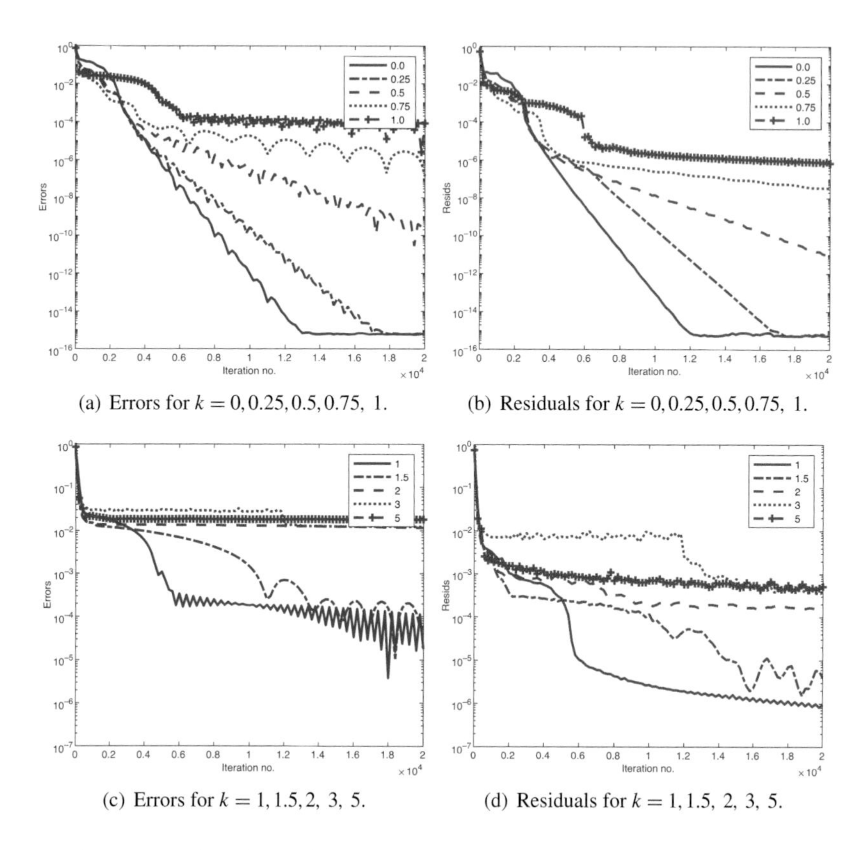

(a) Errors for $k = 0, 0.25, 0.5, 0.75,\ 1$.

(b) Residuals for $k = 0, 0.25, 0.5, 0.75,\ 1$.

(c) Errors for $k = 1, 1.5, 2,\ 3,\ 5$.

(d) Residuals for $k = 1, 1.5,\ 2,\ 3,\ 5$.

Figure 9.1 The effect of smoothness on the behavior of a difference map algorithm defined by $D_{\mathbb{A}_a B_+}$ is revealed by plots of the errors and residuals for 10,000 iterates of this map. Note that in (a, b), the y-axis ranges from 10^{-16} to 10^0 whereas in (c, d) it ranges from 10^{-7} to 10^0. The x-axis ranges from 0 to 2×10^4 iterates in all plots.

Example 9.3 As noted earlier, for generic, nonnegative images, f_0, the intersections in $\mathbb{A}_{\boldsymbol{a}} \cap B_+$ are transversal in the sense defined above; that is, $T_{f_0}\mathbb{A}_{\boldsymbol{a}} \cap \partial B_+ = \{f_0\}$. We showed in Section 4.5 that, for an image defined by convolving another image with an inversion symmetric function, this is no longer the case. Indeed, Table 4.1 shows that the dimensions of these intersections increase with the smoothness of the image. This fact gives us an opportunity to explore the effects of nontransversality on the behavior of an algorithm based on iterating $D_{\mathbb{A}_{\boldsymbol{a}} B_+}$.

For this experiment we compare the behavior of the algorithm on an image of the form

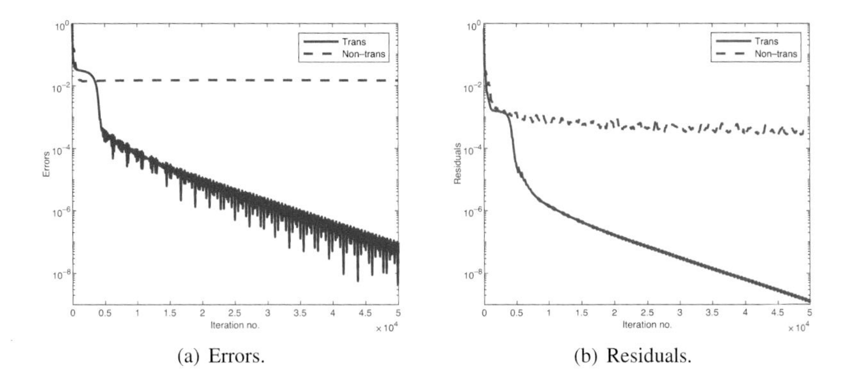

(a) Errors. (b) Residuals.

Figure 9.2 A comparison of the behavior of the 50,000 iterates of $D_{\mathbb{A}_a B_+}$ on two images with the same smoothness, but so that for one (solid line) the intersection $\mathbb{A}_a \cap B_+$ is transversal and for the other (dashed line) it is not.

$$f_{0j}^k = \sum_{m=1}^{M} a_m \psi_k \left(\frac{\|\boldsymbol{j} - \boldsymbol{c}_m\|}{r_m} \right), \tag{9.11}$$

where $\psi_k(\boldsymbol{x}) = (1 - \|\boldsymbol{x}\|^2)^k \chi_{[0,1]}(\|\boldsymbol{x}\|)$, with an image, formed by convolving ψ_1 with $\boldsymbol{f}_0^0$, to obtain

$$f_{1j}^1 = \sum_{\boldsymbol{i} \in J} f_{0(\boldsymbol{j}-\boldsymbol{i})}^0 \psi_1 \left(\frac{\boldsymbol{i}}{P} \right). \tag{9.12}$$

In all cases, the random magnitudes, centers and radii, $\{(a_m, \boldsymbol{c}_m, r_m)\}$, are the same. In order for these images to have the same inherent smoothness, we choose k so that the asymptotic behavior of the 2D-Fourier transform of $\psi_k(\boldsymbol{x})$ matches that of $\psi_1 * \psi_0(\boldsymbol{x})$. An elementary calculation shows that

$$\widehat{\psi_k}(\boldsymbol{\xi}) = c_k \frac{J_{k+1}(\|\boldsymbol{\xi}\|)}{\|\boldsymbol{\xi}\|^{k+1}}. \tag{9.13}$$

If $k = 5/2$, then both Fourier transforms decay like $\|\boldsymbol{\xi}\|^{-4}$.

Figure 9.2 shows the results of running 50,000 iterates of the map $D_{\mathbb{A}_a B_+}$ on $\boldsymbol{f}_0^{\frac{5}{2}}$ and $\boldsymbol{f}_1^1$. The iterates using the convolved image appear to be converging rapidly for about 200 iterates after which they appear to have largely stagnated. The fact that the residual is slowly decreasing suggests that the iterates may be in an attracting basin defined by a nonzero critical point of $d_{\mathbb{A}_a B_+}$. On the other hand, the iterates in the experiment using the generic image appear not to have stagnated but seem to be slowly spiraling in toward a center manifold defined

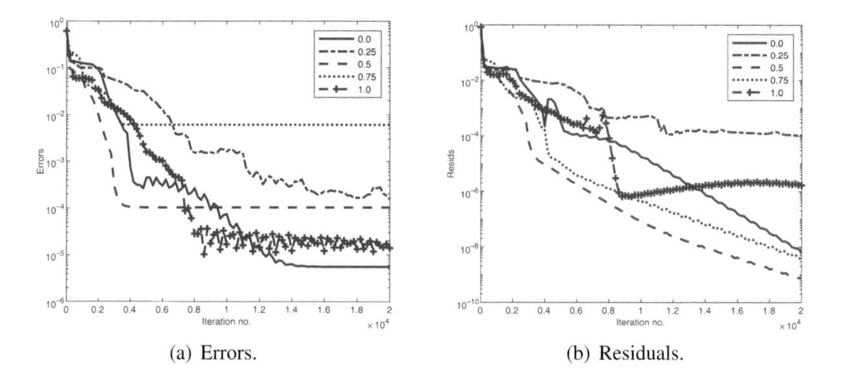

(a) Errors.

(b) Residuals.

Figure 9.3 A comparison of the behavior of the 20,000 iterates of $D_{\mathbb{A}_a B_+^{\mathbb{C}}}$ on complex images with $k \in \{0.0,\ 0.25,\ 0.5,\ 0.75,\ 1.0\}$. The values of these images lie in the sector $0 \leq \arg z < \frac{\pi}{250}$.

by an exact intersection. The difference in these trajectories is almost certainly explained by the fact that one intersection is transversal and the other is not.

The final example in this section examines the behavior of an algorithm, based on $D_{\mathbb{A}_a B_+^{\mathbb{C}}}$, on complex images, $\boldsymbol{f} \in \mathbb{C}^J$, that belong to the set $B_+^{\mathbb{C}}$:

$$B_+^{\mathbb{C}} \stackrel{d}{=} \{\boldsymbol{f} \in \mathbb{C}^J : 0 \leq \arg f_{\boldsymbol{j}} < \frac{\pi}{2} \quad \text{for all } \boldsymbol{j} \in J\}. \tag{9.14}$$

In these experiments we use 256×256 images.

Example 9.4 For the first experiment we consider complex images with $k \in \{0.0,\ 0.25, 0.5,\ 0.75,\ 1.0\}$. The imaginary part of image is much smaller than the real part and is restricted to have its support contained in the support of the real part, and, in these examples the arguments actually lie in the sector $0 \leq \arg x \leq \frac{\pi}{250}$. The results of running $D_{\mathbb{A}_a B_+^{\mathbb{C}}}$ for 20,000 iterates for examples of this type are shown in Figure 9.3. Initial conditions are selected for each value of k by choosing the initial phases that give the best result out of 10 trials. While this algorithm works somewhat better on real images with the same degree of smoothness, its performance on these complex images is quite acceptable. The iterates are still converging for $k = 0.25, 1.0$ and appear to have stagnated for the remaining values of k; nonetheless, the final errors are less than about 10^{-4} for all but $k = 0.75$, for which it is less than 10^{-2}.

In the final experiment of this example we assess the effects of a having a larger imaginary part. For these experiments we consider a collection of $k = 0$, 256×256 images where the $0 \leq \arg f_{\boldsymbol{j}} < m\pi$ for

$$m \in \{0.01,\ 0.02,\ 0.04,\ 0.08,\ 0.16,\ 0.21,\ 0.27\}.$$

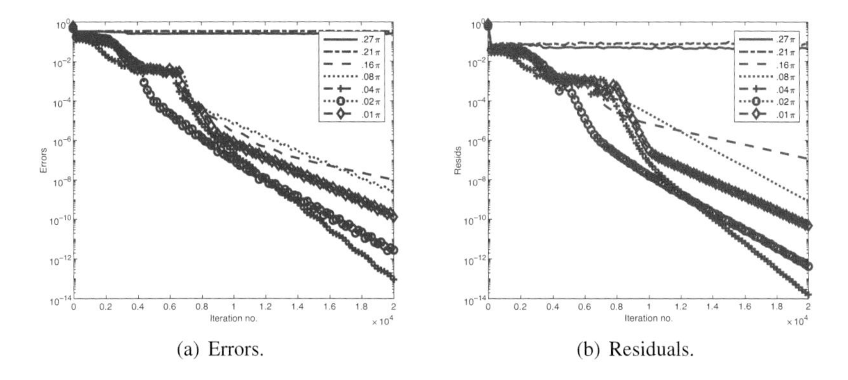

(a) Errors. (b) Residuals.

Figure 9.4 A comparison of the behavior of 20,000 iterates of $D_{\mathbb{A}_a B_+^{\mathbb{C}}}$ on complex images with $k = 0$, whose values images lie in the sectors $0 \leq \arg f_j < m\pi$, where m belongs to the set $\{0.01, 0.02, 0.04, 0.08, 0.16, 0.21, 0.27\}$. The legends on the plots identify the argument estimate used for each example.

For these plots we use the initialization, selected from 10 trials, that provides the best outcome. The results are shown in Figure 9.4, and as expected, smaller angles generally produce better results. Only the $m = 0.27\pi$ and 0.21π cases seem to have stagnated. A reasonable error is attained even for a maximum argument a little greater than $\pi/8$.

These experiments show that the complex nonnegativity constraint provides a very effective reconstruction method for a complex image, f, when it is known that the coordinates $\{f_j\}$ belong to a sector with opening less than about $\pi/8$. For the uniqueness results proved in Chapter 4 to apply, the values of the image are required to lie in a sector with opening less than $\pi/2$.

9.3 Algorithms Based on Minimization in the 1-Norm

If the unknown image is known to be nonnegative, then, as shown in Section 4.3, the ℓ_1-norm is minimized on the magnitude torus, $\mathbb{A}_a$, at points with nonnegative (or nonpositive) coordinates. This observation offers another approach to defining a hybrid iterative map. As noted earlier, for all $\boldsymbol{f} \in \mathbb{A}_a$, the magnitude of the zeroth DFT coefficient satisfies $|\widehat{f}_{\mathbf{0}}| \leq \|\boldsymbol{f}\|_1$, with equality for images of a single sign.

Hence, the magnitude DFT data contains the radius, $r_1 = |\widehat{f}_{\mathbf{0}}|$, of the ℓ_1-ball that intersects $\mathbb{A}_a$ in a finite, nonempty set of points. This set generically consists of the trivial associates of a single solution to the phase retrieval problem. We denote this ball by $B^1_{r_1}$, and the map to the nearest point on $\partial B^1_{r_1}$, in the ℓ_2-norm, by $P_{B^1_{r_1}}$. Because the $\partial B^1_{r_1}$ is a piecewise affine set, this nonlinear map has a very efficient implementation, which is described in Appendix 9.A. This fact makes $D_{\mathbb{A}_a B^1_{r_1}}$ a viable candidate for defining a new algorithm for phase retrieval.

In this section we compare the results of using the map

$$D_{\mathbb{A}_a B^1_{r_1}}(f) = f + P_{\mathbb{A}_a} \circ R_{B^1_{r_1}}(f) - P_{B^1_{r_1}}(f), \tag{9.15}$$

with those obtained using $D_{\mathbb{A}_a B_+}$. For these experiments we introduce a more systematic protocol for comparing two algorithms, which will be extensively used in Part III. Statistics that allow for an assessment of the relative merits of two algorithms are the residuals, the true errors, and the *correlation* between these two quantities. The correlation is important as only the residual is observable in a real experimental setting. As before, the true error is the distance from the nearest exact intersection to reconstructed image and the residual is defined as

$$\|f^{(n)} - f^{(n+1)}\| = \begin{cases} \|P_{\mathbb{A}_a} \circ R_{B^1_{r_1}}(f^{(n)}) - P_{B^1_{r_1}}(f^{(n)})\| & \ell_1 - \text{constraint}, \\ \|P_{\mathbb{A}_a} \circ R_{B_+}(f^{(n)}) - P_{B_+}(f^{(n)})\| & \text{nonneg.-constraint}. \end{cases}$$

The experiment below compares these statistics for these two algorithms on images with smoothness $k = 0, 2, 4$. For each level of smoothness, we run the two algorithms, on a single image, with the same 720 different choices of random initial data. We show a 2D scatter plot with the $\log_{10}$-true errors along the x-axis and the $\log_{10}$-residuals along the y-axis. These 2D plots give a clear appraisal of the ranges of values assumed by these two statistics and the extent of their correlation.

In many signal and image processing algorithms, an ℓ_1-penalty term is used to enforce a "sparsity" constraint. In the present context, this is completely justified if the image is known, a priori, to be nonnegative. Simple numerical experiments show that, for images not satisfying this hypothesis, the minimum ℓ_1-norm is often *not* attained at a sparse image lying on the magnitude torus. It should be noted that 2D images, with small support in phase retrieval, are nonzero at about $1/4$ of the available indices, which would not be considered sparse in most signal or image processing applications.

Example 9.5 The 256×256 images used in this experiment are shown in Figure 9.5. In all cases the intersections $\mathbb{A}_a \cap B_+$ and $\mathbb{A}_a \cap B^1_{r_1}$ are transversal, so these algorithms can be expected to behave rather similarly. In fact, individual runs of the two algorithms, starting with the same choice

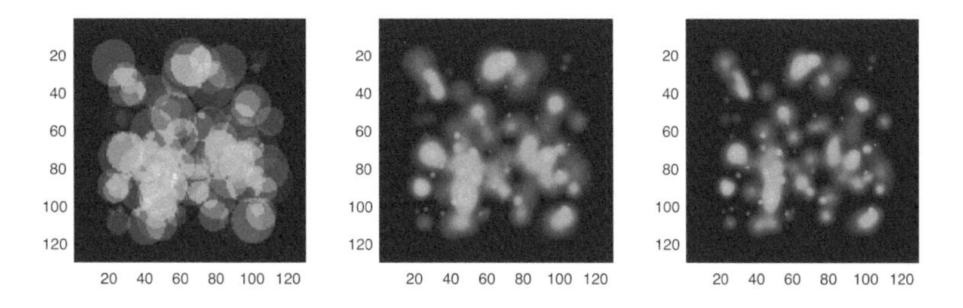

Figure 9.5 Images, with smoothness $k = 0, 2, 4$, used to compare $D_{\mathbb{A}_a B_+}$ to $D_{\mathbb{A}_a B^1_{r_1}}$.

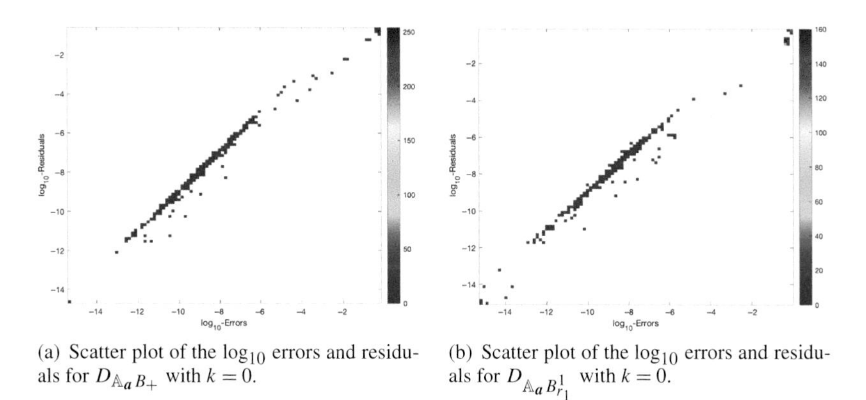

(a) Scatter plot of the $\log_{10}$ errors and residuals for $D_{\mathbb{A}_a B_+}$ with $k = 0$.

(b) Scatter plot of the $\log_{10}$ errors and residuals for $D_{\mathbb{A}_a B^1_{r_1}}$ with $k = 0$.

Figure 9.6 Scatter plots of the $\log_{10}$ errors and residuals produced by 720 trials of algorithms based on 5,000 iterates of either $D_{\mathbb{A}_a B_+}$ or $D_{\mathbb{A}_a B^1_{r_1}}$ on a nonnegative, $k = 0$ image. The number of pairs in each bin is indicated by the color.

of random phases, can actually be quite different; though, on average their behavior is similar.

Figures 9.6–9.8 shows scatter plots, after 5,000 iterates using either $D_{\mathbb{A}_a B_+}$ or $D_{\mathbb{A}_a B^1_{r_1}}$ of the $\log_{10}$ of the errors, on x-axis, versus the $\log_{10}$ of the residuals. These algorithms behave very similarly. We summarize the results of this experiment in Table 9.1. The ranges of errors and residuals are very close in all cases. The mean errors ($\overline{\text{err}}$) and mean residuals ($\overline{\text{rsd}}$) are also quite similar.

The only notable difference is in the mean errors and residuals for the $k = 2$ example. The algorithm that used the ℓ_1-norm as auxiliary data has larger mean errors and residuals. Examining the plot in Figure 9.7, we see that this algorithm had a fairly large number of "failures" with a final error greater than 10^{-1}, which the other algorithm did not experience. The plots in

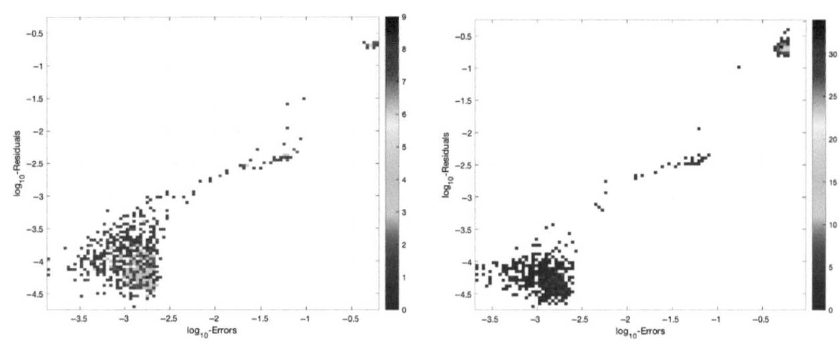

(a) Scatter plot of the $\log_{10}$ errors and residuals for $D_{\mathbb{A}_a B_+}$ with $k = 2$.

(b) Scatter plot of the $\log_{10}$ errors and residuals for $D_{\mathbb{A}_a B^1_{r_1}}$ with $k = 2$.

Figure 9.7 Scatter plots of the $\log_{10}$ errors and residuals produced by 720 trials of algorithms based on 5,000 iterates of either $D_{\mathbb{A}_a B_+}$ or $D_{\mathbb{A}_a B^1_{r_1}}$ on a nonnegative, $k = 2$ image. The number of pairs in each bin is indicated by the color.

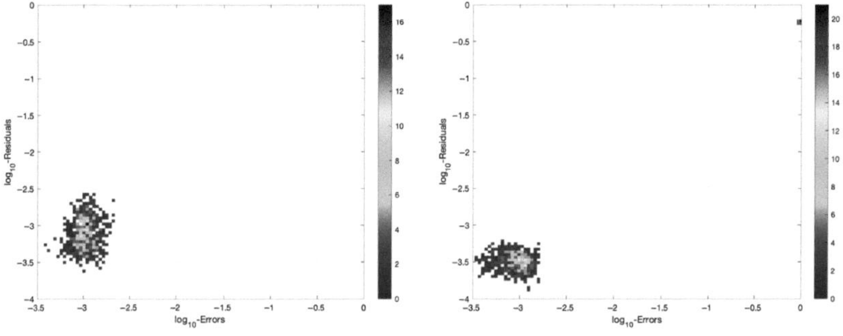

(a) Scatter plot of the $\log_{10}$ errors and residuals for $D_{\mathbb{A}_a B_+}$ with $k = 4$.

(b) Scatter plot of the $\log_{10}$ errors and residuals for $D_{\mathbb{A}_a B^1_{r_1}}$ with $k = 4$.

Figure 9.8 Scatter plots of the $\log_{10}$ errors and residuals produced by 720 trials of algorithms based on 5,000 iterates of either $D_{\mathbb{A}_a B_+}$ or $D_{\mathbb{A}_a B^1_{r_1}}$ on a nonnegative, $k = 4$ image. The number of pairs in each bin is indicated by the color.

the $k = 4$ case reveal that the ℓ_1-based algorithm also had about 35 failures in this case, whereas the nonnegativity-based algorithm had none. Finally, both algorithms exhibit a wide range of results on the $k = 0$ example, with the errors ranging from 10^{-1} to 10^{-15}. Indeed, both had a large number of failures with an error around 10^{-1}. Nonetheless, the errors and residuals for both algorithms are highly correlated.

Table 9.1. *Statistics comparing the hybrid iterative maps* $D_{\mathbb{A}_a B_+}$ *and* $D_{\mathbb{A}_a B^1_{r_1}}$. *The correlations,* corr_+, corr_{ℓ_1} *are the correlations between the* $\log_{10}$*- errors and residuals for each algorithm separately. Also shown are the averages of the* $\log_{10}$ *of the errors and residuals after 5,000 iterates*

k	corr_+	corr_{ℓ_1}	$\overline{\mathrm{err}}_+$	$\overline{\mathrm{err}}_{\ell_1}$	$\overline{\mathrm{rsd}}_+$	$\overline{\mathrm{rsd}}_{\ell_1}$
0	0.99	0.91	−5.33	−5.43	−4.95	−5.03
2	0.99	0.95	−2.76	−1.97	−3.82	−2.96
4	0.82	0.99	−2.98	−2.9	−3.11	−3.34

9.A Appendix: An Efficient Method for Projection onto a Ball in the 1-Norm

In this appendix, we describe an efficient algorithm for finding the point that minimizes the ℓ_2-distance to the boundary of an ℓ_1-ball. For this discussion, we assume that the ambient space is $\mathbb{R}^N$.

Let $B^1_r = \{\boldsymbol{x} : \|\boldsymbol{x}\|_1 \leq r\}$, let $\boldsymbol{x}^0 = (x^0_1, \dots, x^0_N)$ be an initial point, and set

$$\check{\boldsymbol{y}} = \arg\min_{\boldsymbol{y} \in \partial B^1_r} \|\boldsymbol{x}^0 - \boldsymbol{y}\|_2. \tag{9.16}$$

In the discussion that follows we only consider the case $\|\boldsymbol{x}^0\|_1 > r$.

We start with several observations about the minimizer $\check{\boldsymbol{y}}$

(i) If, for an index i, $x^0_i = 0$, then $\check{y}_i = 0$ as well.
(ii) More generally, $\operatorname{sign} x^0_i = \operatorname{sign} \check{y}_i$ for nonzero coordinates.

The proofs are elementary. It therefore suffices to consider the case that $x^0_i > 0$ for every i. With the coordinates reordered so that $x^0_i \geq x^0_{i+1}$, consider the following algorithm:

(i) Set $l = N$.
(ii) Let $c = \frac{x^0_1 + \cdots + x^0_l - r}{l}$.
(iii) If $c < x^0_l$, go to step (iv). Otherwise, set $l = l - 1$, and return to (ii).
(iv) Set $y_i = 0$ for $i > l$ and $y_i = x^0_i - c$ for $i \leq l$.

If the algorithm does not terminate for an $l \geq 1$, then we see that

$$x^0_1 + \cdots + x^0_l - r \geq l x^0_{l+1} > 0. \tag{9.17}$$

Hence, if it terminates at the next step, then

$$c \geq x^0_{l+1} \geq x^0_{l+2} \geq \cdots \geq x^0_n. \tag{9.18}$$

If the algorithm has not terminated when we reach $l = 1$, then it is always true that $x^0_1 - r < x^0_1$, and in this case $y = (r, 0, \ldots, 0)$. In general, $y_i > 0$ for $1 \leq i \leq l$, and 0 otherwise, and therefore,

$$\|\boldsymbol{y}\|_1 = \sum_{i=1}^{l} y_i = r. \tag{9.19}$$

To show that $\boldsymbol{y}$ gives the minimum distance, we see that any other possible minimizer would take the form $\tilde{\boldsymbol{y}} = \boldsymbol{y} + \Delta \boldsymbol{y}$, where

$$y_i + \Delta y_i \geq 0 \text{ for } 1 \leq i \leq N \text{ and } \sum_{i=1}^{N} \Delta y_i = 0. \tag{9.20}$$

The identity follows from the inequalities, as the minimizer must lie on ∂B^1_r. A calculation, using the identity in (9.20), shows that

$$\|\boldsymbol{x}^0 - (\boldsymbol{y} + \Delta \boldsymbol{y})\|_2^2 = \|\boldsymbol{x}^0 - \boldsymbol{y}\|_2^2 + \|\Delta \boldsymbol{y}\|_2^2 + 2 \sum_{i=l+1}^{N} (c - x^0_i)\Delta y_i. \tag{9.21}$$

From the definition of l and equation (9.18) it follows that every term in the last summation is nonnegative, and therefore,

$$\|\boldsymbol{x}^0 - (\boldsymbol{y} + \Delta \boldsymbol{y})\|_2^2 \geq \|\boldsymbol{x}^0 - \boldsymbol{y}\|_2^2, \tag{9.22}$$

with equality if and only if $\Delta \boldsymbol{y} = \boldsymbol{0}$. This demonstrates that our algorithm finds a global minimizer. Letting $\check{\boldsymbol{y}}$ denote $\boldsymbol{y}$ with the coordinates returned to their original order and signs, we have that

$$P_{B^1_{r_1}}(\boldsymbol{x}) = \check{\boldsymbol{y}}. \tag{9.23}$$

A similar discussion applies if $\|\boldsymbol{x}^0\|_1 < r$.

10
Asymptotics of Hybrid Iterative Maps

We close this part of the book by considering the behavior of hybrid iterative maps after large numbers of iterations. The content of this chapter is rather speculative, consisting mostly of examples that illustrate various experimental phenomena. It is motivated by the observation that, except under very specific circumstances, (i.e., transversality of $\mathbb{A}_{\boldsymbol{a}}$ and B at intersection points or a simple $d_{\mathbb{A}_{\boldsymbol{a}}B}$-landscape) the iterates of hybrid iterative maps do not converge. Rather, stagnation seems to occur with very high probability. The discussion in this chapter is not intended to suggest new algorithms, but rather to elucidate the extraordinary range, and beauty, of the dynamics that underlie stagnation. After a fairly small number of iterations, very little additional information about the unknown object is gained by continuing to run these algorithms. An analysis of the information available after an algorithm has stagnated is studied in Chapter 12.

The linearizations of hybrid iterative maps at attracting points on the center manifold appear to have somewhat unusual properties: they are often not contractions, are very far from normal operators, and have complex eigenvalues, many with absolute values very close to 1. Stagnation, wherein the distance to $\mathbb{A}_{\boldsymbol{a}} \cap B_S$ becomes essentially constant and the differences $\{\boldsymbol{f}^{(n+1)} - \boldsymbol{f}^{(n)}\}$ do not go to 0, is completely consistent with these observations and the fact that there are multiple attracting basins in close proximity to one another. Examining the behavior of hybrid iterative maps after a very large number of iterates provides strong evidence that, in the "stable" parts of their phase spaces, these maps stagnate without converging, with the orbits accumulating on lower-dimensional subsets.

10.1 Stagnation

At first there would seem to be three distinct situations that could produce stagnation:

(i) Global phase stagnation: The iterates of the algorithms never find an attracting basin defined by a center manifold of an exact intersection, hence stagnation occurs in the global phase.
(ii) Local phase stagnation: The iterates fall into an attracting basin defined by the center manifold(s) of one, or several, exact intersection points. If $\dim T_{f_0}\mathbb{A}_{\boldsymbol{a}} \cap B \neq 0$ or $T_{f_0}\mathbb{A}_{\boldsymbol{a}}$ and B make a very small angle over a high-dimensional subspace, then the attraction is too weak to pull the iterates into close enough proximity to a single center manifold for them to converge.
(iii) Stagnation caused by the attracting basins of generalized center manifolds, which are defined by nonzero critical points of d_{AB}. These are attracting basins not directly connected to intersections of A and B and contain no fixed points.

In experiments, the third possibility seems to occur quite often, especially with smoother objects, which, in our units, have $k > 1$. Examples of this phenomenon are shown in Section 7.2.1 and Section 8.1.3. This may indeed be a circumstance which produces chaotic dynamics on a positive dimensional invariant set, and it probably arises with real (noisy) data for which the set $A \cap B$ is, in fact, empty. It is difficult to analyze in the phase retrieval problem, as it is hard to locate these "near-miss" local minima. Empirically, a good way to find them is to run a hybrid map algorithm for many iterates. In several examples below, entrainment to the attracting basin of such a generalized center manifold does seem to be the best explanation for the observed phenomena.

As shown in Section 8.2, the linearization of a hybrid iterative map at an attracting fixed point on the center manifold may fail to be a contraction. This produces orbits that lie on slowly shrinking, elongated ellipsoids. This can easily be imagined to produce stable, but nonconvergent orbits transiting among several center manifolds, or generalized center manifolds.

Global phase stagnation does not seem to be a significant phenomenon. While it may require a large number of iterations, an attracting basin, of one sort or another, seems always to be found. A long delay in locating a basin of attraction could be related to the existence of local minima of the distance function d_{AB} at very small, but positive, distances, or to an object and support mask that are invariant under the inversion map. On the other hand, local phase stagnation is often observed, and it is stable over *millions* of iterates.

Remark 10.1 A detailed study of stagnation phenomena connected to hybrid input output (HIO) algorithms was carried out in Fienup and Wackerman (1986). The emphasis in this paper is on transient phenomena observed after several hundred iterates, and methods to remove the attendant artifacts. When this research was done, in the mid-1980s, the computational power to perform hundreds of thousands, or even millions, of iterations did not exist.

In the experiments below, we examine various aspects of the asymptotic behavior of these algorithms. We employ several devices to illustrate what is happening: the simplest information, which we have used many times already, consists of semilog plots of the absolute error, which is the distance to the nearest exact intersection point, $\{\|\boldsymbol{r}^{(n)} - \boldsymbol{f}_0\|_2\}$, along with a semilog plot of the residuals, which for hybrid iterative maps is taken to equal $\{\|\boldsymbol{f}^{(n+1)} - \boldsymbol{f}^{(n)}\|_2\}$. Once an algorithm has found an attracting basin, the iterates either converge more or less geometrically or stagnate, with the absolute errors and residual remaining essentially constant. The directions with higher contraction rates converge more quickly and decisively, leaving only those with the slowest contraction rates. Indeed, the rate of convergence can be so slow as to be practically indistinguishable from stagnation.

Stagnation appears to occur along sets consisting largely of common tangent directions, which is revealed when $\|\boldsymbol{f}^{(n+1)} - \boldsymbol{f}^{(n)}\|_2 \propto \|\boldsymbol{r}^{(n)} - \boldsymbol{f}_0\|_2^2$, see Example 8.12. Indeed, we sometimes see even higher powers appear after many iterates. The explanation for these higher powers may be that the iterates have found points near to curves of higher-order contact between $\mathbb{A}_{\boldsymbol{a}}$ and B_S, as described in Remark 2.12. Decomposing $\boldsymbol{r}^{(n)} = \boldsymbol{f}_0 + \boldsymbol{\tau}^{(n)} + \boldsymbol{\nu}^{(n)}$, where $\boldsymbol{\tau}^{(n)} \in T^0_{\boldsymbol{f}_0}\mathbb{A}_{\boldsymbol{a}}$ and $\boldsymbol{\nu}^{(n)} \in N^0_{\boldsymbol{f}_0}\mathbb{A}_{\boldsymbol{a}}$, we see that $\|\boldsymbol{\nu}^{(n)}\|_2 \approx \|\boldsymbol{\tau}^{(n)}\|_2^2$, and that both vectors satisfy the support condition to many digits, though not exactly. That such vectors might exist is not too surprising as

$$\dim T^0_{\boldsymbol{f}_0}\mathbb{A}_{\boldsymbol{a}} \cap B_S \approx \dim N^0_{\boldsymbol{f}_0}\mathbb{A}_{\boldsymbol{a}} \cap B_S, \tag{10.1}$$

and both dimensions can be fairly large. In fact the normal component, $\boldsymbol{\nu}^{(n)}$, carries $\boldsymbol{f}_0 + \boldsymbol{\tau}^{(n)}$ very close to $P_{\mathbb{A}_{\boldsymbol{a}}}(\boldsymbol{f}_0 + \boldsymbol{\tau}^{(n)})$, the nearest point on $\mathbb{A}_{\boldsymbol{a}}$. It is important to realize that, for most choices of $\boldsymbol{\tau} \in T^0_{\boldsymbol{f}_0}\mathbb{A}_{\boldsymbol{a}}$, largely supported in S, there is no vector $\boldsymbol{\nu}$, almost supported in S, such that

$$\|\boldsymbol{f}_0 + \boldsymbol{\tau} + \boldsymbol{\nu} - P_{\mathbb{A}_{\boldsymbol{a}}}(\boldsymbol{f}_0 + \boldsymbol{\tau})\|_2 \ll \|\boldsymbol{f}_0 + \boldsymbol{\tau} - P_{\mathbb{A}_{\boldsymbol{a}}}(\boldsymbol{f}_0 + \boldsymbol{\tau})\|_2. \tag{10.2}$$

Thus, the points $\{\boldsymbol{r}^{(n)}\}$ found by a hybrid iterative map are special points, which appear to be very close to local minima of $d_{\mathbb{A}_{\boldsymbol{a}} B_S}$, where this function assumes very small values. As indicated in Section 8.1.3, it is expected that quasi-attracting basins are associated with such points. It is difficult to give

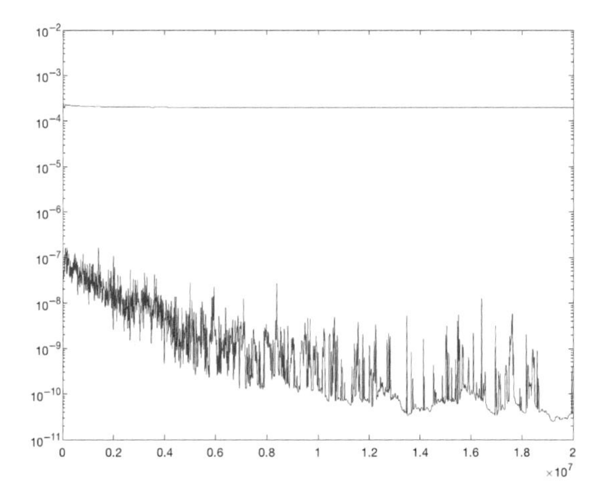

Figure 10.1 The $\log_{10}$ of the errors (blue) and residuals (red) for 20 million iterates of $D_{\mathbb{A}_a B_{S_2}}$ on a $k = 4$ 256×256 image defined by convolution with a Gaussian.

a definitive explanation for this phenomenon as the distances themselves, $\{\|\boldsymbol{r}^{(n)} - \boldsymbol{f}_0\|_2\}$, can remain large.

Figure 10.1 shows what is very likely an example of this phenomenon. This plot shows the errors and residuals from 20 million iterates of the hybrid map, based on $D_{\mathbb{A}_a B_{S_2}}$ for a 256×256 image with $k = 4$, defined by convolving with a Gaussian and a 2-pixel support neighborhood. As the Gaussian is inversion symmetric, this produces a large intersection between $T_{f_0}\mathbb{A}_{\boldsymbol{a}}$ and B_S, as well as between $N_{f_0}\mathbb{A}_{\boldsymbol{a}}$ and B_S. The algorithm is started very close to an exact intersection, with random noise added to the phases producing an initial error of about 10^{-3}. The residual, which is initially slowly decreasing, on average, remains very jittery, eventually oscillating between about 10^{-10} and 10^{-8}. This indicates approach along a direction where B_S makes second-order contact with $\mathbb{A}_{\boldsymbol{a}}$. The error itself is not perceptibly changed after the 4,000,000th iterate.

After many iterates, the changes in the errors usually become quite small, and no longer provide a good indicator of the later behavior of the algorithm. The iterates though are often jittery on what appears to be a characteristic length scale. To get more insight into what is happening, we use several other tools. The simplest is a sequence of histograms of the angles of the phase errors. Though it's a bit imprecise, we call these *phase error histograms*. An example is shown in Figure 10.2. Even after the errors and residuals have largely stagnated, sometimes these plots show the phase errors are continuing

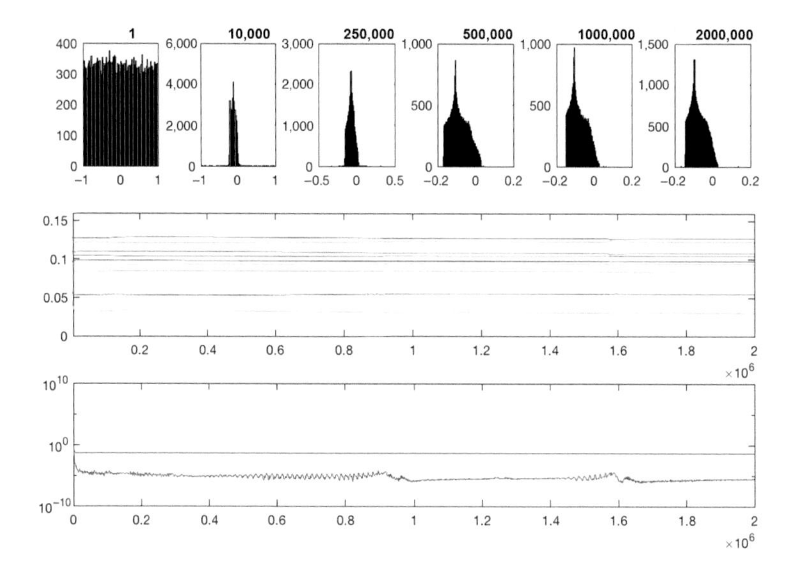

Figure 10.2 Two million iterates of $D_{\mathbb{A}_a B_{S_1}}$ on a $k = 2$ object, with a 1-pixel neighborhood. The top row are histograms of the angles of the phase errors, with iteration numbers indicated at the top. The x-axes of these histograms are in multiples of π. The second row shows distances to the nine nearest 1-pixel translates, and the bottom plots show the errors (blue) and residuals (red) on a semilog-scale.

to diminish, up to a point. This may not be reflected by noticeable changes in the errors because the corresponding discrete Fourier transform (DFT) coefficients have very small magnitudes. This phenomenon seems to occur when the algorithms may still be converging albeit at a very, very slow rate. The rate of convergence is so slow that it is not, in any practical sense, different from stagnation. Nonetheless, continued narrowing around 0 of the phase error histograms allows us to distinguish this case from a case of true stagnation.

As shown in Example 8.5, the distances between trivial associates are often comparable to, or even smaller than, the residuals $\{\|\boldsymbol{f}^{(n+1)} - \boldsymbol{f}^{(n)}\|_2\}$. In this circumstance one expects that the nearest exact intersection point might actually cycle among a small group of nearby trivial associates. To investigate this question, we plot the distances to the nine nearest exact solutions (the 1-pixel neighborhood). In cases where the phase error histograms showed continued improvement, there is typically one exact intersection point that is an order of magnitude closer than the others. This point appears to remain the closest intersection point essentially ad infinitum. In cases where the phase error histograms show no improvement, the nine closest points are almost equally distant from the iterates, and the one which is closest appears to cycle among them, again essentially ad infinitum. The latter case seems to represent true stagnation.

A final graphical device that we employ entails projecting the iterates onto a lower-dimensional subspace. To define this subspace, we pick an object and random initial phases and then proceed for a million iterates. The lower-dimensional subspace is defined by finding the first three principal components that describe the main sources of variation in the next 500,000 iterates. We then run the same algorithm for 5,000,000 additional iterates and plot the projections into the subspace defined by the three principal components found in the previous step. These plots again fall into two classes: when the residuals are very small, the iterates appear to lie on a very structured set, whereas, when the residuals remain large, the iterates are rather chaotic and fill out a volume in the three-dimensional space of the projection.

In the examples that follow, we use these graphical devices to tease out the behavior of hybrid iterative algorithms after they have stagnated. These examples together with those given earlier in the book provide a clear qualitative picture of the possible asymptotic behaviors of these algorithms, and what determines which case one is in. There are essentially three possibilities, with one case having two subcases:

(i) Geometric convergence to a true intersection point.
(ii) Stagnation in the attracting basin defined by a single intersection point
  (a) $d(\boldsymbol{r}^{(n)}, P_{\mathbb{A}_a}(\boldsymbol{r}^{(n)})) \approx d(\boldsymbol{r}^{(n)}, \boldsymbol{f}_0)^2$ and
  (b) $d(\boldsymbol{r}^{(n)}, P_{\mathbb{A}_a}(\boldsymbol{r}^{(n)})) \ll d(\boldsymbol{r}^{(n)}, \boldsymbol{f}_0)^2$.
(iii) Stagnation in the attracting basin defined by several nearby intersection points.

Case 1: *True convergence* most often happens when the iterates find the attracting basin defined by a point in the set $\mathbb{A}_{\boldsymbol{a}} \cap B$ where the intersection is transversal, or almost transversal. With the support condition, this only appears to happen if the object is not too smooth ($k \leq 1$), and the iterates find an attracting basin along the boundary of those allowed by support condition. When using nonnegativity, one can tolerate smoother objects, provided the intersection with ∂B_+ is transversal. The rate of convergence is tied to the smoothness of the object (see Figure 9.1).

Case 2: *Single-point stagnation* arises when the iterates find the attracting basin defined by a single point $\boldsymbol{f}_0 \in \mathbb{A}_{\boldsymbol{a}} \cap B$ for which $\dim T_{\boldsymbol{f}_0}\mathbb{A}_{\boldsymbol{a}} \cap B$ is positive, but small, and the singular values of the matrix H are not too close to 1. In this case the iterates remain closest to a single true intersection point, though this distance remains rather large (10^{-1} to 10^{-2}). There are two subcases to consider, either (a) $d(\boldsymbol{r}^{(n)}, P_{\mathbb{A}_a}(\boldsymbol{r}^{(n)})) \approx d(\boldsymbol{r}^{(n)}, \boldsymbol{f}_0)^2$ or (b) $d(\boldsymbol{r}^{(n)}, P_{\mathbb{A}_a}(\boldsymbol{r}^{(n)})) \ll d(\boldsymbol{r}^{(n)}, \boldsymbol{f}_0)^2$, see Example 8.12. In the former case, the iterates seem to move rather chaotically, but at a scale determined by $d(\boldsymbol{r}^{(n)}, \boldsymbol{f}_0)^2$. In the latter case, the iterates seem to follow very regular trajectories, and it seems quite plausible that they are actually entrained to a quasi-attractive basin defined by a very small, local minimum of $d_{\mathbb{A}_a B_S}$, as described above.

Case 2(a) also arises when we start the iteration very close to a true, but nontransversal, intersection, as shown in Figure 10.1. In this case, it appears that points in $T_{\boldsymbol{f}_0}\mathbb{A}_{\boldsymbol{a}} \cap B_S$ are also "quasi-fixed" points. The dynamics of the map on this set are nonconvergent, on a scale proportional to $d(\boldsymbol{r}^{(n)}, \boldsymbol{f}_0)^2$. The stagnation in Case 2(a) appears to result from the fact that the attraction toward the center manifold along directions lying in $T_{\boldsymbol{f}_0}\mathbb{A}_{\boldsymbol{a}} \cap B$ is very weak and *diminishes* as one approaches the center manifold itself.

Case 3: *Multipoint* stagnation arises when $\dim T_{\boldsymbol{f}_0}\mathbb{A}_{\boldsymbol{a}} \cap B$ is large and/or the matrix H has many singular values very close to 1. This is the case if the object is smooth ($k > 1$), or the support condition is not sufficiently precise for a nonsmooth object. In this case the iterates remain more or less equidistant from the center manifolds defined by a few nearby true intersection points. The closest intersection point appears to cycle among these points, and the residuals remain comparable to the distances between these points. This sort of stagnation appears to be the result of the competing weak attractions exerted by the center manifolds defined by these almost equidistant intersection points, or perhaps nearby generalized center manifolds.

10.2 Numerical Examples

In the examples below, we use both the support and nonnegativity conditions as auxiliary information, and only explore circumstances where the iterates have stagnated. The conditions under which the iterates actually converge are, by this time, quite clear. Convergence is revealed in an experiment by the fact that the residual is converging more or less geometrically toward 0. In this case, the true error also converges, at the same rate, to zero, at least in an averaged sense. Since the linearization at the attracting fixed point is often not a contraction, the iterates follow trajectories lying on shrinking, but highly eccentric ellipsoids.

As the examples below show, the dynamics of stagnation can take many different forms. In Example 10.3 we see that, when the residuals are quite small, the iterates seem to follow very structured trajectories, lying along curves. When the residuals remain large, however, the trajectories look quite chaotic and fill a three-dimensional volume. This is essentially the dichotomy expressed above between stagnation along a tangent direction of approach to a single center manifold, and stagnation in a region defined by the center manifolds of several nearby trivial associates. In this section, the smooth examples are defined by convolving with a Gaussian, with the attendant rather high-dimensional intersections between $T_{f_0}\mathbb{A}_{\boldsymbol{a}}$ and B_S, or B_+ (see Section 2.2).

Example 10.2 In these experiments, we study the "long-time" behavior of algorithms based on hybrid iterative maps, illustrating the different sorts of stagnation that we have observed. The top rows of these plots show histograms of the angles of the phase errors at iterates 1; 10,000; 250,000; 500,000; 1,000,000; and 2,000,000. The x-axes of these histograms are in multiples of π. The second rows are plots of the distances from $\{\boldsymbol{r}^{(n)}\}$ to the nine nearest exact intersection points. The third rows are the usual semilog-plots of the exact errors and residuals.

Figure 10.2 shows two million iterates of the map $D_{\mathbb{A}_{\boldsymbol{a}}B_{S_1}}$ acting on a 256×256 image, $\boldsymbol{f}_0$, with $k = 2$ and a 1-pixel support neighborhood. The asymptotic distance to the closest point is about 0.03. The plot shows that the height of the peak around 0 in the phase error histogram increased markedly between the millionth and two millionth iterates. A numerical computation shows that the dimension of the intersection $\dim T_{\boldsymbol{f}_0^{[0,-1]}}\mathbb{A}_{\boldsymbol{a}} \cap B_{S_3} = 10$. In this example the residuals remain quite jittery, with an average approximately the square of the error.

In Figure 10.3, we see an example of stagnation where the iterates are in an attracting basin defined by a collection of nearby translates. In this example, $k = 0$ and we use a 3-pixel neighborhood. The closest translate, at least at the end of the experiment, is $\boldsymbol{f}_0^{(-2,-2)}$, though the distance, 0.14, to this point is considerably larger than in the previous experiment. The iterates are about equally close to a collection of 1-pixel translates, and these distances remain quite jittery. Both the errors and residuals vary over a very small range, with the residual very close to the square of the error. Nonetheless, the phase error histograms do show a little bit of sharpening around 0 up to about the 500,000th iterate.

For our final experiment, shown in Figure 10.4, we use the nonnegativity constraint on a $k = 4$ image. In this case there are two translates that are almost equally distant from the iterates, with an asymptotic distance of about 0.04. These errors are essentially unchanged after the 200,000th iterate, while the residual continues to decrease, a little, until about the 1,000,000th iterate.

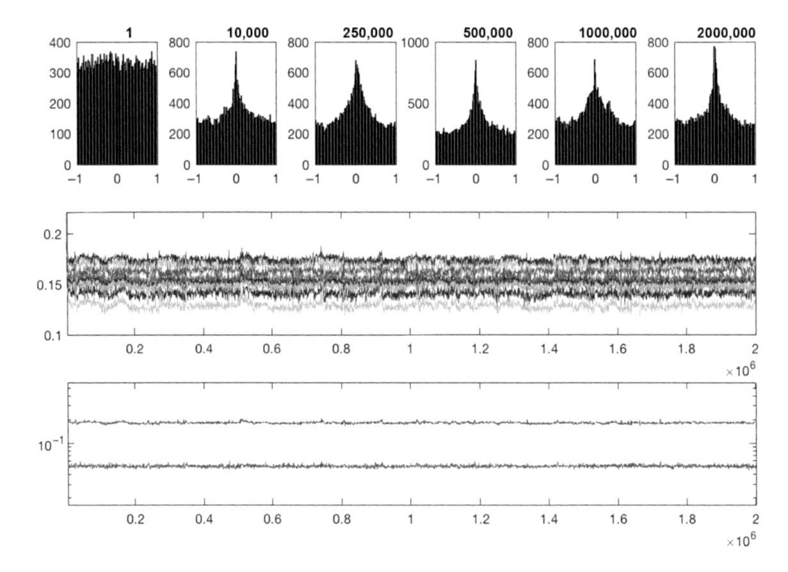

Figure 10.3 Two million iterates of $D_{\mathbb{A}_a B_{S_3}}$ on a $k = 0$ object, with a 3-pixel neighborhood. The top row are histograms of the angles of the phase errors, with iteration numbers indicated at the top, the second row are distances to the nine nearest 1-pixel translates, the bottom plots show the errors and residuals.

Asymptotically, the error squared is about 10 times the residual. The phase errors change very little after the 10,000th iterate. Between the 500,000th iterate and 1,000,000th iterate the histogram appears to have become a little fatter at the base and not as sharply peaked. There is almost no change between the 1,000,000th and 2,000,000th iterate. It should be emphasized that the smoothness of this image is a result of convolution with a Gaussian, which causes a nontransverse intersection of $\mathbb{A}_a$ and B_+. (The examples in Section 9.2 do not have this additional difficulty.) This example demonstrates that, while the nonnegativity constraint generally produces better results than the support constraint, if the intersection is nontransversal, then it eventually succumbs to the same fate.

The main lesson here is that the dynamics of hybrid iterative maps are typically very complicated when the intersection between $\mathbb{A}_a$ and B is nontransversal. In the previous example, we have teased out what look like two rather distinct regimes in the world of stagnating iterations; nonetheless, it is very difficult to have any intuitive feel for the actual behavior of iterates in these nonconvergent cases. Of course, the main factor underlying this difficulty is the very high dimensionality of both the ambient space and the subsets A,

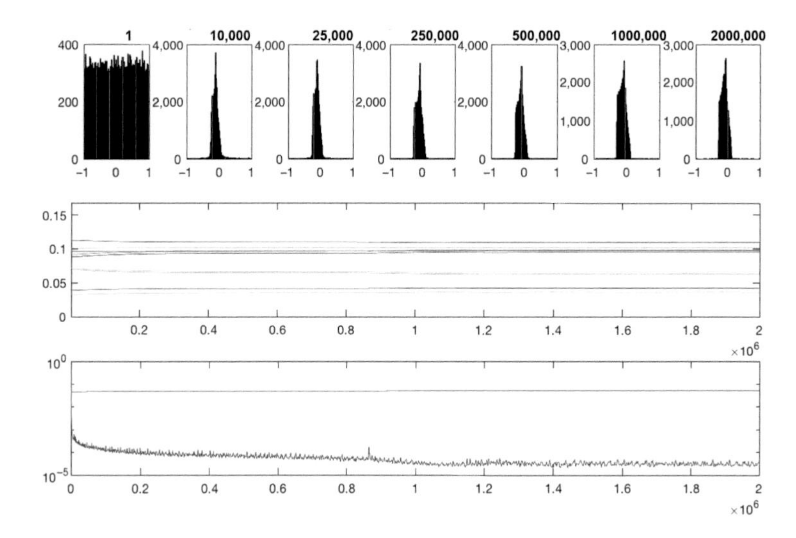

Figure 10.4 Two million iterates of $D_{\mathbb{A}_a B_+}$ on a $k = 4$ object. The top row are histograms of the angles of the phase errors, with iteration numbers indicated at the top, the second row are distances to the nine nearest 1-pixel translates, the bottom plots show the errors and residuals.

B. Our last collection of experiments seeks to give some insight into the range of possibilities for these algorithms. To that end we project the iterates of hybrid map algorithms into a three-dimensional space. The three-dimensional projection is obtained by using principal component analysis, and therefore we are confined to working with fairly small images. In the examples that follow, we work with a variety of 64×64 images.

The subspace into which we project iterates is selected as follows: we obtain magnitude DFT data, $\boldsymbol{a}$, from a 64×64 image, $\boldsymbol{f}_0$, and initialize an algorithm, using a hybrid iterative map $D_{\mathbb{A}_a B}$, by selecting random phases to obtain an initial image, $\boldsymbol{f}^{(0)}$. The algorithm is run for one million iterates. We then collect samples of the next 500,000 iterates $\{\boldsymbol{f}^{(n_j)} : n_j = 10^6 + (j-1)100 \text{ for } j = 1, \ldots, 5001\}$, and perform principal component analysis on this collection of images, obtaining the first 3 principal component directions $(\boldsymbol{v}_1, \boldsymbol{v}_2, \boldsymbol{v}_3)$. In the experiment below, we then run the algorithm for an additional five million steps, and display the following projected points

$$\begin{aligned}\{(\langle \boldsymbol{f}^{(m_j)}, \boldsymbol{v}_1\rangle, \langle \boldsymbol{f}^{(m_j)}, \boldsymbol{v}_2\rangle, \langle \boldsymbol{f}^{(m_j)}, \boldsymbol{v}_3\rangle) :\\ \text{for } m_j = 1.5 \times 10^6 + 50j \text{ with } j = 1, \ldots, 10^5\},\end{aligned} \tag{10.3}$$

in a scatter plot on the left-hand side of the figures below. On the right-hand side we use a similar procedure to obtain principal components for a reduced

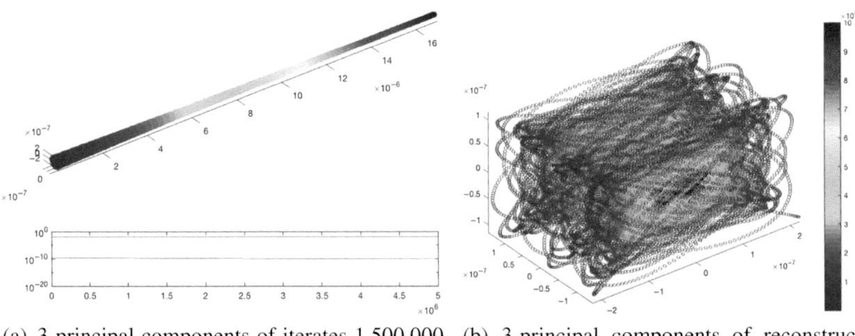

(a) 3-principal components of iterates 1,500,000 to 6,500,000.

(b) 3-principal components of reconstructions 1,500,000 to 6,500,000. The iterates are ordered from blue to red.

Figure 10.5 An example with $k = 0,\ p = 3$, with closest exact intersection $\boldsymbol{f}_0^{[0,0]}$.

dimensional representations of the approximate reconstructions, defined as $\{\boldsymbol{r}^{(n)} = P_B(\boldsymbol{f}^{(n)})\}$. These plots show a remarkable range of behaviors, with varying degrees of structure and chaos. The iterates in these plots are ordered from blue to red.

As a final note, it should be observed that the objects used in these examples are 64×64 images with only $1/16$th of the unknowns present in 256×256 images studied in most of the book. It is reasonable to expect that the properties observed after 1.5 million iterates for such a small example, might not be manifest for larger objects until one has performed many more iterates.

Example 10.3 In the first five experiments of this series, we use the support constraint as auxiliary information, and in the last we use non-negativity. The first three experiments use a piecewise constant image and a 3-pixel neighborhood, with different random starting points. The 3D plot in Figure 10.5(a) shows the projections of iterates 1,500,000 to 6,500,000, and that in Figure 10.5(b) shows their projections onto B_{S_3}. Below the 3D plot in 10.5(a) are plots of the errors (in blue) and residuals (in red). The errors are constant and about 10^{-2}, whereas the residuals are close to constant and about 10^{-10}. The iterates in this experiment seems to be entrained to a local minimum of $d_{\mathbb{A}_{\boldsymbol{a}} B_{S_3}}$. The figure on the right is a projection, eliminating the central manifold directions, of that on the left. It would seem that the long dimension of the left trajectory lies along the (perhaps generalized) center manifold, where the map is essentially translating by a fixed distance. Its projection to B_{S_3} is following some sort of spiraling trajectory, suggesting a linearization with complex eigenvalues. The closest point in $\mathbb{A}_{\boldsymbol{a}} \cap B_{S_3}$ is $\boldsymbol{f}_0^{[0,0]}$ for which $\dim T_{\boldsymbol{f}_0^{[0,0]}} \mathbb{A}_{\boldsymbol{a}} \cap B_{S_3} = 24$.

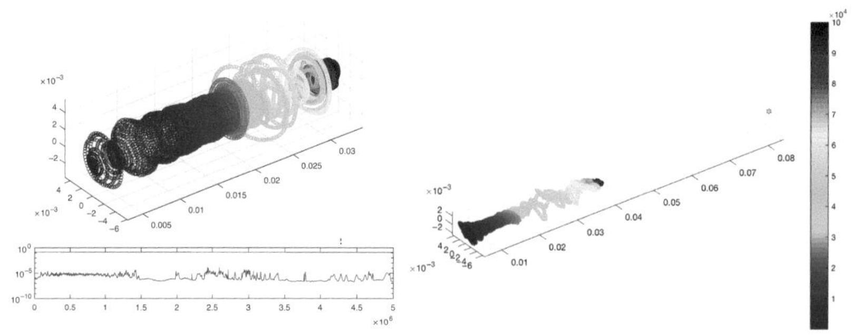

(a) 3-principal components of iterates 1,500,000 to 6,500,000.

(b) 3-principal components of reconstructions 1,500,000 to 6,500,000.

Figure 10.6 Another example with $k = 0,\ p = 3,$ with closest exact intersection $f_0^{[0,0]}$.

The image used to create Figure 10.6 is the same as that used to create Figure 10.5, but with a different choice of initial phases. Once again, the closest point is $f_0^{[0,0]}$; nonetheless, the trajectory is quite different. The light blue star in the upper right-hand side of Figure 10.6(b) is the projection of $f_0^{[0,0]}$. The reconstructions appear to lie on an almost planar, two-dimensional surface, whereas the iterates seem to lie on a curve lying on a surface of revolution in three-dimensional space. The errors are about 10^{-1} and the residuals are about 10^{-6}, indicating that the iterates in this experiment are also entrained to a local minimum of $d_{\mathbb{A}_a B_{S_3}}$.

The plots shown in Figure 10.7 come from the same object as that used in the previous two experiments. This time, the closest exact intersection is $f_0^{[-2,-2]}$, for which $\dim T_{f_0^{[0,0]}} \mathbb{A}_a \cap B_{S_3} = 2$. The iterates and reconstructions seem to eventually lie very close to lines. The projection of the exact intersection is shown as a light blue star in the upper right of Figure 10.7(b). The projections of the reconstructions, in this three-dimensional space, seem to be converging to this point. Indeed, the true error is also decreasing, though at a very slow rate.

In Figure 10.8 we use an object with $k = 2$ and a 1-pixel neighborhood. The nearest exact intersection is $f_0^{[0,0]}$, for which $\dim T_{f_0^{[0,0]}} \mathbb{A}_a \cap B_{S_1} = 22$. The reconstructions appear to be essentially a projection of the iterates onto the plane where $x = y$. The errors, which are constant, are about 10^{-2} and the residuals, which are decreasing, are about 10^{-10} by the end of the experiment. This again indicates that the iterates in this experiment are entrained to a local minimum of $d_{\mathbb{A}_a B_{S_3}}$.

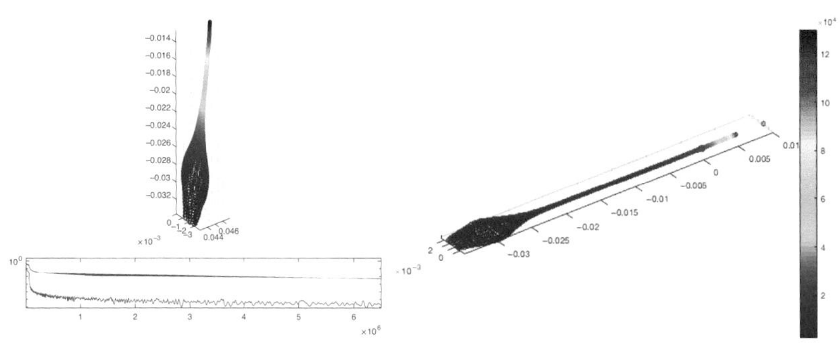

(a) 3-principal components of iterates 1,500,000 to 6,500,000.

(b) 3-principal components of reconstructions 1,500,000 to 6,500,000.

Figure 10.7 Example with $k = 0,\ p = 3$, with closest exact intersection $\boldsymbol{f}_0^{[-2,-2]}$.

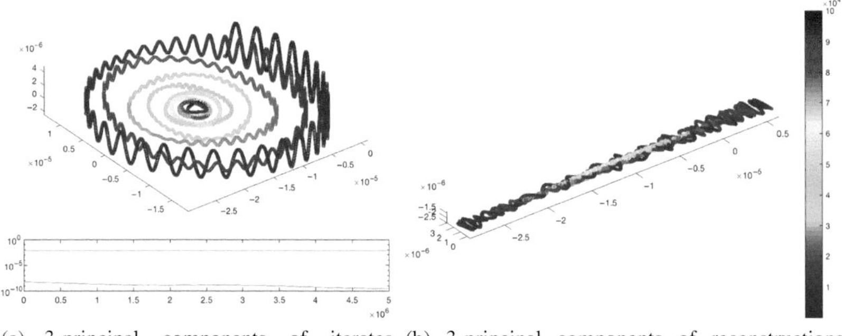

(a) 3-principal components of iterates 1,500,000 to 6,500,000.

(b) 3-principal components of reconstructions 1,500,000 to 6,500,000.

Figure 10.8 Example with $k = 2,\ p = 1$, with closest exact intersection $\boldsymbol{f}_0^{[0,0]}$.

It is quite remarkable how regular the trajectories in these four examples appear (at least in this projection view). Despite the fact that the residual is nearly constant throughout the last 5,000,000 iterates, the variability in these projected trajectories appears to diminish significantly toward the ends of these experiments. A likely explanation for this is that the directions of principal variation for the latter iterates are quite different from those found using iterates 1,000,000 to 1,500,000.

In Figure 10.9 we use an object with $k = 4$ and a 3-pixel neighborhood. The nearest, exact intersection is $\boldsymbol{f}_0^{[-1,-1]}$. A calculation shows that the dimension of the intersection, $\dim T_{\boldsymbol{f}_0^{[0,0]}}\mathbb{A}_{\boldsymbol{a}} \cap B_{S_1}$, is around 40. The errors are about 10^{-1} and the residuals are about 10^{-2}. This would appear to be a situation where

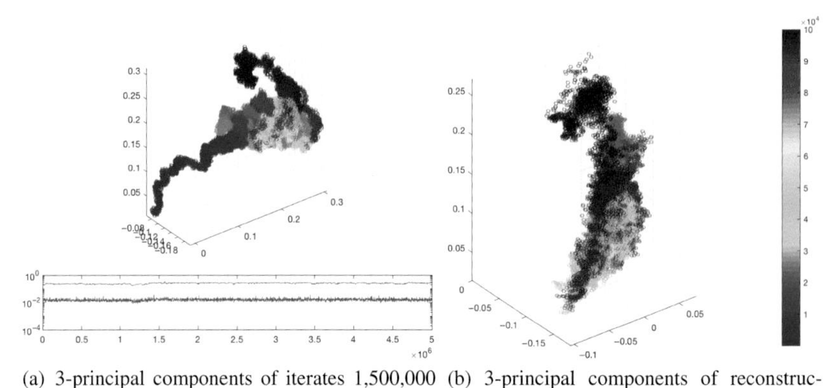

(a) 3-principal components of iterates 1,500,000 to 6,500,000.

(b) 3-principal components of reconstructions 1,500,000 to 6,500,000.

Figure 10.9 Example with $k = 4,\ p = 3$, with closest exact intersection $\boldsymbol{f}_0^{[-1,-1]}$.

(a) 3-principal components of iterates 1,500,000 to 6,500,000.

(b) 3-principal components of reconstructions 1,500,000 to 6,500,000.

Figure 10.10 Example with $k = 6$, where we have used the nonnegativity constraint to define the algorithm.

the iterates have stagnated and are in an attracting basin defined by several nearby trivial associates. The utter lack of structure in these trajectories is quite a striking departure from the previous examples.

In the final example, which is shown in Figure 10.10, we have used the nonnegativity constraint, instead of support, to define the hybrid iterative map. This trajectory appears to be confined to a smooth curve. Its projection into B_+ appears to lie, more or less, along a line, but this may be an artifact of how the axes are scaled. The closest exact intersection is shown in Figure 10.10(b) as a light blue star in the upper right. The errors are very close to 10^{-1} whereas the residuals are about 10^{-7}, again indicating entrainment to a local minimum of $d_{\mathbb{A}_a B_+}$.

The figures shown in these experiments are just a tiny sample of the different sorts of behavior that we have observed. There seems to be a dichotomy between highly structured trajectories, and random trajectories which is highly correlated with the size of the residual, which, in turn seems to be related to whether the iterates have found a common tangent direction defined by a single, nearest intersection point, or if the iterates are being pulled by a collection of nearby trivial associates. While nothing about these experiments gives clues as to how better reconstructions might be obtained, they provide powerful evidence of the rich structure hidden in the dynamics of hybrid iterative maps.

PART III

Further Properties of Hybrid Iterative Algorithms and Suggestions for Improvement

11
Introduction to Part III

In the analysis and numerical examples considered thus far, we have shown that the problem of phase retrieval is bedeviled by the failure of the magnitude torus $\mathbb{A}_a$ and the subset B, defined by the auxiliary information, to meet transversally. This, along with small angles between the tangent spaces to $\mathbb{A}_a$ and B, at intersection points, often render the center manifold weakly attracting in a high-dimensional subspace of normal directions. These facts, coupled with the existence of multiple attracting basins caused by the existence of trivial associates, and critical points of $d_{\mathbb{A}_a B}$ appear to make it very difficult for standard iterative algorithms to converge, *to machine precision,* in all but the most favorable circumstances.

From our analysis, these favorable circumstances arise when the object has a fairly hard outer edge (its interior can be smooth), and either the support is known very precisely or the object is known to be nonnegative. In the absence of these conditions, algorithms can be expected to stagnate in proximity to an exact intersection, but not approach it very closely, or to converge at an extremely slow rate. In this part of the book, we examine several proposals for modifying the measurement process to break the symmetries that lead to nontransversality and the existence of trivial associates. In addition, we also examine the statistical properties of the stochastic process obtained by starting with a random choice of phases and running a hybrid iterative map-based algorithm for many iterates. Averaging such an ensemble of reconstructions is very often a step used in practical reconstruction algorithms. Our experiments indicate that this can produce small improvements in the reconstructed image, but only if done judiciously.

We begin by considering the statistics of the hybrid iterative maps, which are themselves quite striking. We then consider two approaches to breaking the symmetries inherent in this problem. For the first we imagine that we have a soft object, represented as samples of a relatively smooth function. Even with a very precise estimate of the support of such an object we have seen that standard iterative algorithms fail to reconstruct it from magnitude discrete Fourier transform (DFT) data. Here we examine what happens if the object is cut off sharply along a closed curve. If the location of this curve is accurately known, or the object is nonnegative, then it can be very accurately reconstructed using hybrid iterative map algorithms (consistent with our earlier experiments).

We also consider the effect of placing a hard object in a position external to the object of interest. This is sometimes called external "holography," and it has been explored elsewhere (see Maretzke and Hohage 2016; Barmherzig et al. 2019a, 2019b; Jacobsen 2019). Depending upon where the hard object is placed there are three different reconstruction methods. The distances are calibrated by the diameter, d, of the support of the object of interest.

(i) If the hard object is very far (more than $3d$) from the object one is attempting to image, then one can simply use the inverse Fourier transform to reconstruct the autocorrelation of the sum of the two objects. The autocorrelation function contains two isolated copies of the object of interest convolved with the hard external object (Maretzke and Hohage 2016; Jacobsen 2019). In this approach to holography, it is not necessary to know the location of the external object.

(ii) If the hard object is placed at about $d/2$ from the object of interest, then, provided the shape and location of the external object are both precisely known, a standard hybrid iterative algorithm works very well to reconstruct the object of interest (Fienup 1987). If only the shape is known, then the standard hybrid iterative algorithm can be modified to simultaneously determine the location of the external object and reconstruct the image.

(iii) If a known hard object is placed very close to the object of interest, then the noniterative holographic Hilbert transform (HHT) method, see Section 7.4, works well to reconstruct the object of interest.

The HHT method is a limiting case of external holography: the hard object is very small and placed very close to the object of interest. In Section 13.4 we discuss the implementation of the HHT method introduced in Section 7.4 and give the results of a variety of numerical experiments. From these experiments it is evident that there are circumstances where this method may be applied, so long as a small, hard object of known size and shape can be created and placed near the unknown object. A phase retrieval algorithm using the

Hilbert transform was explored in a series of papers by Nakajima and Asakura (see Nakajima and Asakura 1985, 1986; Nakajima 1995). They did not try to couple this approach to reconstruction with a holographic measurement protocol, and the method was not too successful.

Remark 11.1 As in earlier parts of the book (initially stated in Remark 1.10), the support of an image, $\boldsymbol{f}$, used in a numerical example in this part is defined to be the set $\{\boldsymbol{j} : |f_{\boldsymbol{j}}| \geq 10^{-14}\}$.

12
Statistics of Algorithms

In the bulk of this book we consider discrete images parameterized by vectors $\boldsymbol{f} \in \mathbb{R}^J$. The problem of phase retrieval is phrased as that of finding points in the intersection $\mathbb{A}_{\boldsymbol{a}} \cap B$, where $\mathbb{A}_{\boldsymbol{a}}$ is the magnitude torus defined by (idealized) measurements and $B \subset \mathbb{R}^J$ is a set defined by "auxiliary" information. Most of the algorithms we consider for solving this problem are defined by iterating a map $F : \mathbb{R}^J \to \mathbb{R}^J$, and looking for fixed points. We have seen that whether or not such fixed points can be accurately found, depends largely on the geometry near points $\boldsymbol{f} \in \mathbb{A}_{\boldsymbol{a}} \cap B$; in particular: is such an intersection transversal and are there small angles between $T_{\boldsymbol{f}}\mathbb{A}_{\boldsymbol{a}}$ and B? We have shown that transverse intersections are unusual, at least when using a support constraint, and have argued that, for the most part, these algorithms stagnate, and display several distinct kinds of nonconvergent behavior. The smooth examples ($k > 0$) in this chapter are defined by convolving a piecewise constant image with a Gaussian. As noted in Section 2.2 this inevitably leads to nontransversal intersections between $\mathbb{A}_{\boldsymbol{a}}$ and B_S. It is also important to recall that while a larger value of k correspond to smoothing with a wider Gaussians, the value of k does not correspond, in any simple way, to a measure of ordinary differentiability.

In this chapter, we consider the statistical properties of these sorts of algorithms in cases where they appear to stagnate. To that end, we fix a number of iterates, m, and consider the statistical properties of sets of the form

$$F^m(\mathscr{R}) = \{F^m(\boldsymbol{x}) : \boldsymbol{x} \in \mathscr{R}\}, \tag{12.1}$$

where $\mathscr{R} \subset \mathbb{A}_{\boldsymbol{a}}$ is a "uniformly distributed" subset of the magnitude torus. Recall that the iterative algorithms we consider are initialized by randomly assigning a phase to each discrete Fourier transform (DFT) coefficient. More precisely, the initial phases $\{e^{i\theta_{\boldsymbol{k}}} : \boldsymbol{k} \in J\}$ are independent random variables, each uniformly distributed on the unit circle. A uniformly distributed subset $\mathscr{R}$ arises by independently making many such initial choices. Various statistics

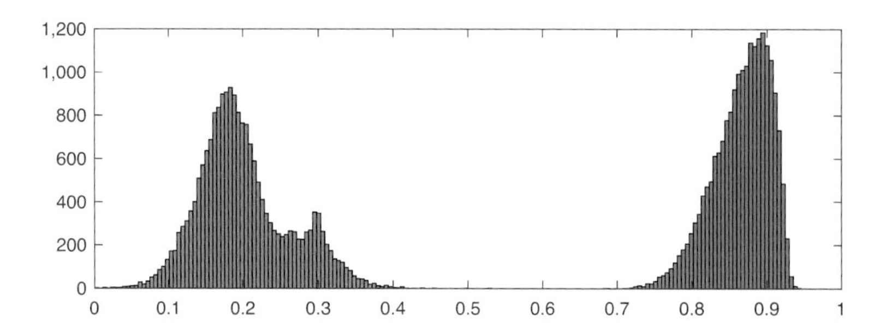

Figure 12.1 Histograms of the distances to the set of trivial associates of a reference image, before (right) and after (left) applying $D^{20000}_{\mathbb{A}_a B_{S_1}}$ to 20,000 random points on $\mathbb{A}_a$.

associated to $F^m(\mathscr{R})$ give further evidence that, even though they are non-convergent, the trajectories of these maps fall into several distinct classes. Figure 12.1 illustrates this idea in a simple example. For this experiment $\boldsymbol{f}_{\text{ref}}$ is a 64×64 image of ℓ^2-norm 1, with $k = 2$, and $\mathscr{R}$ consists of 20,000 randomly selected points on the magnitude torus defined by $\boldsymbol{f}_{\text{ref}}$. The figure shows two histograms, the one on the right is the distribution of ℓ^2-distances from the points in $\mathscr{R}$ to the set of trivial associates of $\boldsymbol{f}_{\text{ref}}$ (the initial errors), while that on the left is the same statistic, but for the set of points $\{P_{\mathbb{A}_a} \circ R_{B_{S_1}} \circ D^{20000}_{\mathbb{A}_a B_{S_1}}(\boldsymbol{x}) : \boldsymbol{x} \in \mathscr{R}\}$, the errors after 20,000 iterates. Evidently the set of trivial associates of $\boldsymbol{f}_{\text{ref}}$ defines an attracting set for $D^{20000}_{\mathbb{A}_a B_{S_1}}$, but it is also clear that there are several classes of trajectories and few have gotten very close to this attracting set.

Given that these iterations rarely converge, a question of clear import is whether (or when) the ensemble $F^m(\mathscr{R})$ contains more information about the unknown phases than a single trajectory, and how to identify and extract that information. The second question we consider is whether some sort of averaging procedure leads to improved reconstructions. Many reconstruction methods described in the literature employ an averaging step.

The set $\mathscr{R}$ is a finite sample of the probability space defined by the normalized product measure,

$$d\mu_{\mathbb{T}^N} = \frac{d\theta_1 \cdots d\theta_N}{(2\pi)^N},$$

on the torus $\mathbb{T}^N = S^1 \times \cdots \times S^1$, with the angles $-\pi < \theta_j \leq \pi$. We use these angles to define a point on $\mathbb{A}_a$:

$$(\theta_1, \ldots, \theta_N) \mapsto (a_1 e^{i\theta_1}, \ldots, a_N e^{i\theta_N}). \tag{12.2}$$

As noted earlier, we use the term *phase* to refer to points on the unit circle, i.e., points of the form $e^{i\theta}$, with $\theta \in \mathbb{R}$; in the Engineering literature these are often called phasors. With this convention, the phase of a DFT coefficient is well defined, even though the angle, θ, is only defined mod 2π. As the volume of the magnitude torus is $(2\pi)^N \prod_{i=1}^N a_i$, the natural probability measure on the magnitude torus is $d\mu_{\mathbb{T}^N}$ as well.

Pushing this measure forward with the map F^m defines a new probability measure on the torus. For a large m this push-forward measure can be expected to encode properties of the asymptotic dynamics of the map F. We do not attempt to do a theoretical analysis of the properties of this measure, but instead study the empirical statistics of the sets $F^m(\mathscr{R})$ defined by four different images, with different levels of smoothness, and/or different sized support neighborhoods. In all cases $|\mathscr{R}| = 20,000$, $F = D_{\mathbb{A}_a B_{S_p}}$, and we take $m = 20,000$. The four examples we study show some striking phenomena that suggest that a theory for $F^m_* d\mu_{\mathbb{T}^N}$ (or perhaps $\lim_{m\to\infty} F^m_* d\mu_{\mathbb{T}^N}$) may well exist and be accessible to the sufficiently determined researcher.

12.1 Statistics of Phases

As noted above, we represent the phases in question as points on the unit circle; it is well known that there is no globally defined, reasonable notion of a continuous circle-valued mean, i.e., for an $n > 1$ there does not exist a continuous map

$$A : [S^1]^n \to S^1,$$

such that

(i) $A(e^{i\theta}, \ldots, e^{i\theta}) = e^{i\theta}$.
(ii) The value of $A(e^{i\theta_1}, \ldots, e^{i\theta_n})$ is unchanged if the arguments are permuted.

(See Weinberger 2004.) There are, nonetheless, two obvious choices for defining an empirical mean: Let $\Theta = \{e^{i\theta^{(q)}} : 1 \le q \le Q\}$ be a collection of points on the unit circle. The mean can be defined as

$$e^{i\bar{\theta}} = \frac{\sum_{q=1}^{Q} e^{i\theta^{(q)}}}{\| \sum_{q=1}^{Q} e^{i\theta^{(q)}} \|_2}, \tag{12.3}$$

when the sum in the denominator does not vanish. If we normalize $\theta^{(q)}$ to lie in $(-\pi, \pi]$, then we can also define the mean angle to be

$$\hat{\theta} = \frac{1}{Q} \sum_{j=1}^{Q} \theta^{(q)} \mod (-\pi, \pi]. \tag{12.4}$$

Neither definition defines a continuous map from $[S^1]^Q \to S^1$: if the points in Θ are very uniformly distributed on S^1, then the denominator in (12.3) will be very close to 0 and the point, $e^{i\overline{\theta}}$, is therefore essentially random; whereas $\hat{\theta}$ will be very close to 0 and so $e^{i\hat{\theta}} \approx 1$. Neither definition works well in this case, but if the angles are largely clustered near to a single value, then both approaches reliably cluster near that value.

For most of the experiments in this section we use the definition in (12.3) to compute the mean. There is an invariant definition of an empirical variance, for points on S^1 given by

$$\sigma_{S^1}^2 = \frac{1}{2N(N-1)} \sum_{q,p=1}^{Q} \text{dist}_{S^1}^2 (e^{i\theta^{(q)}}, e^{i\theta^{(p)}}). \tag{12.5}$$

Here

$$\text{dist}_{S^1}(e^{i\theta^{(q)}}, e^{i\theta^{(p)}}) = \min\{|\theta^{(q)} - \theta^{(p)}|, 2\pi - |\theta^{(q)} - \theta^{(p)}|\},$$

is the geodesic distance on the circle. Unfortunately, $\sigma_{S^1}^2$ is too computationally demanding for the examples in this section and we therefore use the traditional formula

$$\sigma^2 = \frac{1}{2N(N-1)} \sum_{q,p=1}^{Q} |\theta^{(q)} - \theta^{(p)}|^2. \tag{12.6}$$

Of course, $\sigma_{S^1}^2 \leq \sigma^2$, and for points concentrated in an interval of length π, properly centered, these two formulæ give the same answer. For points uniformly distributed on S^1, $\sigma^2 \approx 2\sigma_{S^1}^2$. A circumstance where $\sigma_{S^1}^2$ is small, but σ^2 is large can arise when the points on S^1 are uniformly clustered near to -1.

In our applications the phases that arise are phase ratios, that result from angular differences

$$e^{i\Delta\theta_{\boldsymbol{k}}^{(q,m)}} = e^{i\theta_{\boldsymbol{k}}^{(q,m)}} e^{-i\theta_{\boldsymbol{k}}^{\text{ref}}}, \tag{12.7}$$

where $e^{i\theta_{\boldsymbol{k}}^{(q,m)}}$ is the phase of the $\boldsymbol{k}$th DFT coefficient found by iterating a map m-times starting with initial phases $\{e^{i\theta_{\boldsymbol{k}}^{(q)}} : \boldsymbol{k} \in J\}$, and $e^{i\theta_{\boldsymbol{k}}^{\text{ref}}}$ is the phase of the $\boldsymbol{k}$th DFT coefficient of a reference image. Following standard practice in the imaging literature, we often refer to the angular differences $\{\Delta\theta_{\boldsymbol{k}}^{(q,m)}\}$, as *phase errors*. The overall sign of each reconstructed image is selected to largely agree with the sign of the reference image. At least for larger magnitude DFT coefficients, this makes it unlikely for $e^{i\Delta\theta_k^{(q,m)}} \approx -1$. With this in mind it is reasonable to expect that a small value for the variance predicts phases that have strongly clustered around a single value, and a large variance is indicative of rather uniform distribution of the phase errors. In fact, we will see that in regions of the J-grid where the variance is large, the mean phase strongly resembles a uniformly distributed random variable, as predicted by the discussion above.

By examining the empirical mean and variance of the phase error, one can see rather clearly that the phases of certain DFT coefficients are determined with a high degree of certainty by iterative maps, whereas essentially no information is available about the phases of others. Unsurprisingly, the phases of coefficients with larger magnitudes are determined with much greater certainty than those with smaller magnitudes. What is somewhat surprising is that the transition between those with small variance and those with large variance is quite abrupt.

12.2 Statistics of Ensembles

For the experiments in this section, we choose a random image, $\boldsymbol{x}_0$, of a reasonably small size and compute its DFT magnitude data, $\boldsymbol{a}$. The images in this section are usually 64×64, as it is very time consuming to run tens of thousands of trials of a hybrid iterative map algorithm for several thousand iterates each on a larger image. The set $\mathscr{R}$ is defined by choosing Q random points $\{\boldsymbol{x}^{(q)} \in \mathbb{A}_{\boldsymbol{a}} : q = 1, \ldots, Q\}$. The dimension of the torus $\mathbb{A}_{\boldsymbol{a}}$ is 2,046, so for practical values of Q, say between 10^4 and 10^5, the set $\mathscr{R}$ remains a very sparse sample of $\mathbb{A}_{\boldsymbol{a}}$. As a point of comparison, there are $2^{2046} \approx 10^{616}$ points on the torus with phases $e^{i\theta_k} \in \{1, -1\}$.

Nonetheless the behavior we observe for ensembles in the size range $10^4 - 10^5$ is very suggestive, reasonable, and reproducible. In Figure 12.2 we see the distribution of distances to a fixed reference point for 40,000 randomly selected points on a magnitude torus defined by a 128×128 image. These look normally distributed. While it is difficult to compute the expected distance of a randomly selected point to a fixed reference point, $\boldsymbol{f}_{\text{ref}}$, it is easy to compute the expected value of $d(\boldsymbol{f}_{\text{ref}}, \boldsymbol{f})^2$, where the distance is measured along the torus. As a reference point we use $\boldsymbol{f}_{\text{ref}}$, for which $\widehat{f}_{\text{ref},\boldsymbol{k}} \in [0, \infty)$, that is, we set all the phases to 1, or the angles to 0. If the radii of the circles are $\{r_1, \ldots, r_N\}$, then

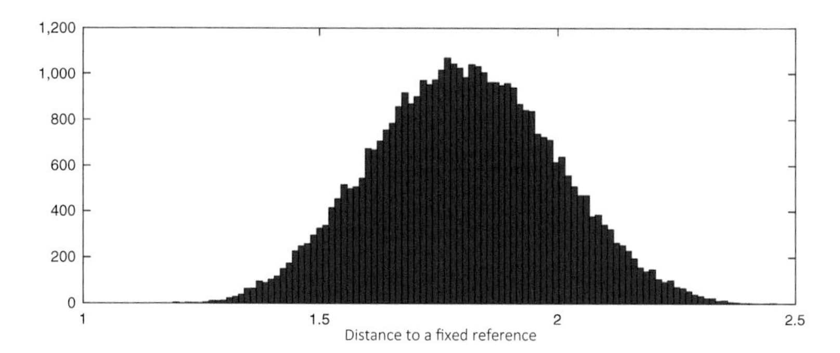

Figure 12.2 Histogram of the distances to a fixed reference point for 40,000 randomly selected points on a magnitude torus $\mathbb{A}_{\boldsymbol{a}} \subset \mathbb{R}^{128\times 128}$.

$$\text{mean}[d(\boldsymbol{f}_{\text{ref}}, \boldsymbol{f})^2] = \int_{-\pi}^{\pi} \cdots \int_{-\pi}^{\pi} \left[\sum_{i=1}^{N} r_i^2 \theta_i^2\right] \frac{d\theta_1 \cdots d\theta_N}{(2\pi)^N} = \frac{\pi^2}{3} \sum_{i=1}^{N} r_i^2. \tag{12.8}$$

Similarly the variance of the $d(\boldsymbol{f}_{\text{ref}}, \boldsymbol{f})^2$ can be calculated; it is

$$\text{var}[d(\boldsymbol{f}_{\text{ref}}, \boldsymbol{f})^2] = \frac{8\pi^4}{45} \left[\sum_{i=1}^{N} r_i^4\right], \tag{12.9}$$

which is also readily computed for a specific example.

For the example used to produce Figure 12.2, we take a torus so that its image under the (unitary) DFT is a product of circles with radii $\{r_i\}$, satisfying $\sum_{i=1}^{N} r_i^2 = 1$. The empirical mean of the squared distance is $0.9995\frac{\pi^2}{3}$, which is a very good approximation to the correct answer. The predicted variance is 0.389 and the empirical variance is 0.401, which is, again, a good estimate. These errors in the mean and variance agree with the expected size of the error of about $1/\sqrt{Q}$, which is typical for Monte Carlo integration using Q points. The example indicates that the squared distance, for even a relatively small sample obtained using the random number generator in MATLAB, has the correct first and second moments for a uniformly distributed sample.

Remark 12.1 There is a similar discussion of the statistics of reconstructed phases in Chapman et al. (2006).

12.2.1 Errors and Residuals

In the experiments that follow, we use the hybrid iterative maps $D_{\mathbb{A}_{\boldsymbol{a}} B_{S_p}}$, employing a p-pixel support constraint. We let $\boldsymbol{x}^{(q,n)}$ denote the nth iterate of this map starting at $\boldsymbol{x}^{(q)}$ and let $\boldsymbol{r}^{(q,n)} = P_{\mathbb{A}_{\boldsymbol{a}}} \circ R_{B_{S_p}}(\boldsymbol{x}^{(q,n)})$ be a projection of this iterate onto the magnitude torus. The DFT coefficients of $\boldsymbol{r}^{(q,n)}$ are given by $\hat{r}_{\boldsymbol{k}}^{(q,n)} = a_{\boldsymbol{k}} e^{i\theta_{\boldsymbol{k}}^{(q,n)}}$, where we normalize so that $\theta_{\boldsymbol{k}}^{(q,n)} \in (-\pi, \pi]$.

Recall that we define the residual at the $(n+1)$st iterate to be

$$\text{resid}^{(q,n+1)} = \frac{\|P_{\mathbb{A}_{\boldsymbol{a}}} \circ R_{B_{S_p}}(\boldsymbol{x}^{(q,n)}) - P_{B_{S_p}}(\boldsymbol{x}^{(q,n)})\|_2}{\|\boldsymbol{x}_0\|_2} = \frac{\|\boldsymbol{x}^{(q,n+1)} - \boldsymbol{x}^{(q,n)}\|_2}{\|\boldsymbol{x}_0\|_2}; \tag{12.10}$$

it is a measure of the failure of $\boldsymbol{r}^{(q,n)}$ to belong to $\mathbb{A}_{\boldsymbol{a}} \cap B_{S_p}$, as well as measure of the extent to which the algorithm has converged. The true error is the distance to closest trivial associate of the reference image:

$$\text{err}^{(q,n)} = \min\left\{ \frac{\|\boldsymbol{x}^{(q,n)} - \boldsymbol{x}_{\text{ref}}^{(\boldsymbol{v})}\|_2}{\|\boldsymbol{x}_0\|_2}, \frac{\|\boldsymbol{x}^{(q,n)} - \check{\boldsymbol{x}}_{\text{ref}}^{(\boldsymbol{v})}\|_2}{\|\boldsymbol{x}_0\|_2} : \boldsymbol{v} \in J \right\}. \tag{12.11}$$

With synthetic data, both the residual and true error are available. In a real experiment, only the residual is available. The simplest statistic to assess the properties of the sets $D^m_{\mathbb{A}_{\boldsymbol{a}} B_{S_p}}(\mathscr{R})$, and thereby the statistical properties of the algorithm, is the distribution of the values $\{\text{resid}^{(q,m)} : q = 1, \ldots, Q\}$. The utility of this quantity as a measure of success can be explored by considering the correlation between this set and the true errors, $\{\text{err}^{(q,m)} : q = 1, \ldots, Q\}$.

Figure 12.3 shows histograms of the $\log_{10}$ of the residuals for the 20,000th iterate of $D_{\mathbb{A}_{\boldsymbol{a}} B_{S_p}}$ with 20,000 random starting points for 64×64 images with different smoothness levels, and support neighborhoods: $k = 1, p = 1$; $k = 2, p = 1$; $k = 4, p = 1$; and $k = 4$, and $p = 2$. With this type of data, we have already seen that a hybrid iterative map generally fails to converge. The residuals are not unimodally distributed as might be expected for a convergent algorithm, but rather appear to follow multimodal distributions. This is consistent with the discussion and examples in Chapter 10, which indicate that there are several distinct classes of nonconvergent trajectories for hybrid iterative maps.

In Figure 12.3(a–c) we use a 1-pixel neighborhood, and in (d) a 2-pixel neighborhood. In (a) there would appear to be at least three distinct peaks, each about one order of magnitude apart. It is plausible that some of the gaps in this histogram might fill in with a larger collection of samples, leaving just two peaks about two orders of magnitude apart. In (b) there are four clear peaks and in (c) there are three. In (d) there are two distinct peaks, but the two distributions have significant overlap.

The leftmost peaks likely correspond to trajectories that have found an attracting basin defined by a single exact intersection point, whereas the

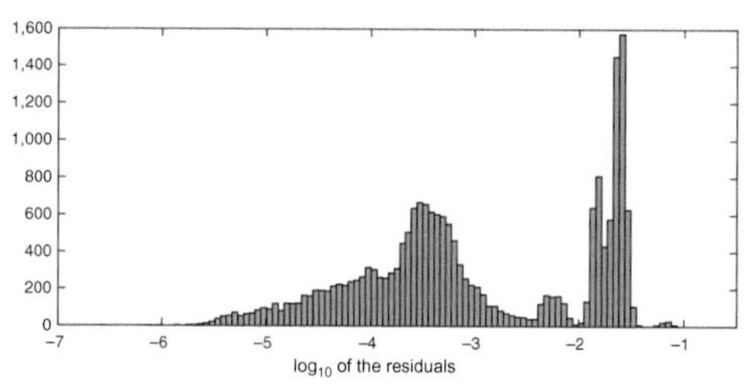

(a) Results using a $k = 1$ image and a 1-pixel ($p = 1$) support neighborhood.

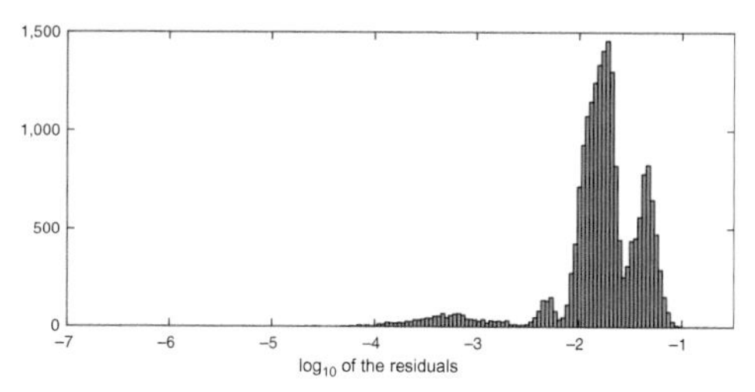

(b) Results using a $k = 2$ image and a 1-pixel ($p = 1$) support neighborhood.

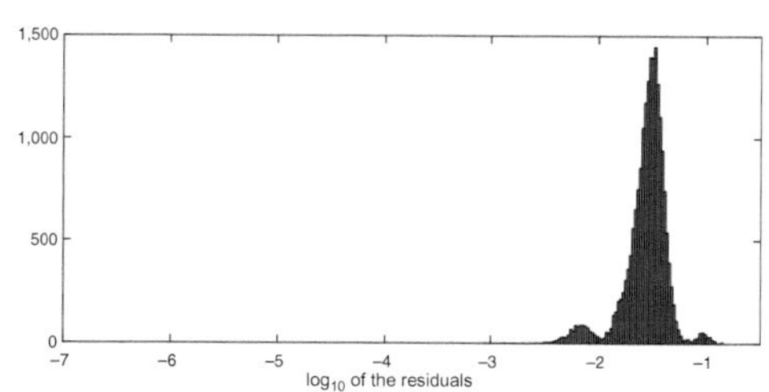

(c) Results using a $k = 4$ image and a 1-pixel ($p = 1$) support neighborhood.

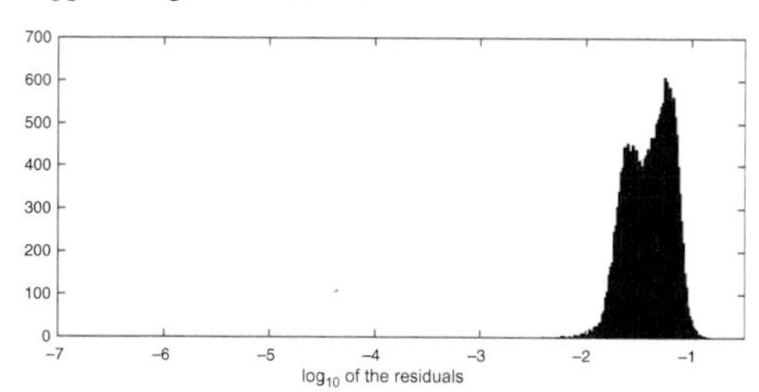

(d) Results using a $k = 4$ image and a 2-pixel ($p = 2$) support neighborhood.

Figure 12.3 Histograms of the $\log_{10}(\text{resid}^{(q,m)})$, after 20,000 iterates of $D_{\mathbb{A}_a B_{S_p}}$ for 20,000 different initial choices of random phases using images with three levels of smoothness and two different support neighborhoods.

peaks to the right may represent the more chaotic behavior caused by the continuing influence of several distinct center manifolds. In Section 12.3.1 we see that, for the purposes of assessing the information available in the ensemble, $D^m_{\mathbb{A}_a B_{S_p}}(\mathscr{R})$, it is better to discard the data from all trajectories, but those unambiguously contributing to the left most peaks. While averaging the results of all the experiments is a bad idea, it is sometimes possible to get an improved reconstruction by averaging reconstructions drawn from trajectories that contribute to the leftmost peak.

Figure 12.4 contains 2D-scatter plots of the $\log_{10}$ of the true errors versus the $\log_{10}$ of the residuals. The correlation between the error and residual is generally high, but the range of patterns evident in these plots shows that the story is rather complex. The residuals do tend to be about the squares of

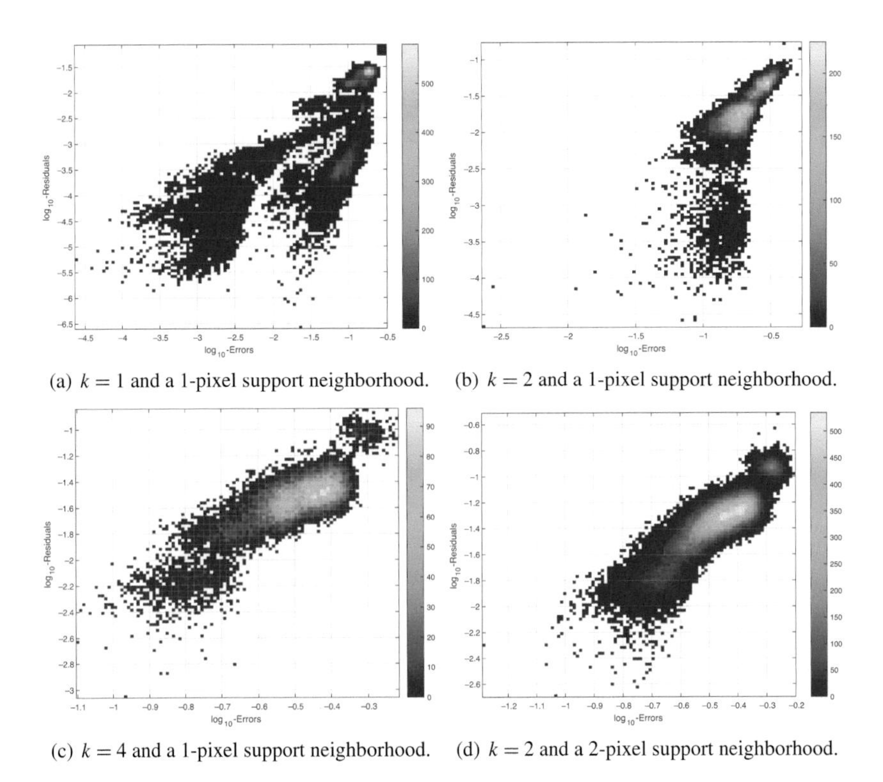

(a) $k = 1$ and a 1-pixel support neighborhood. (b) $k = 2$ and a 1-pixel support neighborhood.

(c) $k = 4$ and a 1-pixel support neighborhood. (d) $k = 2$ and a 2-pixel support neighborhood.

Figure 12.4 Scatter plots of the $\log_{10}(\text{err}^{(q,m)})$ versus $\log_{10}(\text{resid}^{(q,m)})$ after 20,000 iterates for several different initial images and support neighborhoods; here $m = 20,000$. The density of points in these plots is indicated by the individual color bars.

the errors, which is predicted by the observation that the iterates of hybrid maps (eventually) tend to be confined to directions in $T_{f^{(v)}_{\rm ref}}\mathbb{A} \cap B_{S_p}$, see Example 8.12. However, there are also distinct clusters where there are other relationships, or no relationship between these two quantities.

In Figure 12.4(a), which arises from the 1-pixel neighborhood of the $k = 1$ image, there appear to be at least four distinct families of trajectories: There is a large group of trajectories in the upper right corner that have large residuals and large errors. These trajectories may not yet have settled into common tangent directions. Below these are three clusters each with a different linear relation between the $\log_{10}$-data. There is a small cluster where the slope is nearly 0, then a much more extensive, though sparse collection where the slope is about $\frac{3}{2}$. This group contains the trajectories with the smallest errors, though not those with the smallest residuals. This group might be composed of trajectories that are very slowly converging to an exact intersection. Finally, there is a large cluster where the slope is about $\frac{5}{2}$. Indeed, the trajectory with the smallest residual is in this family, with the residual about $10^{-6.5}$ and the true error about $10^{-1.6}$.

The plot in Figure 12.4(b) arises from a 1-pixel neighborhood of the $k = 2$ image. There is a large cluster (with two peaks) in the upper right corner where the residuals are about the squares of the errors, but the errors are larger than 10^{-1}. This is clearly the "typical" behavior and indicates stagnation quite far from a true intersection. Below this cluster is a smaller, rather diffuse, cluster where the errors and residuals are not well correlated. The errors are between $10^{-0.5}$ and 10^{-1}, whereas the residuals range from $10^{-2.7}$ to $10^{-4.5}$. These trajectories appear to have found normal directions with supports nearly contained in S_1 that carry the image much closer to the torus than would be possible along a tangent direction alone. This is a possibility discussed in Section 10.1. In the lower left there are a few trajectories that may also be slowly converging. This group contains the trajectory with the smallest residual, about $10^{-4.6}$, which also has the smallest true error, about $10^{-2.6}$.

In both Figures 12.4(c) ($k = 4, p = 1$) and (d) ($k = 4, p = 2$) there is essentially a single cluster with several regions of higher density. Throughout these clusters the residuals are strongly correlated to the squares of the errors. In (c) the image with the smallest residual, about $10^{-3.1}$, has almost the smallest error, about $10^{-0.97}$. There is an image with residual $10^{-2.63}$ and error $10^{-1.09}$. The plot in (d) tells a similar story. Overall, we see that the residual is a useful predictor of the error, though it is sometimes much smaller than the square of the error, and often the minimum residual image is not the minimum error image. Indeed, in all four plots we see that the group of images with very small residuals have errors that span a rather substantial range. Nonetheless, in the experiments that follow we often use the minimum residual image as the reference image. As we will see, from a statistical perspective, using this image as the reference produces almost the same phase error statistics as we would obtain if we were to use the true image itself.

12.2.2 The Mean and Standard Deviation of the Phase Errors

We next consider the mean and standard deviation of the phase errors produced by running the algorithm $D_{\mathbb{A}_a B_{S_p}}$ on an ensemble, $\mathscr{R}$, of uniformly sampled initial phases. As noted above, the phase errors lie on the unit circle; for such a random variable there does not exist a notion of mean that is continuous and satisfies other desirable properties (Weinberger 2004). Nonetheless, the empirical mean defined in (12.3), provides useful information, especially when the empirical standard deviation, whose square is defined in (12.6), is small.

The notion of "phase error" presupposes a reference image. A reason that we need such a reference is that the different reconstructions differ mostly by rigid translations. Therefore, to compare the phases of individual DFT coefficients in a collection of reconstructed images defined by the set of iterates, $\{D^m_{\mathbb{A}_a B_{S_p}}(\mathscr{R})\}$, we first need to register the images to a reference image. It also makes it possible to make a good choice of overall sign, that is between an image $\boldsymbol{f}$ and $-\boldsymbol{f}$. This discourages phases errors near to -1, where the empirical mean of the angles, that is (12.4), is discontinuous. For this purpose, the minimum residual image provides a reasonable choice. As we see, the mean errors do not change qualitatively if we replace the minimum residual image with the exact reconstruction as the reference image, and the two computations of the standard deviation are nearly identical. Moreover, since the mean is not a continuous operation (at -1 in our normalization) it is desirable to make the quantities being averaged lie as close to 1 as possible. Again, for phases with small standard deviation this is facilitated by using a reasonable choice of reference image to define phase errors, rather than working with the phases themselves.

Our experiments consider the per-frequency means and standard deviations computed from the 20,000 trials, where each trial uses 20,000 iterates of the maps $D_{\mathbb{A}_a B_{S_p}}$ acting on the four 64×64 images used above. These are the same images and trials used in Section 12.2.1. In this experiment we compare the per-frequency statistics obtained by registering the reconstruction to the exact reference image versus those obtained by registering to the minimum residual image. Let $\{\boldsymbol{r}^{(q)} : 1 \leq q \leq 20{,}000\}$, denote the reconstructions obtained after the 20,000th iterate, and $\{\hat{r}_{\boldsymbol{k}}^{(q)} : 1 \leq q \leq 20,000,\, \boldsymbol{k} \in J\}$, their DFT coefficients. If $\boldsymbol{f}^{\text{ref}}$ denotes a reference image, and $\{\widehat{f}_{\boldsymbol{k}}^{\text{ref}}\}$, its DFT coefficients, then the phase errors for trial q used in the statistical analysis are defined to be

$$e^{i\Delta\theta_{\boldsymbol{k}}^{(q)}} = \frac{\hat{r}_{\boldsymbol{k}}^{(q)}\overline{\widehat{f}_{\boldsymbol{k}}^{\text{ref}}}}{|\hat{r}_{\boldsymbol{k}}^{(q)}\widehat{f}_{\boldsymbol{k}}^{\text{ref}}|} \quad \text{for } \boldsymbol{k} \in J. \tag{12.12}$$

As noted, the mean of $\{e^{i\Delta\theta_{k}^{(q)}} : 1 \leq q \leq 20,000\}$ is computed using (12.3) and the square of the standard deviation using (12.6). That is, these statistics are computed using the angles $\{\Delta\theta_{\boldsymbol{k}}^{(q)}\}$; we, nonetheless, continue to refer to them as phase errors. These angles lie in the range $(-\pi,\pi]$, which determines the scales in these plots.

The first rows of subplots (that is (a–d)), in Figure 12.5 show the standard deviation and the second rows show the mean, with $\boldsymbol{f}^{\text{ref}}$ the exact reconstruction in the left columns and $\boldsymbol{f}^{\text{ref}}$ the minimum residual image found among the 20,000 trials in the right columns. The most striking features of these plots are:

(i) the rapid transition, in all cases, from a very low standard deviation to essentially the maximal standard deviation;
(ii) the very small effect of the choice of reference on the standard deviation images; and
(iii) the strong qualitative similarities of the mean error images between the two choices of reference.

One would not expect the mean errors to be pointwise independent of the choice of reference, but we do see a large overlap in the regions where the mean phase errors are close to 0. These largely coincide with the region where the variance is very small.

From these experiments we see that there is a well-defined collection of DFT coefficients whose phases are reliably determined, in an averaged sense, by running a hybrid iterative map many times. These are the coefficients for which the phase error standard deviation is small. As expected, this region shrinks with increasing smoothness (compare (a) to (b) to (c)), as well as with a less precise support constraint (compare (c) to (d)). In all cases this region can be reliably identified *without* knowledge of the exact reconstruction. Note that the range of the mean errors in (a) is much smaller than in the other three plots.

From the plots in Figure 12.5 it is evident that the phase retrieval algorithm does a much better job, at least in a statistical sense, with DFT coefficients corresponding to frequencies near to the center of $\boldsymbol{k}$-space. These frequencies tend to have larger DFT coefficients; indeed, one can experimentally show that

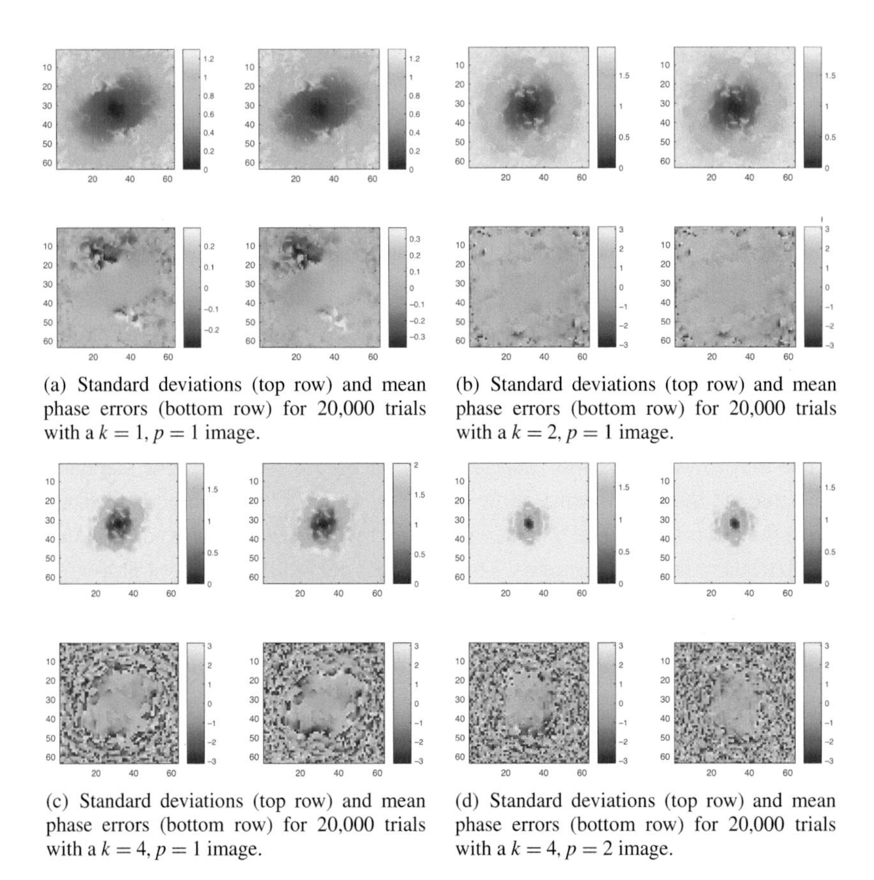

(a) Standard deviations (top row) and mean phase errors (bottom row) for 20,000 trials with a $k = 1, p = 1$ image.

(b) Standard deviations (top row) and mean phase errors (bottom row) for 20,000 trials with a $k = 2, p = 1$ image.

(c) Standard deviations (top row) and mean phase errors (bottom row) for 20,000 trials with a $k = 4, p = 1$ image.

(d) Standard deviations (top row) and mean phase errors (bottom row) for 20,000 trials with a $k = 4, p = 2$ image.

Figure 12.5 Comparison of the per-pixel standard deviations and means of the phase errors computed using different reference images. In each subfigure the results using the exact reconstruction appear on the left, and those using the minimum residual reconstruction on the right. All images are centered on $\boldsymbol{k} = (0, 0)$. The scales shown in the color bars are in radians.

there is strong relationship between the magnitude of a DFT coefficient and how well its phase can be determined. Above a certain threshold the phase can be reliably determined, while below, it cannot be.

The plots in Figure 12.5 give the mean and standard deviation when all 20,000 trials are considered. As we observed in Section 12.2.1, the residuals cluster the trajectories into several different classes. This and the correlation between errors and residuals suggests that one perform a similar analysis for the cluster of images produced by the smaller residual trajectories. In Section 12.3 we consider how this affects the statistical properties of the phase errors.

12.3 Averaging to Improve Reconstructions

In applications to practical phase retrieval, a reconstruction algorithm is typically run for many independent trials. A final reconstructed image is then obtained using an average, or weighted average, of the "best" reconstructions found among these independent reconstructions (see Fienup and Wackerman 1986; Chapman et al. 2006). In Section 12.2 we saw that there is a principled method for deciding when an ensemble of reconstructions provides a strong consensus value for the phase of any particular DFT coefficient. We also noted that the reconstructions are naturally clustered by their residuals into several groups.

In this section, we first examine how the statistics of the phase errors change if we replace the full set of reconstructions with a subset having small residuals. We then compare the outcomes of averaging all the reconstructions to that of averaging only the images having small residuals. One can either average registered reconstructions in the "image domain," or average the phases themselves in the "frequency domain." These two approaches give different averaged images, with rather different qualitative properties, though neither approach produces a consistently superior result. Throughout this section, the minimum residual reconstruction is used as a gold standard "reference image." By using a carefully selected subset of images to average, one can usually obtain an improvement in the image quality, as measured by ℓ^2-error, over the minimum residual reconstruction. Sometimes one can obtain a markedly better reconstruction.

In practical settings, and in the absence of additional information about the sample, the only meaningful criteria one has for choosing "good" reconstructions is some measure like the size of the residual. We use the ℓ_2-norm to measure these errors, but other choices are certainly possible and may, in some contexts, be superior. Examining the correlations between the residuals and the true errors, as shown in Figure 12.4, it seems likely that this procedure will produce a better reconstruction for all cases, but the $k = 2, p = 1$ case. In this case, the correlation between the error and the residual is quite small among reconstructions with small residuals. A criterion that chooses images to average based on the residual will therefore produce a set of images containing many reconstructions with (relatively) large errors. In this example, the smallest residual reconstruction happens to also have the smallest true error, by a large margin, so it will be difficult to beat. For the other cases, it seems conceivable that averaging some subset of small residual images will lead to better reconstructions.

Unfortunately, the plots in Figure 12.4 are based on the true error, which is unavailable in practice. Thus, it is difficult to predict if averaging will produce a better reconstruction. While Figure 12.4(b) is quite different from Figures 12.4(a, c, d), there is no obvious qualitative difference that distinguishes Figure 12.3(b) from Figures 12.3(a, c). To complicate matters further, even when the averaged image has a smaller ℓ^2-error than the minimum residual image, it often has a larger residual. This is a common occurrence for ill-conditioned problems, when many of the reconstructions have achieved spuriously small residuals – often much smaller than the square of the ℓ^2-error.

12.3.1 Statistics of the Small Residual Reconstructions

We first consider how the phase error statistics change if we consider the subset of images with "small" residuals. The histograms in Figure 12.3 give clear indications which reconstructions have significantly smaller residuals. To illustrate the effects of choosing subsets with smaller residuals we consider the case $k = 4, p = 1$. Figure 12.5(c) shows the mean phase errors and standard deviations for all 20,000 trials in this case. From the histogram in Figure 12.3(c) we see that the small residual cluster ends at 10^{-2}.

Figure 12.6 shows the statistics of the full data set along with the subsets with the residuals bounded above by $10^{-2}, 10^{-2.2}$, and $10^{-2.4}$. These subsets contain 20,000, 1044, 421, and 55 reconstructions respectively. The region of $\boldsymbol{k}$-space where the standard deviation is small clearly expands as we place stronger restrictions on the residual, as does the region of $\boldsymbol{k}$-space where the mean phase error is close to 0. We emphasize that the plots shown in Figure 12.6 only use data that would be available in a real experimental situation.

One might suspect that the reduction in standard deviation is simply a consequence of using a smaller data set. To show that this is not the case, we show, in Figure 12.6(c), the mean phase errors and standard deviations for a subset of 68 reconstructions with residuals between $10^{-1.1}$ and $10^{-1.05}$. Even though this is a small set, the regions where the mean phase errors are close to 0 and the standard deviations are small are considerably smaller than those obtained using the full data set.

12.3.2 Averaged Images

The experiments in Section 12.3.1 suggest that averaging over subsets of reconstructions with small residuals might give improvements in the final image. To make a precise statement we need to posit a "best guess"

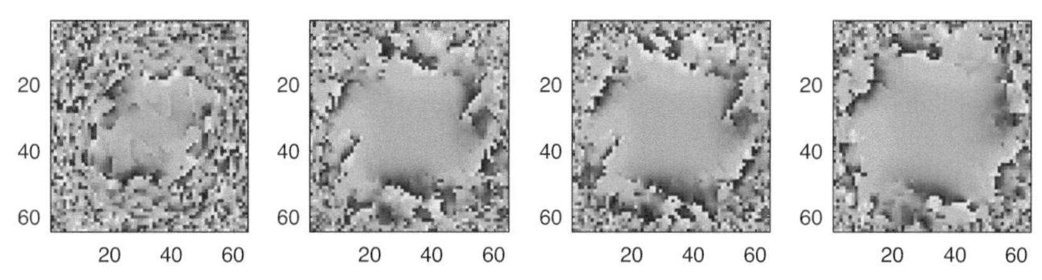

(a) Mean phase errors for subsets of trials of 20,000 iterates of $D_{\mathbb{A}_a S_{B_1}}$ for a $k = 4, p = 1$ image. The leftmost image uses the full data set; after that, from left to right, the maximum residuals are $10^{-2}, 10^{-2.2}$, and $10^{-2.4}$.

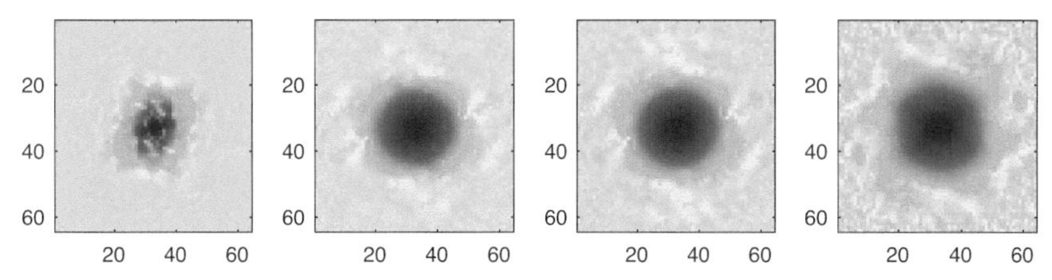

(b) Standard deviations for subset of trials of 20,000 iterates of $D_{\mathbb{A}_a S_{B_1}}$ for a $k = 4, p = 1$ image. The leftmost image uses the full data set; after that, from left to right, the maximum residuals are $10^{-2}, 10^{-2.2}$, and $10^{-2.4}$.

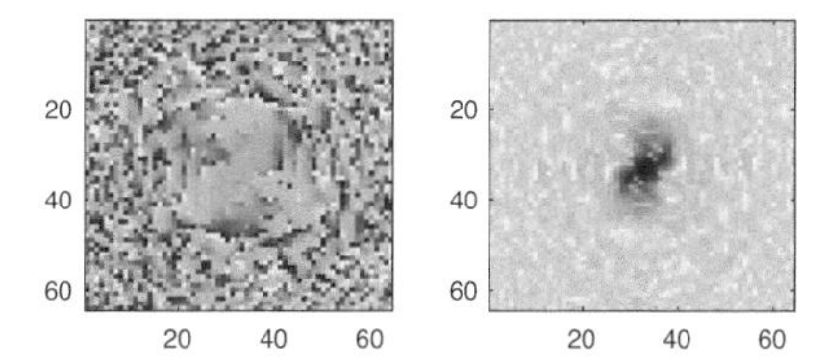

(c) Means phase errors and standard deviations for the subset of trials of 20,000 iterates of $D_{\mathbb{A}_a S_{B_1}}$ for a $k = 4, p = 1$ image with residuals in $[10^{-1.1}$ and $10^{-1.05}]$.

Figure 12.6 Plots showing how the statistics of the phase errors change as we consider subsets of reconstructions defined by bounds on the residual. While no color bars are shown, all images in a given row use the same color map. All images are centered on $\boldsymbol{k} = (0, 0)$.

reconstruction to compare with. As noted before, the only quantitative measure of convergence available in real experiments is the residual, so the natural choice would be the image with the minimum residual. From Figure 12.4 we see that the correlation between the residual and the error is not that strong for less smooth objects, and, in all cases, there is a large dispersion of true errors among images with small residuals.

For the $k = 1, p = 1$ case, the minimum residual image has a relatively large true error, whereas in the $k = 2, p = 1$ case it has an anomalously small error.

This suggests that averaging in the $k = 1, p = 1$ case has a good chance of producing an image with a smaller ℓ^2-error, whereas, this is unlikely to happen in the $k = 2, p = 1$ case. As we shall see, these expectations are borne out in the experiments below.

It should be remembered that the information in Figure 12.4 is, in principle, unavailable in a real imaging experiment. These plots make clear that the statistical relationships between the residuals and true errors are highly example dependent. In a real imaging application, it suggests the creation of a very realistic phantom object on which one could experiment to learn the statistical properties of the available reconstruction algorithms on the type of objects one is actually attempting to image. Such synthetic experimental results could then inform the decision as to whether averaging some collection of reconstructions is likely to provide a better reconstruction, or if minimum residual reconstruction is a better bet.

As a practical matter there are two distinct approaches to averaging a collection of reconstructions, $\{\boldsymbol{r}^{(q)} : 1 \leq q \leq Q\}$. In this discussion we assume that the images in this collection have been registered to a reference image. One can average either in the image domain, or in the frequency domain. The image-domain average is

$$\overline{\boldsymbol{r}}^i = \frac{1}{Q}\sum_{q=1}^{Q} \boldsymbol{r}^{(q)}. \tag{12.13}$$

Let $\{\hat{\boldsymbol{r}}^{(q)} : 1 \leq q \leq Q\}$ denote the DFT data for these reconstructions; the individual coefficients can be written as

$$\hat{r}_{\boldsymbol{k}}^{(q)} = a_{\boldsymbol{k}} e^{i\theta_{\boldsymbol{k}}^{(q)}}, \tag{12.14}$$

where the magnitudes $\{a_{\boldsymbol{k}}\}$ are regarded as measured data. The phases can then be averaged using either (12.3) or (12.4). Our experiments show that the differences between these two approaches to averaging phases are small and neither is consistently better than the other. In the experiments below we use (12.4). The frequency-domain average is then defined to be the image, $\overline{\boldsymbol{r}}^f$, with DFT data $\widehat{\overline{\boldsymbol{r}}}_{\boldsymbol{k}}^f = a_{\boldsymbol{k}} e^{i\bar{\theta}_{\boldsymbol{k}}}$. Once again, there does not seem to be a way to predict whether $\overline{\boldsymbol{r}}^i$ or $\overline{\boldsymbol{r}}^f$ has the smaller true error, but they are qualitatively somewhat different: the image-space average does a better job of suppressing noise, whereas the frequency-space average is noisier but appears to have somewhat higher resolution. These effects are difficult to quantify.

To conclude this discussion, we show the results of averaging different subsets of reconstructions for some of the examples considered earlier in this

Table 12.1. *Errors and residuals for the $k = 1, p = 1$ example.*

		$N = 3,846$		$N = 55$	
	Min. residual	Frequency-domain avg.	Image-domain avg.	Frequency-domain avg.	Image-domain avg.
error	2.4×10^{-2}	1.9×10^{-3}	2.0×10^{-3}	1.27×10^{-3}	1.27×10^{-3}
residual	1.65×10^{-6}	6×10^{-7}	5.3×10^{-4}	4.7×10^{-7}	1.23×10^{-4}

chapter. We consider the outcome of averaging all of the reconstructions in the smallest residual cluster as well as those coming from the leftmost tail of this cluster, see Figure 12.3. This cluster is fairly well defined, except for the $k = 4, p = 2$ case, where we take two subsets of images from the left side of the tail of the smaller residual cluster. In all cases we are drawing from the 20,000 trials used to produce the images in Section 12.2.

We begin with the $k = 1, p = 1$ case where the minimum residual image has a residual of 1.65×10^{-6} and an error of 2.4×10^{-2}; the residual is considerably smaller than the square of the error, 5.76×10^{-4}. In Figure 12.3(a) the clump to the left seems to be composed of two separate clusters; those with the smallest residuals lie to the left of about 10^{-4}.

Table 12.1 compares the errors and residuals for this collection of averaged reconstructions to one another, and to the minimum residual image. We see that images produced by averaging 3,846 reconstructions have an error of about 1.9×10^{-3}, and those obtained by averaging 55 reconstructions have an error of about 1.27×10^{-3}, which are both smaller than the error of the minimum residual image. The two methods of averaging produce images with the same sized errors, but the images obtained by averaging in the frequency domain usually have much smaller residuals.

We next turn to the $k = 2, p = 1$ case; here the minimum residual image happens to attain a particularly small value of 2.41×10^{-5} and an error of 2.35×10^{-3}; the residual is a little larger than the square of the error. In Figure 12.3(b), the cluster with smallest residuals is clearly delineated and lies to the left of about $10^{-2.4}$

Table 12.2 compares the errors and residuals for this collection of averaged reconstructions to one another, and to the minimum residual image. We see that all of the images produced by averaging have larger errors than the minimum residual by almost an order of magnitude. As before, averaging over the smaller subset of reconstructions with very small residuals produces a better result. It is interesting to note that the relationship between residual and

Table 12.2. *Errors and residuals for the $k = 2, p = 1$ example.*

		$N = 1{,}240$		$N = 55$	
	Min. residual	Frequency-domain avg.	Image-domain avg.	Frequency-domain avg.	Image-domain avg.
error	2.35×10^{-3}	1.72×10^{-2}	3.58×10^{-2}	1.71×10^{-2}	2.6×10^{-2}
residual	2.41×10^{-5}	1.34×10^{-4}	3.27×10^{-2}	2.84×10^{-5}	2.0×10^{-2}

Table 12.3. *Errors and residuals for the $k = 4, p = 1$ example.*

		$N = 1{,}044$		$N = 152$	
	Min. residual	Frequency-domain avg.	Image-domain avg.	Frequency-domain avg.	Image-domain avg.
error	1.08×10^{-1}	8.51×10^{-2}	8.0×10^{-2}	8.7×10^{-2}	7.9×10^{-2}
residual	1.33×10^{-3}	1.0×10^{-2}	6.4×10^{-2}	7.51×10^{-3}	5.0×10^{-2}

error for the frequency-space averages is very similar to that obtained directly from the hybrid iterative map, one is about the square of the other, whereas for image-space averages these two numbers are almost the same.

For the next case, we take $k = 4, p = 1$; the minimum residual image has a residual of 1.33×10^{-3} and an error of 1.08×10^{-1}; the residual is a little smaller than the square of the error. In Figure 12.3(c) there is a well-defined cluster of reconstructions with the smallest residuals, lying to the left of about 10^{-2}. In Figure 12.7, (a) shows the exact reconstruction and minimum residual reconstruction, (b) shows the results of averaging, in both the image and frequency domain, the 1,044 reconstructions with residuals less than 10^{-2}, and (c) shows the results of averaging the 152 reconstructions with residuals less than 5×10^{-3}. The minimum residual image in this case has an error comparable to that of the smallest errors encountered in the 20,000 trials.

Table 12.3 compares the errors and residuals for this collection of averaged reconstructions to one another and to the minimum residual image. All methods of averaging lead to images with about the same error 8×10^{-2}, which is a little smaller than the error of the minimum residual image. The difference between using all 1,044 reconstructions in the smallest residual cluster versus only those 152 reconstructions with smallest residuals is negligible.

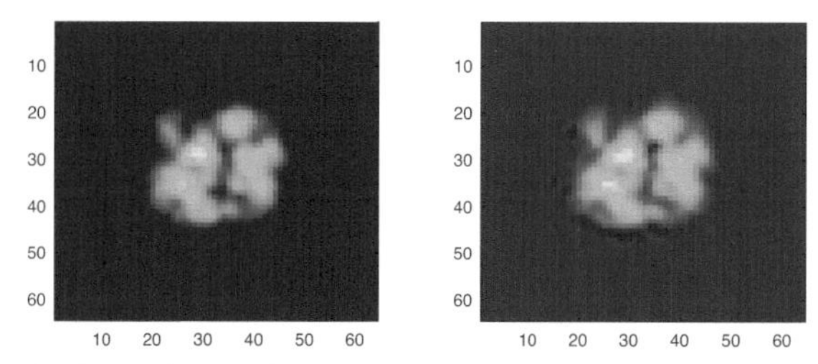

(a) The reference image on the left and reconstruction with minimum residual, from the sample of 20,000 reconstructions, on the right.

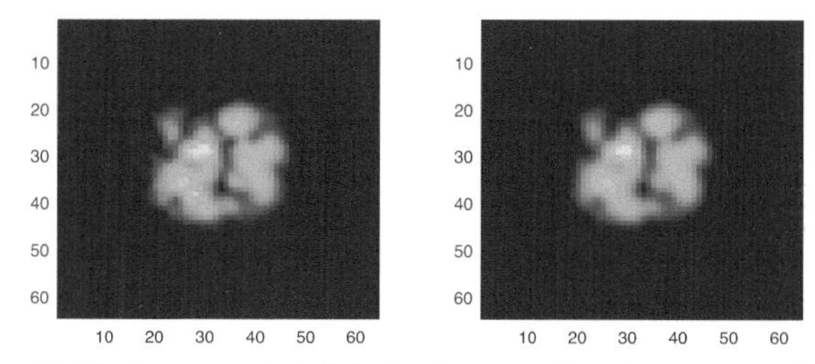

(b) The image on the left is the frequency-domain average, and the image on the right is the image-domain average of the 1,044 reconstructions with residual less than 10^{-2}.

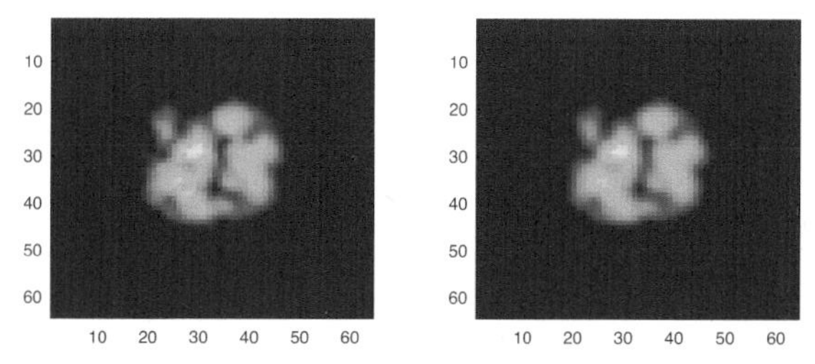

(c) The image on the left is the frequency-domain average, and the image on the right is the image-domain average of the 152 reconstructions with residual less than 5×10^{-3}.

Figure 12.7 The reference image and various reconstructions and averages of reconstructions for a $k = 4, p = 1$ image using 20,000 reconstructions.

For the final case we take $k = 4, p = 2$; the minimum residual image has a residual of 3.56×10^{-3} and an error of 2.34×10^{-1}; the residual is a little smaller than the square of the error. In Figure 12.3(d) there are only two clusters of reconstructions, with considerable overlap. In order to avoid reconstructions that cannot be confidently assigned to one cluster or the other, we have used parts of the leftward tail of the cluster with the smaller residuals. In Figure 12.8, (a) shows the exact reconstruction and

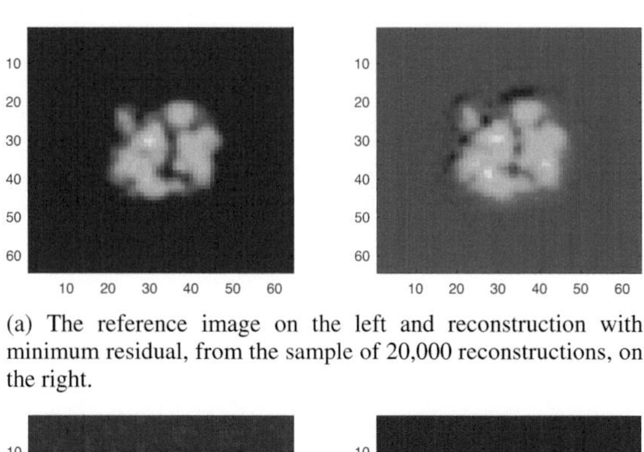

(a) The reference image on the left and reconstruction with minimum residual, from the sample of 20,000 reconstructions, on the right.

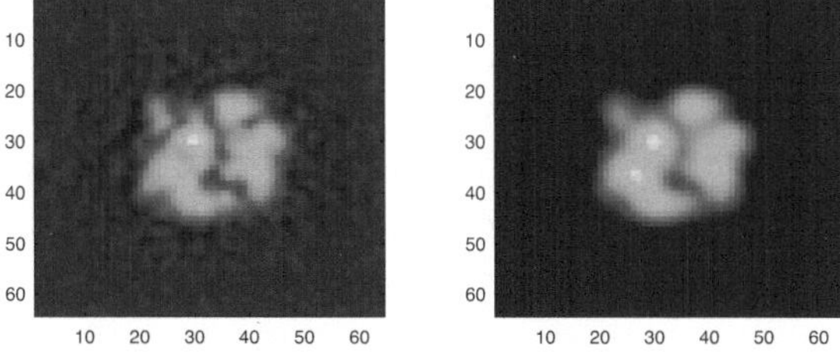

(b) The image on the left is the frequency-domain average, and the image on the right is the image-domain average of the 850 reconstructions with residual less than 1.78×10^{-2}.

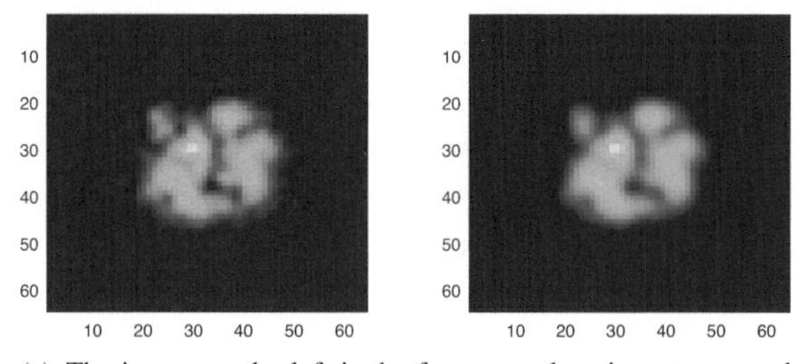

(c) The image on the left is the frequency-domain average, and the image on the right is the image-domain average of the 61 reconstructions with residual less than 10^{-2}.

Figure 12.8 The reference image and various reconstructions and averages of reconstructions for a $k = 4, p = 2$ image using 20,000 reconstructions.

minimum residual reconstruction, (b) shows the results of averaging, in both the image and frequency domain, the 850 reconstructions with residuals less than 1.78×10^{-2}, and (c) shows the results of averaging the 61 reconstructions with residuals less than 10^{-2}. The minimum residual image in this case has an error comparable to that of the smallest errors encountered in the 20,000 trials.

Table 12.4 compares the errors and residuals for this collection of averaged reconstructions to one another and to the minimum residual image. All

Table 12.4. *Errors and residuals for the $k = 4, p = 2$ example.*

		$N = 850$		$N = 61$	
	Min. residual	Frequency-domain avg.	Image-domain avg.	Frequency-domain avg.	Image-domain avg.
error	2.34×10^{-1}	1.89×10^{-1}	1.87×10^{-1}	1.78×10^{-1}	1.62×10^{-1}
residual	3.56×10^{-3}	2.94×10^{-1}	1.65×10^{-1}	2.04×10^{-2}	9.66×10^{-2}

methods of averaging lead to images with about the same error 1.8×10^{-1}, which is a little smaller than the error of the minimum residual image. The difference between using all 850 reconstructions in the smallest residual cluster versus only those 61 reconstructions with smallest residuals is again negligible. As was the case with the previous example, the residuals of the averaged images are much larger than the squares of the errors, indicating that these reconstructions are, in some ways, different from those obtained directly from a hybrid iterative map algorithm.

The per-frequency statistics of the mean phase errors and standard deviations for the cluster with the smallest residuals indicate that this subset provides better information about a larger subset of the unknown phases than the entire ensemble. Many of the reconstructions above look qualitatively correct. The reconstructions in Figures 12.7–12.8, with errors of size $\sim 10^{-1}$, however, do not stand up to careful scrutiny. Except for the $k = 2, p = 1$ case, averaging the reconstructions with very small residuals *does* produce an image with a slightly smaller ℓ^2-error, though often with a larger residual. Careful examination of the averaged reconstructions shows that image space averaging suppresses the noise rather well. This is especially evident in the background, outside of the object itself. On the other hand, small features visible in the reference image are somewhat better resolved in the frequency space averages.

12.4 Some Conclusions

Hybrid iterative maps display certain very interesting statistical regularities despite the fact that, in general, they are not convergent. The residuals naturally group different trials of these algorithms into a small group of clusters. While the logs of the residuals are correlated with the logs of the true errors, in many cases there is still a considerable range of true errors occurring among the cluster of reconstructions with the smallest residuals. We have shown that, for the purposes of studying the statistics of phase errors, the minimum residual reconstruction is an adequate reference for registering reconstructions prior to statistical analysis of these quantities.

The empirical statistics of the phase errors clearly delineate a subset of phases for which an ensemble of runs determines a reasonably accurate consensus value for the phase. This is evident in Figures 12.5 and 12.6: the frequencies where the standard deviation is less than about 0.5 coincide with frequencies where the mean phase error is close to 0. It is important to emphasize that the per-frequency standard deviations of the phases are reliably computed without access to an exact reconstruction by registering the reconstructions obtained in a series of independent trials to a minimum residual reconstruction.

While in the interest of brevity we have omitted experiments that demonstrate it, there is a strong correlation between the magnitude of DFT coefficients and the accuracy of the consensus value determined for the phase. Coefficients with magnitude above a certain threshold are, on average, correctly phased, whereas even many runs of an algorithm will fail to the determine the phases of coefficients with magnitudes below this threshold. To some extent, this result can be inferred from Figures 12.5 and 12.6 as, generally, the magnitudes of the DFT coefficients fall off as $\|\boldsymbol{k}\|_2$ increases. This gives another explanation for the superior performance of hybrid iterative map-based algorithms on less smooth images: a greater proportion of their DFT coefficients have large moduli.

Finally, we observe that the subset of reconstructions with residuals in the leftmost cluster has rather different statistical properties from the complete set of trials. The standard deviations of the phases in this set are small over a larger region of $\boldsymbol{k}$-space, and the mean phase errors are also smaller over a similarly larger set of frequencies. Averaging subsets of these small residual reconstructions, either in the image domain or frequency domain, often leads to reconstructions with somewhat smaller ℓ^2-error than that of the minimal residual reconstruction.

13

Suggestions for Improvements

In this chapter, we consider several ideas that might lead to improved reconstructions in coherent diffraction imaging (CDI). As has been made amply clear by the preceding pages, three geometric facts underlie the difficulty of the phase retrieval problem:

(i) The failure of the intersection between the magnitude torus, $\mathbb{A}_{\boldsymbol{a}}$, and the set, B, defined by the auxiliary information, to be transversal.
(ii) At points in $\mathbb{A}_{\boldsymbol{a}} \cap B$ there may be many directions where the angles between the tangent space to the magnitude torus and that of the set, B, defined by the auxiliary data, are very small.
(iii) The existence of multiple basins of attractions due to existence of trivial associates, as well as nonzero critical points of the distance function, $d_{\mathbb{A}_{\boldsymbol{a}}B}$.

To avoid the conditions that render the inverse problem ill conditioned, and the usual iterative algorithms nonconvergent, we consider approaches that entail (i) a modification of the sample preparation, (ii) a new experimental modality, which uses the holographic Hilbert transform (HHT) method described in Section 7.4 to reconstruct the image, or (iii) a Newton-type algorithm.

For the first approach, we assume that a soft object has been cut, creating a sharp edge and leaving an object of known shape and size. For the second approach we assume that a hard object, of known shape and size, has been added to the field of view outside of the object that we are trying to image (a form of external holography).

In Section 13.3 we then discuss a new algorithm that can reasonably be described as a geometric "Newton's method" for phase retrieval. To solve the phase retrieval problem, we are trying to find points in $\mathbb{A}_{\boldsymbol{a}} \cap B$. The set $\mathbb{A}_{\boldsymbol{a}}$ is a torus, which is always nonlinear; the set B, defined by auxiliary information, is linear if the auxiliary information is an estimate for the support of the unknown object, and polyhedral if we know the object is nonnegative. In either case, we replace the problem of finding a point in $\mathbb{A}_{\boldsymbol{a}} \cap B$, with the more linear problem of finding points in $T_{\boldsymbol{f}^{(n)}}\mathbb{A}_{\boldsymbol{a}} \cap B$, for a succession of points $\boldsymbol{f}^{(n)} \in \mathbb{A}_{\boldsymbol{a}}$.

The final section describes the detailed implementation of the HHT method introduced in Chapter 7.4. This algorithm is entirely different from the iterative approaches considered in the rest of the book. Other than the fact that the object is compactly supported, no additional information is required. As we will see, however, adequate sampling of the magnitude DFT data (relative to the size of the object) is also important in this context, though for somewhat different reasons.

Remark 13.1 As we have seen above, especially in Chapter 12, the outcome of running a hybrid iterative algorithm depends strongly, and unpredictably, on the initial guess. To compare the performance of iterative algorithms with different experimental setups, or different algorithms, we use the following protocol: We run each reconstruction algorithm for 20,000 iterates using 20,000 independently selected random starting points. From these 20,000 trials we then randomly select 10,000 subsets of 400 trials each. For each of these 400 selected trials, we find the reconstruction with minimum residual, which we regard as the best guess for the optimal reconstruction among these 400 trials. The 10,000 subsets of trials are indexed by $q \in \{1, \ldots, 10{,}000\}$. We then plot histograms of the true errors for the various minimum residual reconstructions. Comparing these plots then gives a systematic basis for comparing the performance of the two experimental setups.

13.1 Use of a Sharp Cutoff

As we have seen in Section 8.3, if the object we seek to image, f, is hard ($k = 0$), and we have a very accurate estimate for its support, then a hybrid iterative map, run for sufficiently many iterates often converges to a highly accurate reconstruction. With a less precise estimate, such an algorithm sometimes converges; this happens when the iterates fall into an attracting basin defined by a trivial associate, $\boldsymbol{f}^{(v)} \in B_S$, for which $\dim T_{\boldsymbol{f}^{(v)}} \cap B_{S_p}$ is very small.

On the other hand, if the object is soft ($k \geq 1$), then it is very unlikely that such an algorithm will ever converge to a highly accurate reconstruction. In this section we explore what happens if a soft object is cut off along a hard edge.

As we see below, such a cutoff can vastly improve the performance of hybrid iterative map algorithms, provided that we also have a precise description of the edge along which the object is cut. The choice that an experimenter has is the shape of the resultant object. If the hard edge is circular, then all that is needed for a precise description of the edge is an accurate estimate of its diameter. On the other hand, if the edge is noncircular then one must also have an accurate estimate for its orientation relative to the detector, which may be difficult to obtain. With a circular edge, the inversion symmetry of the support constraint seems to slightly extend the global phase of an iteration, where the map is searching for an attracting basin, but, with sufficiently many iterates, this problem is usually overcome.

For the experiments in this section, we use a 64×64, $k = 4$ image. We consider the original image, the image cut along a circular contour; and the image cut along a noncircular contour; in all cases the support constraint is a 1-pixel neighborhood of the support of the image, see Figure 13.1. The support of the image in Figure 13.1(a) is defined as the set $\{\boldsymbol{j} : f_{\boldsymbol{j}} > 10^{-10}\}$; in (b) and (c) it is the set where $f_{\boldsymbol{j}} \neq 0$.

We follow the protocol laid out in Remark 13.1; for each $q \in \{1, \ldots, 10,000\}$, we select the minimum residual reconstruction, $\boldsymbol{r}_q$, from the corresponding set of 400 trials, which we regard as the best choice of reconstruction from this set of trials. We then tabulate the true errors, $\text{err}(q)$, for each of these reconstructions. To compare the behavior of this algorithm on the three images in Figure 13.1, we display histograms of the $\log_{10}(\text{err})$ in Figure 13.2. For ease of comparison, the axes in the three histograms are the same.

From these histograms it is apparent that, from among 400 trials, the minimum residual reconstructions for the soft object itself, have a true error of about 10^{-1}. The minimum residual reconstruction found among 400 trials, cutting on a noncircular arc, have true errors reliably less than 10^{-14}. Cutting along a circular arc, the error in minimum residual reconstruction is less than 10^{-14} in the vast majority of trials, and less than 10^{-13} is all cases. It is readily apparent that cutting an object along a sharp edge leads to a well-posed reconstruction problem, provided the shape of the sharp edge is known to high precision (consistent with our earlier analysis). The smoothness of the interior of the object does not interfere with the reconstruction process.

To close this section, we consider the consequences of uncertainty in the shape of the edge along which the soft object is cut – that is, when the support neighborhood cannot be precisely estimated. The theory presented above clearly predicts that there should exist *some* trivial associates of the original image, $\boldsymbol{f}^{(\boldsymbol{v})} \in \mathbb{A}_{\boldsymbol{a}} \cap B_{S_p}$, for which the intersection is transversal, that is

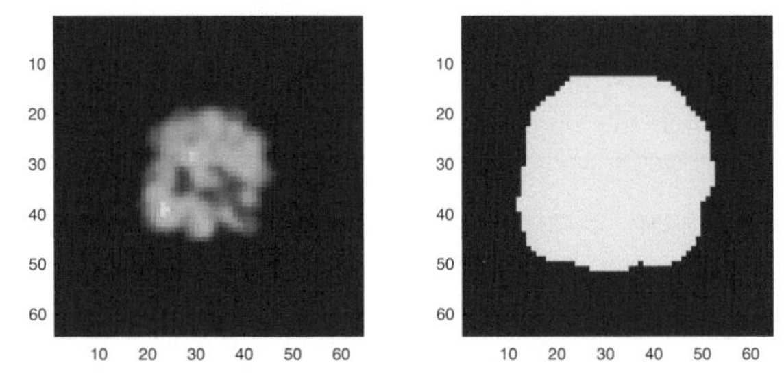

(a) On the left is a 64×64 image with smoothness $k = 4$ and the right panel shows a 1-pixel neighborhood of the support of this image.

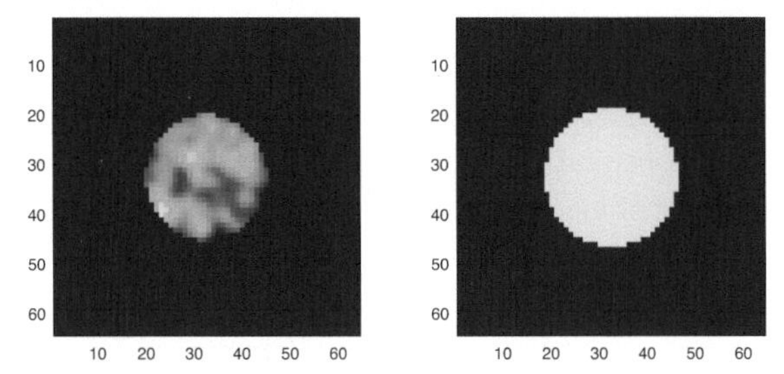

(b) The left panel shows the image in (a) cut along a circular contour and the right panel shows the 1-pixel neighborhood of this region.

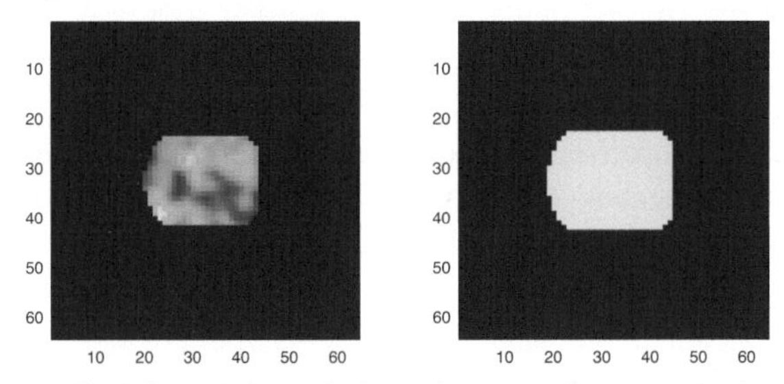

(c) The left panel shows the image in (a) cut along a noncircular contour and the right panel shows the 1-pixel neighborhood of this region.

Figure 13.1 Images and support neighborhoods used for the experiments performed in this section.

$$T_{f^{(v)}} \mathbb{A}_{a} \cap B_{S_p} = f^{(v)}. \tag{13.1}$$

If the iterates of the hybrid iterative map happen to fall in the attracting basin defined by such a trivial associate, then, eventually, the iterates should converge to a point on the center manifold defined by $f^{(v)}$, to machine precision.

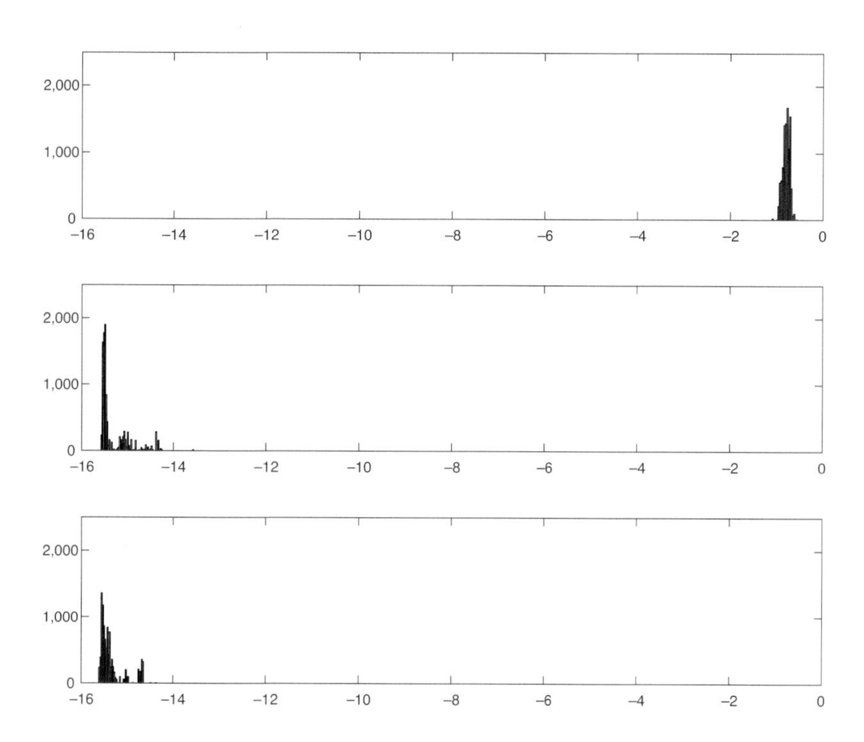

Figure 13.2 Histograms of the errors in the minimum residual reconstruction drawn from 10,000 subsets of 400 randomly selected trials from a collection of 20,000 trials. The plot at the top is for the original soft object; the plot in the middle is for the soft object cut along a circular contour, and the bottom is for the soft object cut along a noncircular contour. The x-axis is the $\log_{10}$ of the true error of the reconstruction. The y-axis is the number of counts in each bin.

The question then is how often does this occur? To investigate, we use the objects shown in Figure 13.1(b, c), which are smooth objects cut off along hard edges, but with a 3-pixel neighborhood of the regions bounded by these edges as auxiliary data for the hybrid iterative map. Note that 3-pixels is about 10% of the diameter of the cut off objects themselves.

Figure 13.3 shows the histograms of the numbers of trials, out of 20,000, using the map $D_{\mathbb{A}_a B_{S_3}}$, that fall into each of the attracting basins defined by $\{\boldsymbol{f}^{(\boldsymbol{v})} : \|\boldsymbol{v}\|_\infty \leq 3\}$. With the circular cutoff, all 20,000 trials found one of these attracting basins, whereas with the noncircular cutoff, 185 trials failed to find any. A somewhat similar experiment is conducted in Elser et al. (2018).

We first consider the circular case: Out of 20,000 trials, 16,131 ended up in the attracting basins defined by $\{\boldsymbol{f}^{(\boldsymbol{v})} : \|\boldsymbol{v}\|_\infty \leq 1\}$. In both cases the only

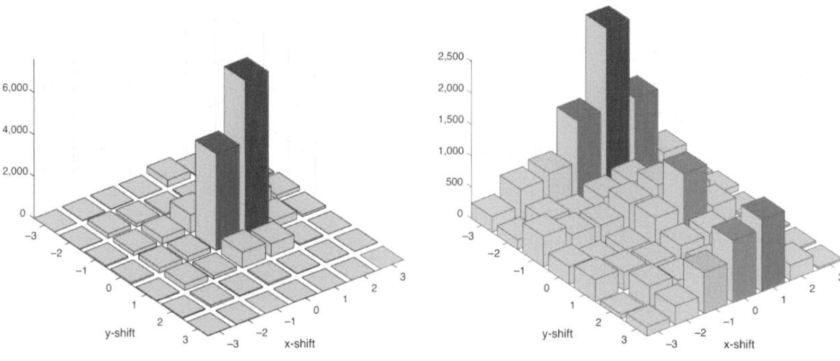

(a) Counts using a circular cutoff. Note that the maximum of the z-axis is 6,000.

(b) Counts using a noncircular cutoff. Note that the maximum of the z-axis is 2,000.

Figure 13.3 A 2D histogram showing the number of trials (out of 20,000) that ended up in each of the 49 attracting basins defined by $\{\boldsymbol{f}^{(\boldsymbol{v})} : \|\boldsymbol{v}\|_\infty \leq 3\}$.

translates for which (13.1) holds are those lying in the corners, i.e., with $\boldsymbol{v} = [\pm 3, \pm 3]$. Out of 20,000 trials, only 135 found a transversal intersection point. For the nonsymmetric case, a much smaller number, 4,057, ended up in the 1-pixel neighborhood, and a larger number, 610, found a transversal intersection. The distribution over the different attracting basins is much more uniform for the asymmetric cutoff than for the symmetric one. The problem of understanding what makes one attracting basin more attractive or less attractive than another is both open and interesting.

To complete this discussion, we show histograms of the errors and residuals for the trials that found attracting basins defined by transversal intersections, as well as those with residuals less than 10^{-5}; these are shown in Figure 13.4. From Figure 13.4(a) it appears that, for the circular cutoff, all of the trials that converged, to machine precision, found transversal intersections, whereas many trials that found transversal intersections remained very far from convergence after 20,000 iterates. For this latter class, the failure to converge could be explained by the existence of many directions where the angles between $T_{\boldsymbol{f}^{(\boldsymbol{v})}}\mathbb{A}_{\boldsymbol{a}}$ and B_{S_3}, though nonzero, are very small. For these trials the plots of residuals and errors are rather similar.

The upper plot in Figure 13.4(a) shows that about 100 of the trials that found transversal intersections ended up with residuals greater than 10^{-2}, and hence did not contribute to the lower plot. From the lower plot it is apparent there are many trials that did not find a transversal intersection, but nonetheless, attained a very small residual, though without a correspondingly small error.

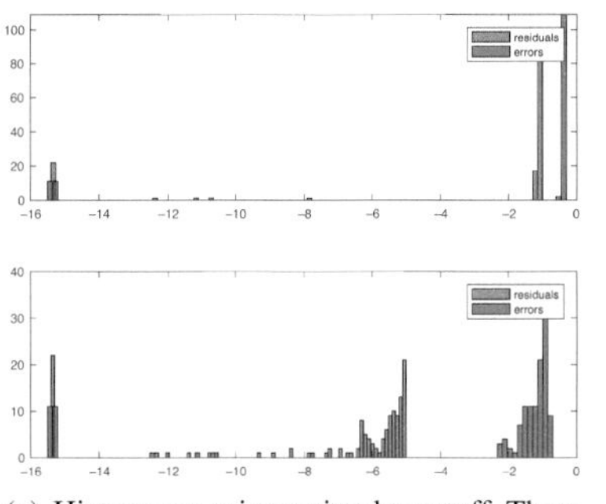

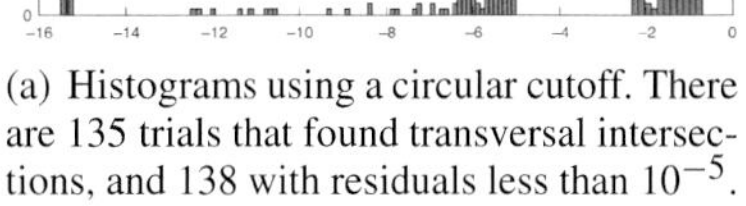

(a) Histograms using a circular cutoff. There are 135 trials that found transversal intersections, and 138 with residuals less than 10^{-5}.

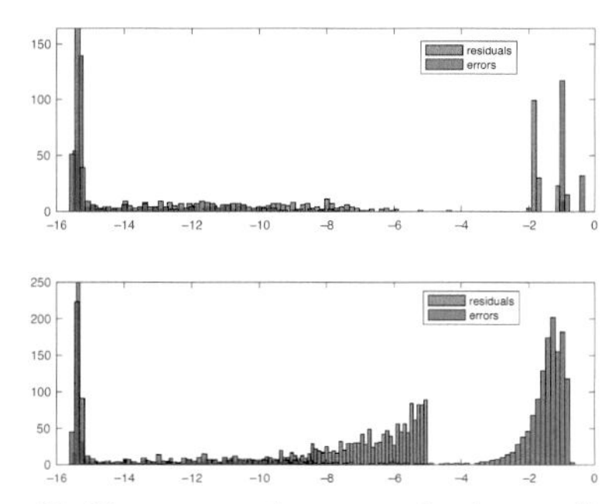

(b) Histograms using a non-circular cutoff. There are 610 trials that found transversal intersections, and 1,959 with residuals less than 10^{-5}.

Figure 13.4 The upper plots shows histograms of the $\log_{10}$ of the errors (blue) and residuals (red) for the trials that found attracting basins defined by transversal intersections. The lower plots show the same data for the trials for which the residuals are less than 10^{-5}.

Indeed, while some of these residuals are extremely small (less than 10^{-10}), *the only trials for which the errors are smaller than* 10^{-3} *actually come from transversal intersections*. This again highlights the inherent ill conditioning of the phase retrieval problem: for perfectly ordinary looking data one can find approximate solutions that have data errors less than 10^{-10}, while at the same time having a true error of about 10^{-2}.

For the noncircular cutoff there are again many trials with small residuals, which did not have small errors. As the number (372) of trials with errors less than 10^{-14} in the lower plot in Figure 13.4(b) is larger than in the upper plot (269), somewhat surprisingly, there appear to be a significant number of cases where the algorithm converged, to machine precision, in attracting basins defined by nontransversal intersections. It is also worth noting that a much smaller proportion of the trials that found a transversal intersection (130/610) failed to converge to machine precision than in the circular cutoff case (100/135). Taken together these observations indicate that, if possible, a soft object should be cutoff along an asymmetric edge.

13.2 External Holography

The method that we call "external holography" consists of placing a hard object with a known shape in the exterior of the object we would like to image. It is easiest to discuss this method in a continuum description of the objects and the measurements. Suppose that the unknown object is described by a function $\rho(\boldsymbol{x})$ with support in a box $[-1,1]^d$, and assume that $s_\rho(\boldsymbol{x})$ is a function, taking values in $\{0,\ 1\}$, which provides an estimate for the support of ρ; that is $\operatorname{supp}\rho \subset \{\boldsymbol{x} : s_\rho(\boldsymbol{x}) = 1\}$. Let $d(\boldsymbol{x})$ be a function with support centered on $\boldsymbol{0}$, and let $s_d(\boldsymbol{x})$ be an estimate for the support of $d(\boldsymbol{x})$. Instead of collecting a diffraction pattern from $\rho(\boldsymbol{x})$ we consider the data obtained by illuminating $\rho(\boldsymbol{x}) + d(\boldsymbol{x} - \boldsymbol{x}_0)$ with coherent X-rays. Here the offset, $\boldsymbol{x}_0$, is a parameter under the control of the experimenter. In the far field, the data collected is modeled as

$$\begin{aligned}\widehat{w}_{\boldsymbol{x}_0}(\boldsymbol{k}) &= |\widehat{\rho}(\boldsymbol{k}) + e^{-i\boldsymbol{x}_0\cdot\boldsymbol{k}}\widehat{d}(\boldsymbol{k})|^2 \\ &= |\widehat{\rho}(\boldsymbol{k})|^2 + 2\operatorname{Re}\left[\widehat{\rho}(\boldsymbol{k})e^{i\boldsymbol{x}_0\cdot\boldsymbol{k}}\overline{\widehat{d}(\boldsymbol{k})}\,\right] + |\widehat{d}(\boldsymbol{k})|^2.\end{aligned} \tag{13.2}$$

The measurements contain modulus data for $\widehat{\rho}$ and $\widehat{d}$, as well as terms reflecting the interference between these two objects. In the physics literature, such a measurement is called *holographic*; since the interference is produced by an object external to the object of interest, we call this *external holography*. Methods of external holography are useful if ρ is smooth and/or s_ρ is not a very tight estimate for $\operatorname{supp}\rho$.

A method of this general type has been considered by Hohage et al. (see Maretzke and Hohage 2016; Jacobsen 2019), where they take the support of d to be very small, compared to that of ρ, and they assume that $\|\boldsymbol{x}_0\|_2$ is more than twice the diameter of the $\operatorname{supp}\rho$. With these assumptions, one can simply invert the measured Fourier magnitude data, to obtain the autocorrelation function

$$\begin{aligned}w_{\boldsymbol{x}_0}(\boldsymbol{x}) = \rho\star\rho(\boldsymbol{x}) + d\star d(\boldsymbol{x}) + \int \overline{d}(\boldsymbol{y}-\boldsymbol{x}_0)\rho(\boldsymbol{y}+\boldsymbol{x})d\boldsymbol{y} \\ + \int d(\boldsymbol{y}-\boldsymbol{x}_0)\overline{\rho}(\boldsymbol{y}-\boldsymbol{x})d\boldsymbol{y}.\end{aligned} \tag{13.3}$$

The first and second terms are supported in a disk of diameter twice the diameter of the $\operatorname{supp}\rho$, whereas the third and fourth terms are supported in disks centered on $\pm\boldsymbol{x}_0$ of the diameter equal to the sum of the diameters of $\operatorname{supp}\rho$ and $\operatorname{supp}d$. Assuming that $\operatorname{supp}d \subset \operatorname{supp}\rho$, it follows that these two integrals have disjoint supports, which are also disjoint from the support of the first two terms. The third term, therefore, provides a smoothed-out version of $\rho(\boldsymbol{x})$, without any need to employ an iterative algorithm. Reconstruction under these hypotheses is then completed by isolating this part of $w_{\boldsymbol{x}_0}$ and then removing, to the extent possible, the effects of convolution by $\overline{d}$. For this method to give a high-resolution image, the function d should therefore have very small support, so that the first zero of its Fourier transform occurs at a large frequency. This method therefore requires a field of view that is a little larger than four times the diameter of the object we seek to image, with a small, hard object placed in the "far" exterior region.

We now consider a different experimental setup with an external object and its effect on hybrid iterative maps. Here, the object does not have to have small support, compared to that of ρ, and a field of view twice the diameter of the $\operatorname{supp} \rho$ often suffices. In this case we use a standard iterative reconstruction algorithm, and therefore require an accurate estimate for the shape of the support of d, though its location can be fixed as part of the reconstruction process. For this to be meaningful, the function d should have a well-defined support, and therefore a fairly sharp edge.

For simplicity we assume that $d(\boldsymbol{x})$ is a radial function with support equal to a disk of radius r centered at $\mathbf{0}$:

$$d(\boldsymbol{x}) = m\chi_{[0,r]}(\|\boldsymbol{x}\|_2)(r^2 - \|\boldsymbol{x}\|_2^2)^{\alpha}. \tag{13.4}$$

The parameter, α, regulates the smoothness of d, with $\alpha = 0$ giving a piecewise constant, discontinuous function. Increasing α leads to smoother functions. In addition to α, there is a collection of parameters whose choice might affect the outcome of this experiment: the radius, r, the placement of the object, $\boldsymbol{x}_0$, and the magnitude, m.

Heuristically, the idea is that having an accurate estimate for the support of *part* of the image breaks the infinitesimal translational symmetry, which renders the reconstruction problem ill posed. It seems reasonable to expect that the more the two objects are "entangled" in the measurement process, the better this heuristic will apply. From the second line in (13.2) we see that the entanglement is a result of the cross term

$$\operatorname{Re}\left[\widehat{\rho}(\boldsymbol{k})e^{\boldsymbol{x}_0\cdot\boldsymbol{k}}\overline{\widehat{d}(\boldsymbol{k})}\right]. \tag{13.5}$$

For the cross term to play an important role it is clear that supports of $\widehat{\rho}$ and $\widehat{d}$ should have a large overlap. If the object described by $d(\boldsymbol{x})$ is oscillatory, so that the support of its Fourier transform is effectively disjoint from that of $\widehat{\rho}$, then the method fails entirely. Moreover, these two functions should be roughly the same "size": Good results are obtained if r is selected to be about a quarter of the side length of the smallest rectangle containing the support of $\rho(\boldsymbol{x})$, and m is selected so that the two functions have comparable sup-norms.

It is less clear how $\boldsymbol{x}_0$ should be selected. As we see below, there is a fairly large range of values where this method works well. The support of the external object should not be too close to that of ρ and it should not be too close to the edge of the field of view. Within these constraints, the value of $\|\boldsymbol{x}_0\|_2$ is not a critical determinant of the success or failure of this approach. What is quite important is that the field of view should be large enough to be able to satisfy both of these requirements. Suppose that the support of the original object sits in a square with side lengths ℓ. For the support constraint to ensure generic uniqueness, the sample spacing in the Fourier domain must define a field of view with side lengths 2ℓ. While the addition of an external object is often effective with the same field of view, a field of view with side lengths 3ℓ seems to provide more reliable results. This is explored in the examples below.

In the continuum, only the $\|\boldsymbol{x}_0\|_2$ would be expected to play a role. In fact, we have found that, for the DFT model we have been analyzing, placing the external object at an angle in the range 30°–45° from a coordinate axis produces favorable and robust results. Outside of this range, the results are rather poor. Given the uniform Cartesian sampling pattern this might be expected to produce the most uniform pattern of entanglement across the sampling grid. In fact, this angular dependence disappears in the continuum problem, even with uniform Cartesian sampling. It is entirely an artifact of the DFT model that we are using. Section 13.2.1 is devoted to numerical experiments that elucidate the effects of the various parameters appearing in the definition of the external object.

13.2.1 Numerical Examples

For the examples in this section, smoother objects are obtained by convolving a piecewise constant image with a Gaussian of various widths. As shown in Section 2.2, this leads to a large value for $\dim T_{\boldsymbol{x}_0}\mathbb{A}_{\boldsymbol{a}} \cap B_S$, and therefore, a more challenging reconstruction problem. Nevertheless, the addition of a hard external object, whose shape is known, leads to a tractable problem. A similar approach can be found in Huang et al. (2016) and Huang and Eldar (2017).

Example 13.2 In this section we show the results of several numerical experiments, which demonstrate, on the one hand, the efficacy of adding an external object, and, on the other hand, the dependence of this approach on the detailed properties of that object. The parameters we consider in defining the external object are its magnitude, radius, and location, m, r, and $\boldsymbol{x}_0$. We also consider the smoothness of the object of interest (the parameter k), and the extent of over-sampling used in the Fourier domain but omit variations in the accuracy of the support neighborhoods of the object of interest and of the external disk. These, too, could have an effect on the performance of this approach to external holography.

Our first experiments demonstrate the efficacy of adding such an object when using hybrid iterative maps that employ a support condition as the auxiliary information. A dramatic improvement in the convergence behavior can often be obtained. First, let us fix the parameters $m, r, \boldsymbol{x}_0$, to values that have been found to work reliably. We vary the smoothness of the object, taking $k = 4, 8, 12$, in the presence or absence of the external object. We also consider external objects with two levels of smoothness, taking $\alpha = 0$ or 0.5 in (13.4).

The parameter m is adjusted so that the magnitude DFT data of the object we are seeking to reconstruct and that of the external object have the same sup-norms. The field of view is normalized to be a square of side length 2, with the object of interest contained in a square of side length 1, and the external object a disk of radius 0.1. We use 256×256 samples of the magnitude DFT data. The external object is placed along a line making a 45° angle with the coordinate axes; its center is placed at a distance of $0.6\sqrt{2}$ from the center of the field of view.

On the left-hand side of Figure 13.5 we show the object of interest with $k = 12$ and the $\alpha = 0$ external disk. For a support mask we use the 1-pixel neighborhood of the external object along with the 4-pixel neighborhood of the object of interest. This is shown on the right-hand side of Figure 13.5. Figure 13.6 shows the errors and residuals for 5,000 iterates of $D_{\mathbb{A}_a B_S}$ with

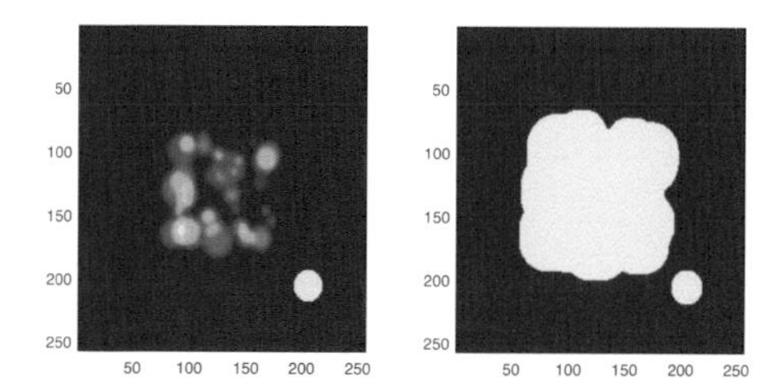

Figure 13.5 The left panel shows the $k = 12$ object of interest with a piecewise constant external disk and right panel shows the support mask used in the experiment. It is a 1-pixel neighborhood of the external object and a 4-pixel neighborhood of the object of interest.

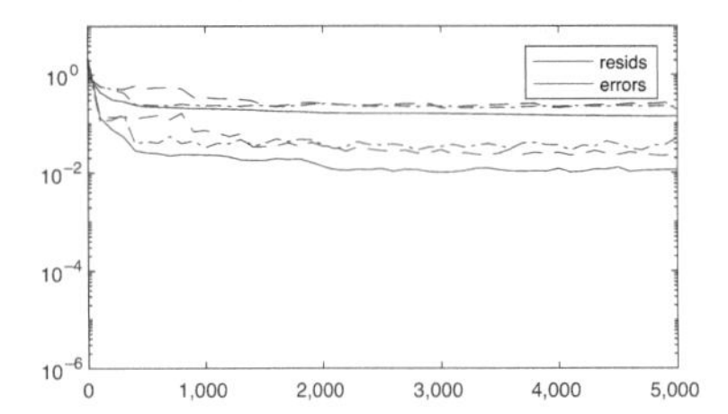

(a) Errors and residuals of 5,000 iterates of $D_{\mathbb{A}_a B_S}$ on objects with $k = 4, 8, 12$ and no external object.

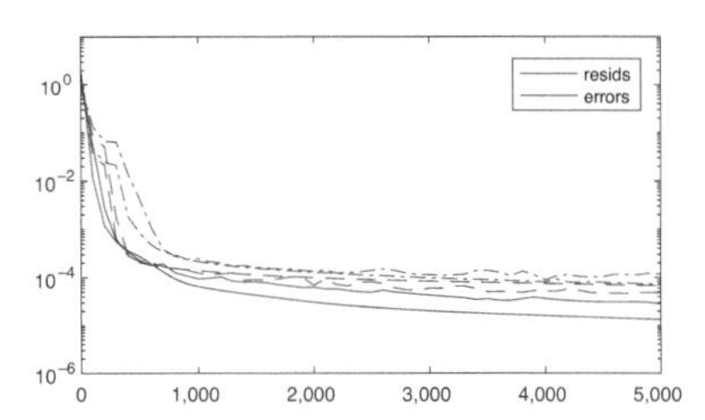

(b) Errors and residuals of 5,000 iterates of $D_{\mathbb{A}_a B_S}$ on objects with $k = 4, 8, 12$ and a piecewise constant external disk.

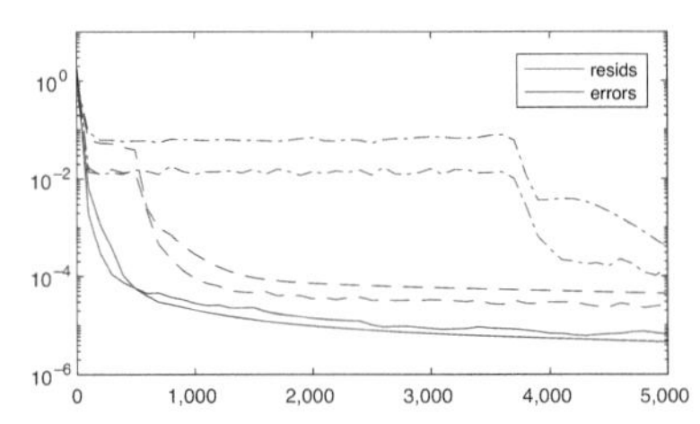

(c) Errors and residuals of 5,000 iterates of $D_{\mathbb{A}_a B_S}$ on objects with $k = 4, 8, 12$ and an $\alpha = 0.5$ external object.

Figure 13.6 These semilog plots show the errors (blue) and residuals (red) of 5,000 iterates of $D_{\mathbb{A}_a B_S}$ on objects with $k = 4, 8, 12$ with (b, c) and without (a) external objects. The solid curves are $k = 4$, the dashed curves are $k = 8$, and the dot-dashed curves are $k = 12$. The support constraint, S, is defined as a 4-pixel neighborhood of the support of the object of interest along with a 1-pixel neighborhood of the support of the external object.

various types of external objects. Clearly, having an external object vastly improves the behavior of this algorithm. Though the $\alpha = 0$ external object works a little faster, both external objects work well, reducing the error by four or five orders of magnitude regardless of the smoothness of the object of interest. While the details of each trial of one of these algorithms is different, these results are typical.

Example 13.3 For the second experiment, we consider the effects of using external objects of different magnitudes relative to the object of interest. For this experiment, we use an object, with $k = 6$, obtained by convolution with a Gaussian; the external object is defined by the parameters

$$\alpha = 0.5, r = 0.1, \ \boldsymbol{x}_0 = (1.4, -1.4), m \in \{0.2,\ 0.6,\ 1, 3,\ 6\}.$$

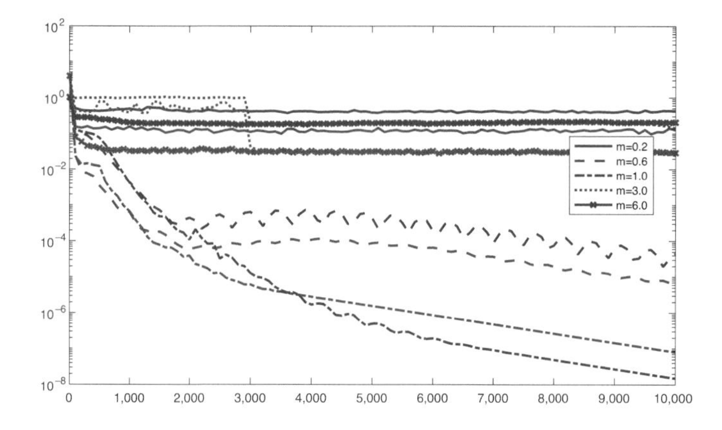

Figure 13.7 Plots of errors and residuals with external objects of various magnitudes.

The plots in Figure 13.7 show the errors and residuals for 10,000 iterates of $D_{\mathbb{A}_a B_S}$ with these choices. We use a 3-pixel neighborhood of the object of interest along with a 1-pixel neighborhood of the external object. The graphs of the errors are blue, and those of the residuals are red. The solid plots have $m = 0.2$, the dashed plots have $m = 0.6$, the dot-dashed plots have $m = 1$, the dotted plots have $m = 3$, and the solid-with-x plots have $m = 6$. From these plots it is clear that the external object is only effective if its magnitude is reasonably close to that of the object of interest, i.e., $m = 0.6$ or 1.

Example 13.4 Our final experiments examine the effects of the parameters r and $\boldsymbol{x}_0$, as well as the extent of oversampling. We use an object with $k = 6$ and a 2-pixel support neighborhood which sits in a 64×64-pixel rectangle, and we examine the effects of an external object on the behavior of a hybrid iterative map algorithm, $D_{\mathbb{A}_a B_S}$, with either double or triple oversampling: that is, 128×128 or 192×192 samples. In Figure 13.8(a) we show the object of interest and its support mask with double oversampling. For purposes of comparison we show, in Figure 13.8(b), the errors and residuals obtained by running 10 trials of 5,000 iterates each of the hybrid iterative map algorithm without an external object. For all of the trials the errors are greater than 10^{-1} and the residuals are about the squares of the errors.

In the experiments below, the coordinates are normalized so that the object of interest lies in the square $R_1 = [-1, 1] \times [-1, 1]$. The external object is a piecewise constant disk ($\alpha = 0$) scaled to have its maximum DFT coefficient equal in magnitude to that of the object of interest, with radius that takes values

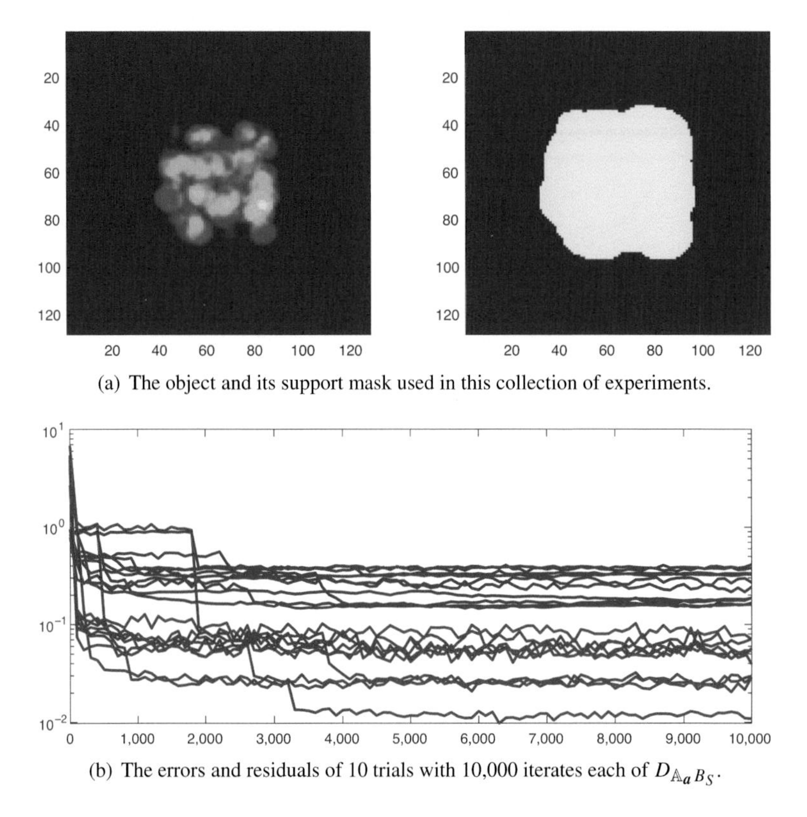

(a) The object and its support mask used in this collection of experiments.

(b) The errors and residuals of 10 trials with 10,000 iterates each of $D_{\mathbb{A}_a B_S}$.

Figure 13.8 The object of interest and support mask used in experiments below. The errors and residuals obtained without an external object.

$2r_j \in \{0.1,\ 0.15,\ 0.2,\ 0.25,\ 0.3,\ 0.35,\ 0.4,\ 0.45\}$. The disk is placed along a ray ℓ_θ, through (0,0), making an angle $\theta = -\pi/4$ with the x-axis. In the double oversampling case the field of view is $R_2 = [-2,\ 2] \times [-2,\ 2]$ and in the triple oversampling case it is $R_3 = [-3,\ 3] \times [-3,\ 3]$.

In the double oversampling case, the center of the disk is placed along the line segment $\ell_\theta \cap R_2 \setminus R_1$. Let d_θ denote the length of this intersection and $p_\theta = \ell_\theta \cap \partial R_1$. The center is placed along this line segment at distances $\mu_k d_\theta$ from p_θ, where $\mu_k \in \{0,\ 0.1,\ 0.2,\ 0.3,\ 0.4,\ 0.5,\ 0.6,\ 0.7\}$. These geometric parameters are illustrated in Figure 13.9. Similar considerations apply using triple oversampling by simply replacing $\ell_\theta \cap R_2 \setminus R_1$ with $\ell_\theta \cap R_3 \setminus R_1$, and d_θ with the length of this intersection.

To get an accurate assessment of the effects of the external object on the behavior of the algorithm, for each choice of parameters $\{r_j,\ \mu_k\}$, we run

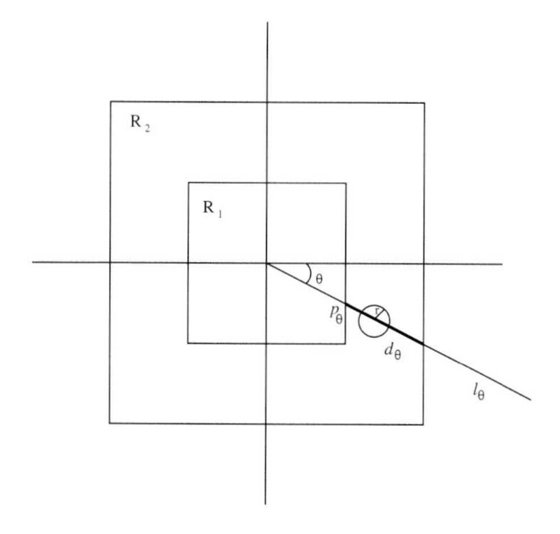

Figure 13.9 Diagram of the geometric parameters used in the experiments in Example 13.4.

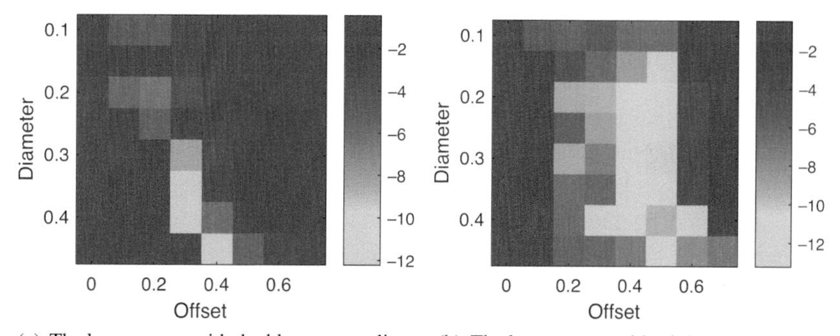

(a) The $\log_{10}$-errors with double oversampling. (b) The $\log_{10}$-errors with triple oversampling

Figure 13.10 Plots showing the $\log_{10}$ of the errors in the minimum residual trials, out of 48, with external objects defined by parameters from the set $\{r_j, \mu_k\}$, with $\theta = -\pi/4$. For the plot in (a) we use a 128×128 field of view and in (b) a 192×192 field of view. In these plots magenta indicates errors of about 10^{-2}, and blue an error of about 10^{-12}.

48 trials of 5,000 iterates each. The "best" reconstruction is then taken to be the one with the minimum residual. Each experiment is run with both 128×128 samples and 192×192 samples, resulting in the plots in Figure 13.10, which show the $\log_{10}$ of the errors for the minimum residual trials. The plot in Figure 13.10(a) employed double oversampling and that in Figure 13.10(b), triple oversampling.

In each plot, the offset distance varies along the x-axis and the diameter varies along the y-axis. In the least effective cases the errors are reduced to about 10^{-2}, that is, by more than an order of magnitude. For the best cases the errors are about 10^{-12}, which is close to machine precision. With triple oversampling, the best results are obtained with a much wider range of parameters, particularly as regards the diameter of the external disk. To summarize, the addition of a "hard" external disk, with known support, offers the possibility of dramatically improved reconstructions obtained using a hybrid iterative map.

Remark 13.5 Crimmins and Fienup (1983) show that having an object with disconnected components, satisfying certain separation conditions, leads, even in one dimension, to a phase retrieval problem that has a unique solution. More generally, they found that strong obstructions to nonuniqueness, like having multiple components, often lead to superior performance of reconstruction algorithms, consistent with the results of this section.

13.3 A Geometric Newton's Method for Phase Retrieval

We next discuss an approach which leads to a geometric Newton-type algorithm and is an application of our analysis of the tangent bundle of $\mathbb{A}_{\boldsymbol{a}}$. As usual, we let $f_0 \in \mathbb{R}^J$ denote the unknown image, and B denote the subset of $\mathbb{R}^J$ specified by the auxiliary constraints; we denote the measured data by $\boldsymbol{a} = \mathscr{M}(f_0)$. The idea is that for $f \in \mathbb{A}_{\boldsymbol{a}}$, the tangent space $T_f\mathbb{A}_{\boldsymbol{a}}$ is the best linear approximation to $\mathbb{A}_{\boldsymbol{a}}$ in a neighborhood of f. It therefore makes sense to replace the search for $\mathbb{A}_{\boldsymbol{a}} \cap B$, with its (partial) linearization: the search for $T_f\mathbb{A}_{\boldsymbol{a}} \cap B$. The extent to which this is actually a linear problem depends on the choice of B; if $B = B_S$, then this is a linear problem. Since $\mathbb{A}_{\boldsymbol{a}}$ is always nonlinear, this is more linear than the original problem. We call this the *geometric* Newton's method for solving the phase retrieval problem. Note that it does not entail the computation of a Hessian, but solving the linearized problems remains computationally demanding. While we do not provide any experiments, these methods should also work well with complex–valued images.

The use of continuous optimization methods is not new in phase retrieval. Gonsalves (1976) uses a Fletcher–Powell gradient search for phase retrieval in the context of wave-front sensing. Fienup (1982) examines approaches to phase retrieval employing conjugate gradient, a Newton–Raphson method, and steepest descent. In Marchesini (2007), hybrid maps were combined with a lower-dimensional subspace saddle point optimization to improve the convergence rate. Pham et al. (2019) considered generalized proximal smoothing methods and found them to be more effective than hybrid maps for biological specimens. Lane investigated the use of conjugate gradient methods in Lane (1991), and Osherovich showed that Newton-type methods can be very successful given good initial approximations for the phases (Osherovich 2011).

The precise form that the geometric Newton's method takes depends on the auxiliary data encoded in B. Because $\dim \mathbb{A}_{\boldsymbol{a}} + \dim B$ is usually much less than $|J|$, it is reasonable to expect that the intersections $T_{\boldsymbol{f}}\mathbb{A}_{\boldsymbol{a}} \cap B$ are usually empty. Instead of trying to find an intersection point, we seek a pair $\widetilde{\boldsymbol{f}} \in T_{\boldsymbol{f}}\mathbb{A}_{\boldsymbol{a}}$, and $\widetilde{\boldsymbol{x}} \in B$ so that $\|\widetilde{\boldsymbol{f}} - \widetilde{\boldsymbol{x}}\|_2$ minimizes $\text{dist}(T_{\boldsymbol{f}}\mathbb{A}_{\boldsymbol{a}}, B)$. If B is convex, then this problem is, in principle, solvable. If $B = B_S$ is the linear subspace defined by a support condition, or $B = B_+$, then either the intersection is nonempty or the minimum distance is positive and is attained for a pair of points $(\widetilde{\boldsymbol{f}}, \widetilde{\boldsymbol{x}}) \in T_{\boldsymbol{f}}\mathbb{A}_{\boldsymbol{a}} \times B$. For the case $B = B_S$, this fact is classical, for $B = B_+$ it follows from:

Theorem 13.6 *Let $A \subset \mathbb{R}^n$ be an affine subspace, and $B_+ \subset \mathbb{R}^n$ the nonnegative orthant. Either $A \cap B_+ \neq \emptyset$, or the $\text{dist}(A, B_+)$ is positive and is attained at a pair of points $\boldsymbol{v}_0 \in A$, $\boldsymbol{x}_0 \in B_+$.*

This theorem is proved in Appendix 13.A.

As an algorithm, the geometric Newton's method is as follows: fix a tolerance $\epsilon_{\text{tol}} > 0$:

(i) Choose a point, $\boldsymbol{f} \in \mathbb{A}_{\boldsymbol{a}}$, and compute the orthogonal projection, $P_{T^0_{\boldsymbol{f}}\mathbb{A}_{\boldsymbol{a}}}$ onto the fiber, $T^0_{\boldsymbol{f}}\mathbb{A}_{\boldsymbol{a}}$ of the tangent bundle to $\mathbb{A}_{\boldsymbol{a}}$ at $\boldsymbol{f}$.

(ii) Find a pair of points $\widetilde{\boldsymbol{f}} \in T_{\boldsymbol{f}}\mathbb{A}_{\boldsymbol{a}}$ and $\widetilde{\boldsymbol{x}} \in B$ so that

$$(\widetilde{\boldsymbol{f}}, \widetilde{\boldsymbol{x}}) = \underset{\varphi \in T_{\boldsymbol{f}}\mathbb{A}_{\boldsymbol{a}}, \xi \in B}{\arg\min} \|\varphi - \xi\|_2. \tag{13.6}$$

(iii) If $\|\widetilde{\boldsymbol{f}} - \widetilde{\boldsymbol{x}}\|_2 > \epsilon_{\text{tol}}$ then return to step (i), with $\boldsymbol{f}$ replaced by $\widetilde{\boldsymbol{f}}$.

For standard auxiliary information the set B is convex, so this method replaces the problem of finding an intersection between the nonconvex set, $\mathbb{A}_{\boldsymbol{a}}$ and B, with that of finding an intersection, or closest pair of points, for a pair of convex sets. While, in principle, this is a solvable problem, in the context of phase retrieval it can be quite challenging. First, the dimension of the ambient space is usually quite large and second, the subsets whose intersections we seek often meet nontransversally, or at very shallow angles. This makes these intersections, or pairs of nearest points, very difficult to find, in practice. For pairs of convex sets, it is known that the alternating projection (AP) algorithm often converges to either an intersection point, or a pair of nearest points.

For the auxiliary data used in phase retrieval the projection, P_B, onto the set B can often be very efficiently implemented. The projection onto the fiber of the tangent bundle of an amplitude torus, $P_{T_f^0 \mathbb{A}_a}$, has an implementation of the form

$$P_{T_f^0 \mathbb{A}_a}(\boldsymbol{g}) = \boldsymbol{g} - \mathscr{F}^{-1} \Lambda^f \left[\operatorname{Re}(\overline{\Lambda^f}(\widehat{\boldsymbol{g}})) \right],$$

where $\mathscr{F}$ is the DFT and Λ^f is a diagonal matrix. See Appendix 2.B.

The AP algorithm, defined by taking $\boldsymbol{g}^{(0)} = \boldsymbol{f}$ and

$$\boldsymbol{g}^{(n+1)} = P_{T_f \mathbb{A}_a} \circ P_B(\boldsymbol{g}^{(n)}), \tag{13.7}$$

can therefore be very efficiently implemented. In Bauschke and Borwein (1996) it is shown that, if B is convex and $T_f \mathbb{A}_a \cap B \neq \emptyset$, then these iterates converge to the intersection point closest to the initial point. If this intersection is empty, but the minimum of $\operatorname{dist}(T_f \mathbb{A}_a, B)$ is positive and attained at pairs of point in $T_f \mathbb{A}_a \times B$, then the iterates, $\{(\boldsymbol{g}^{(n+1)}, P_B(\boldsymbol{g}^{(n)})\}$, converge to such a minimizing pair. While it is unlikely to be a computationally efficient method, the iteration defined in (13.7) can be used whether or not the intersection $T_f \mathbb{A}_a \cap B$ is empty; in the examples below, the Newton iteration step is implemented using the iteration in (13.7).

As with all Newton-type methods, a good initial guess is required for their success. If the initial phases are all within $\frac{\pi}{2} - \epsilon$, for a positive ϵ, of the true phases, then the algorithm usually finds an attracting basin defined by a true intersection. A similar condition is used in Osherovich (2011). The rate of convergence depends, roughly speaking, on the angles between $T_f \mathbb{A}_a$ and the portion of B nearest to $\boldsymbol{f}$, and therefore, such an algorithm is not expected to work well when the true intersections are nontransversal. Implementing the solver for the linearized problem becomes computationally expensive when the images are larger than 64×64. In the examples below we see that using Newton's method as a second step, after running a hybrid iterative algorithm, can quickly provide additional digits of accuracy that would require many more iterates of the hybrid iterative algorithm alone. Indeed, even though the hybrid iterative map may have effectively stagnated, Newton's method might still provide a few extra digits of accuracy.

13.3.1 Numerical Examples for the Geometric Newton's Method

We give some examples of Newton's method with either the support or nonnegativity constraint for 64×64 images. In the first example we demonstrate the behavior of Newton's algorithm alone, starting with random phases that lie within $\pi/2.1$ of the correct phases. In the second we use a few iterates of Newton's method to get several more digits of accuracy after 1,000 iterates of the hybrid map, using either the support or nonnegativity constraint.

Example 13.7 For this experiment we use Newton's method alone to reconstruct a 64×64 image, using either support, or nonnegativity as auxiliary information. The initial image is defined by randomizing the phases with angles lying within intervals of the form $\left[\theta_j - \frac{\pi}{2.1}, \theta_j + \frac{\pi}{2.1}\right]$, where $\{\theta_j : j \in J\}$ are the correct phases. This ensures that the initial image is in an attracting basin for the Newton algorithm.

Figure 13.11 shows the results of running 200 iterates of Newton's method using the smallest enclosing rectangle to define the support constraint. The initial error is about 0.68; after 13 iterates the error has fallen to 4.87×10^{-6}. The remaining iterates show a very slow decrease in the error, with the ratio of the error to the residual equal to about 18. The top row of the figure shows the rectangle defining the support constraint, the reference image, and

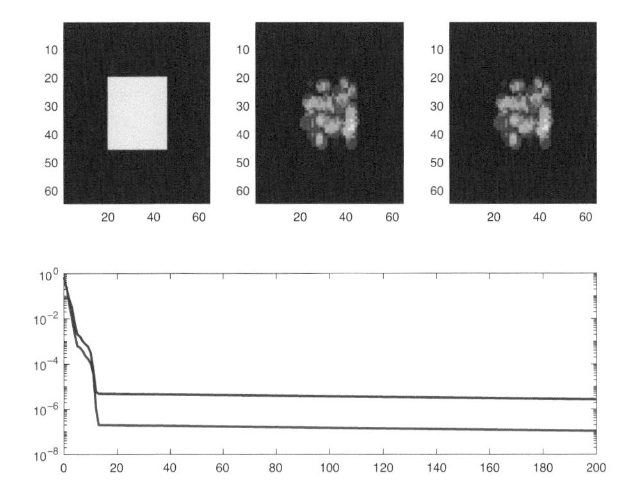

Figure 13.11 200 iterates of Newton's method using a rectangular support constraint with a $k = 0$ example. In the top row of the figure the left panel shows the rectangle defining the support constraint, the middle panel is the reference image, and the right panel is the reconstructed image. The plot on the bottom shows the errors (blue) and the residuals (red).

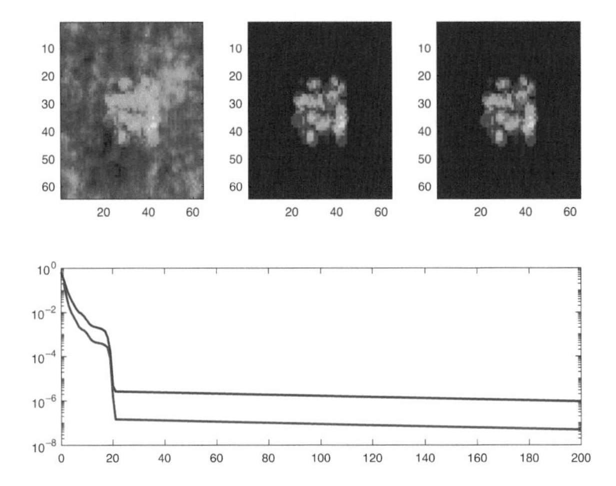

Figure 13.12 200 iterates of Newton's with the nonnegativity constraint on a $k=0$ example. In the top row of the figure the left panel shows the initial image, the middle panel is the reference image, and the right panel is the reconstructed image. The plot on the bottom shows the errors (blue) and the residuals (red).

the reconstructed image. The plot on bottom shows the errors (blue) and the residuals (red).

The next experiment shows the results of applying Newton's method with the nonnegativity constraint on a piecewise constant image. For the initial image we again randomize the phases, with perturbations uniformly distributed in $\left[-\frac{\pi}{2.1}, \frac{\pi}{2.1}\right]$. The initial error is again about 0.68; after 22 iterates the error has fallen to 2.6×10^{-6}. As before, the remaining iterates show a very slow decrease in the error, with the ratio of the error to the residual equal to about 25. The top row in Figure 13.12 shows the initial image, the reference image, and the reconstructed image. The plot on the bottom row shows the errors (blue) and the residuals (red).

Example 13.8 For the second groups of experiments, we use 3 iterates of Newton's method to improve a reconstruction found using 1,000 iterates of a hybrid iterative map algorithm. For these experiments we use random phases to define the initial image.

Figure 13.13 shows the results of this approach on a piecewise constant, 64×64 image using a 1-pixel support constraint as auxiliary information. The initial 1,000 iterates use $D_{\mathbb{A}_a B_{S_1}}$. The iterates of the hybrid iterative map have fallen into the attracting basin defined by the center manifold of $f_0^{([1,1])}$, which is a point where $\mathbb{A}_a$ and B_{S_1} intersect transversally. The error after 1,000

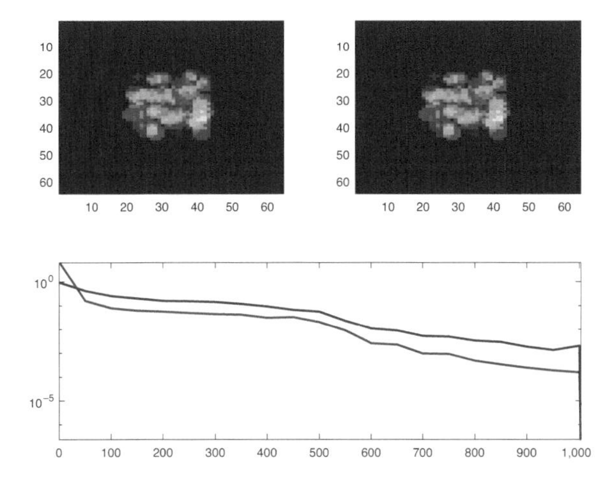

Figure 13.13 1,000 iterates using $D_{\mathbb{A}_a B_{S_1}}$ with a $k = 0$ example, followed by 3 iterates of Newton's method. In the top row of the figure the left panel shows the reconstructed image after the initial 1,000 iterates of $D_{\mathbb{A}_a B_{S_1}}$, and the right panel is the final reconstructed image. In the plot on bottom row is the error (blue) and the residual (red).

iterates of $D_{\mathbb{A}_a B_{S_1}}$ is 2.03×10^{-3}; after 3 iterates of Newton's method, using the support constraint, it is 6.54×10^{-6}.

In the second experiment we use $D_{\mathbb{A}_a B_+}$ for the first 1,000 iterates and then 3 iterates of Newton's method, using nonnegativity, for the last 3 iterates (Figure 13.14). In this experiment we use a 64×64, $k = 2$ image, where the image is defined as in (1.46). After 1,000 iterates of $D_{\mathbb{A}_a B_+}$ the error is 6.93×10^{-3}; the 3 additional iterates of Newton's method reduce the error to 6.35×10^{-5}.

Running additional iterates of Newton's method, with either type of auxiliary information, does not produce a significant decrease in the error. We have also found that using one sort of auxiliary information for the initial iterates and the other sort with Newton's method produces very poor results.

13.4 Implementation of the Holographic Hilbert Transform Method

In Chapter 7.4 we introduce a noniterative reconstruction method that employs a holographic experimental setup and uses the Hilbert Transform to recover the missing phase information. To analyze the implementation of the HHT

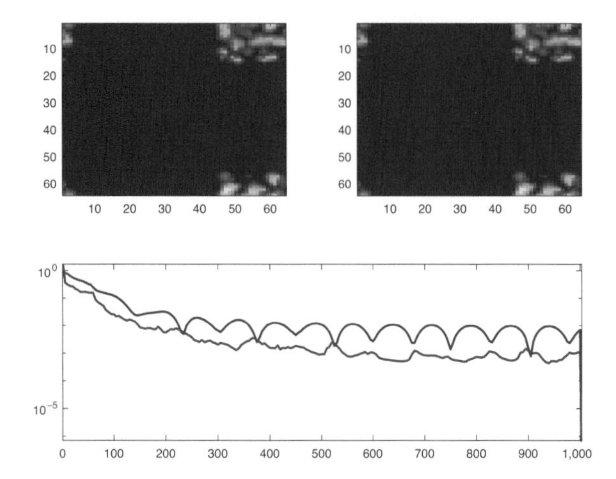

Figure 13.14 1,000 iterates using $D_{\mathbb{A}_a B_+}$ with a $k = 2$ example, followed by 3 iterates of Newton's method. In the top row of the figure the left panel shows the reconstructed image after 1,000 iterates of $D_{\mathbb{A}_a B_+}$, and the right panel is the final reconstructed image. In the plot on bottom row is the error (blue) and the residual (red).

method, on finitely sampled data we begin, as before in the continuum world, assuming that the object being imaged can be modeled as

$$f(\boldsymbol{x}) = M\delta(\boldsymbol{x} - \boldsymbol{c}) + g(\boldsymbol{x}). \tag{13.8}$$

Here, $\boldsymbol{c}$ is a point outside the smallest rectangle containing $\operatorname{supp} g$, and $M > \|g\|_1$. This is another form of external holography, but with a small, highly diffracting object placed just outside the support of the object of interest. The reconstruction method is noniterative and works well with a field of view twice the size of the object of interest.

The Fourier transform of f is

$$\widehat{f}(\boldsymbol{\xi}) = Me^{-2\pi i \boldsymbol{c}\cdot\boldsymbol{\xi}} + \widehat{g}(\boldsymbol{\xi}). \tag{13.9}$$

In this context the measured data consists of

$$\{|\widehat{f}(\boldsymbol{\xi}_{\boldsymbol{k}})|^2 : \boldsymbol{k} \in J\}, \tag{13.10}$$

where $J = [-N : N]^d$ and $\boldsymbol{\xi}_{\boldsymbol{k}} = \boldsymbol{k}\Delta k$. It is assumed that M is known, so that

$$\log|\widehat{f}(\boldsymbol{\xi}_{\boldsymbol{k}})| - \log M = \log\left|1 + \frac{e^{2\pi i \boldsymbol{c}\cdot\boldsymbol{\xi}}\widehat{g}(\boldsymbol{\xi}_{\boldsymbol{k}})}{M}\right| \tag{13.11}$$

can be determined from the data. Using this information we would like to compute an approximation to $\mathscr{H}\left[\log\left|1+\frac{e^{2\pi i c\cdot\xi}\widehat{g}(\xi)}{M}\right|\right]$ in order to recover the harmonic conjugate of $\log|\widehat{f}(\boldsymbol{\xi}_k)|$, namely the *desired phase*. In this section, we consider how the choices of $M, \Delta k$, and N affect the accuracy of this approximation. For simplicity we only do this analysis in the one-dimensional case, with

$$\widehat{h}(\xi) = \log\left|1+\frac{\widehat{G}(\xi)}{M}\right|, \tag{13.12}$$

where

$$\widehat{G}(\xi) = e^{2\pi i c\xi}\widehat{g}(\xi). \tag{13.13}$$

For this analysis, the location, c, of the δ-source is selected so that $G = \mathscr{F}^{-1}(\widehat{G})$, the translate of g, is supported in $[-l, 0]$.

In the following sections we also show the results of numerical experiments in one and two dimensions, using both an exact δ-spike and approximations to such a spike as described in Section 7.4.2. The somewhat surprising outcome of these experiments is that, even for functions φ that are very far from satisfying the hypotheses laid out in Section 7.4.2, the HHT works quite well.

13.4.1 Error Estimates in the 1D case

The transform we need to compute is $\mathscr{H}(\widehat{h}) = -\mathscr{F}[\operatorname{sgn} x \mathscr{F}^{-1}(\widehat{h})]$. The available data are finitely many samples of $\widehat{h}$, $\{\widehat{h}(j\Delta k) : -N \leq j \leq N\}$. We therefore use the approximation

$$\mathscr{H}_a(\widehat{h})(m\Delta k) = -\Delta x \sum_{n=-N}^{N} h_n \operatorname{sgn}(n) e^{-2\pi i m n \Delta k \Delta x}, \tag{13.14}$$

where $m \in [-N : N]$, and the spatial resolution, $\Delta x = (2N\Delta k)^{-1}$, is determined by the highest frequency sampled. The sequence $\{h_n\}$ is defined by

$$h_n = \Delta k \sum_{j=-N}^{N} \widehat{h}(j\Delta k) e^{2\pi i j n \Delta k \Delta x}. \tag{13.15}$$

In a calculation of this type there are two sources of error:

(i) Sampling errors, which result from collecting data on a discrete grid.
(ii) Truncation errors, which result from collecting finitely many data points.

To assess both errors we use the Poisson summation formulæ: suppose that φ is a sufficiently nice function with Fourier transform $\widehat{\varphi}$, then

$$\begin{aligned}\frac{1}{K}\sum_{j=-\infty}^{\infty}\widehat{\varphi}\left(\frac{j}{K}\right)e^{\frac{2\pi ijx}{K}} &= \sum_{j=-\infty}^{\infty}\varphi(x+jK),\\ \frac{1}{K}\sum_{j=-\infty}^{\infty}\varphi\left(\frac{j}{K}\right)e^{-\frac{2\pi ij\xi}{K}} &= \sum_{j=-\infty}^{\infty}\widehat{\varphi}(\xi+jK).\end{aligned} \tag{13.16}$$

A standard hypothesis for the validity of these formulæ is that there is a constant C and a $0 < \delta$ so that

$$|\varphi(x)| + |\widehat{\varphi}(x)| \leq \frac{C}{(1+|x|)^{1+\delta}}. \tag{13.17}$$

We henceforth assume that the data $\widehat{g}(\xi)$ satisfies such an estimate. This implies that g is a continuous function, so strictly speaking, the analysis in this section does not directly apply to piecewise constant functions, or, more generally, to functions with jump discontinuities. No additional assumption is required on $h(x)$ as it decays exponentially; see Lemma 13.9.

M is selected so that $M > 2\|g\|_1$, and therefore, (13.11) implies that

$$\widehat{h}(\xi) = \operatorname{Re}\left[\sum_{j=1}^{\infty}\frac{(-1)^{j+1}}{j}\left(\frac{\widehat{G}(\xi)}{M}\right)^j\right], \tag{13.18}$$

is uniformly and absolutely convergent for $\xi \in \mathbb{R}$. Standard properties of the Fourier transform imply that

$$h(x) = \sum_{j=1}^{\infty}\frac{(-1)^{j+1}}{j}\frac{G^{*j}(x) + \overline{G}^{*j}(-x)}{M^j}, \tag{13.19}$$

where $*_j$ denotes the j-fold convolution product:

$$H^{*j} = H \overset{j-\text{times}}{* \cdots *} H. \tag{13.20}$$

The terms of the sum satisfy

$$\operatorname{supp} G^{*j}(x) \subset [-jl, 0] \text{ and } \overline{G}^{*j}(-x) \subset [0, jl], \tag{13.21}$$

and all of these functions vanish at $x = 0$. A classical estimate for the convolution shows that

$$\|G^{*j}\|_\infty \leq \|G\|_1^{j-1}\|G\|_\infty. \tag{13.22}$$

Using this estimate and (13.21) we easily establish the following estimate for h.

Lemma 13.9 *Let* $a = \log(M/\|G\|_1)/l$. *Then*

$$|h(x)| \leq C_{M,l}\|G\|_\infty e^{-a|x|}, \tag{13.23}$$

where

$$C_{M,l} = \frac{2e^{la}}{M}. \tag{13.24}$$

Proof To prove the lemma we observe that if $lp \leq |x| < l(p+1)$, then (13.21) implies that

$$h(x) = \sum_{j=p+1}^{\infty} \frac{(-1)^{j+1}}{j} \frac{G^{*j}(x) + \overline{G}^{*j}(-x)}{M^j}, \tag{13.25}$$

and therefore, (13.22) gives the estimate

$$\begin{aligned} |h(x)| &\leq \sum_{j=p+1}^{\infty} \frac{\|G\|_1^{j-1}\|G\|_\infty}{jM^j} \\ &\leq \frac{\|G\|_\infty}{(p+1)M} \frac{e^{-\alpha p}}{1-e^{-\alpha}}, \end{aligned} \tag{13.26}$$

where $\alpha = \log(M/\|G\|_1)$. The conclusion follows as $e^{-\alpha} < \frac{1}{2}$ and $|x| < (p+1)l$. □

Remark 13.10 If we let

$$h_p(x) = \sum_{j=1}^{p} \frac{(-1)^{j+1}}{j} \frac{G^{*j}(x) + \overline{G}^{*j}(-x)}{M^j}, \tag{13.27}$$

then the proof of the lemma also shows that

$$|h(x) - h_p(x)| \leq \frac{\|G\|_\infty}{(p+1)M} \frac{e^{-\alpha p}}{1-e^{-\alpha}}, \tag{13.28}$$

for all $x \in \mathbb{R}$.

The first step in the computation of $\mathscr{H}(\widehat{h})$ is the calculation of $h = \mathscr{F}^{-1}(\widehat{h})$; if $N = \infty$, then using the Poisson summation formula, this is approximated by

$$\begin{aligned} h(x) &\approx \Delta k \sum_{j=-\infty}^{\infty} \widehat{h}(j\Delta k)e^{2\pi ij\Delta kx} \\ &= \sum_{j=-\infty}^{\infty} h\left(x + \frac{j}{\Delta k}\right), \end{aligned} \tag{13.29}$$

The error at this step from sampling is, therefore,

$$e_1(x) = \sum_{j\neq 0} h\left(x + \frac{j}{\Delta k}\right). \tag{13.30}$$

This error is easily estimated using Lemma 13.9

$$|e_1(x)| \leq C_{M,l}\|G\|_\infty \frac{e^{-\frac{a}{2\Delta k}}}{1-e^{-a\Delta k^{-1}}} \quad \text{for } x \in \left[\frac{-1}{2\Delta k}, \frac{1}{2\Delta k}\right]. \tag{13.31}$$

By taking Δk small enough, we can make $e^{-a(2\Delta k)^{-1}}$ as small as we like.

The truncation error is

$$e_2(x) = \Delta k \sum_{|j|>N} \widehat{h}(j\Delta k) e^{2\pi i j\Delta k x}. \tag{13.32}$$

To control the truncation error, e_2, we need to take N large enough so that

$$\begin{aligned} |e_2(x)| &\leq 2\frac{\Delta k}{M} \sum_{|j|>N} |\widehat{g}(j\Delta k)| \\ &\approx \frac{1}{M} \int\limits_{N\Delta k<|\xi|} |\widehat{g}(\xi)| d\xi << 1. \end{aligned} \tag{13.33}$$

Taking M large also helps to control this error.

For the next step we compute $-\mathscr{F}[\operatorname{sgn}(x)h]$. If we had the samples $\{h(n\Delta x) : -\infty < l < \infty\}$, then we would set

$$-\mathscr{F}[\operatorname{sgn}(x)h](\xi) \approx -\Delta x \sum_{n=-\infty}^{\infty} h(n\Delta x)\operatorname{sgn}(x) e^{-2\pi i l\Delta x\xi}. \tag{13.34}$$

Using the finitely sampled Fourier data we actually evaluate the expression in (13.14). For the moment assume that we can determine the values $\{h(n\Delta x)\}$.

The sampling error is given by

$$\widetilde{e}_1(\xi) = \int_{\infty}^{-\infty} h(x)e^{-2\pi i x\xi}\operatorname{sgn}(x)dx - \sum_{n=-\infty}^{\infty} h(n\Delta x)e^{-2\pi i n\Delta x\xi}\operatorname{sgn}(x)\Delta x, \tag{13.35}$$

which we estimate below for $\xi \in [-N : N]\Delta k$. Finally, the truncation error is given by

$$\widetilde{e}_2(\xi) = \Delta x \sum_{|n|>N} h(n\Delta x)\operatorname{sgn}(n)e^{-2\pi i n\Delta x\xi}. \tag{13.36}$$

A simple estimate for this sum is given by

$$
\begin{aligned}
|\widetilde{e}_2(\xi)| &\leq \int_{|x|>\frac{1}{2\Delta k}} |h(x)|dx \\
&\leq \frac{C_{M,l}\|G\|_\infty e^{-\frac{a}{2\Delta k}}}{a}.
\end{aligned}
\tag{13.37}
$$

We now use the expansion in (13.19) and Lemma 13.9 to estimate $\widetilde{e}_1(\xi)$ in terms of the data, $\widehat{g}(\xi)$,

$$
\begin{aligned}
|\widetilde{e}_1(\xi)| &= \left| \int_{-\infty}^{\infty} h(x)\,\mathrm{sgn}(x) e^{-2\pi i x\xi} dx - \sum_{n=-\infty}^{\infty} h(n\Delta x) e^{-2\pi i l \Delta x \xi}\,\mathrm{sgn}(x)\Delta x \right| \\
&= \left| \int_{-N\Delta x}^{N\Delta x} h(x) e^{-2\pi i x\xi}\,\mathrm{sgn}(x) dx - \sum_{n=-N}^{N-1} h(n\Delta x)\,\mathrm{sgn}(x) e^{-2\pi i x\xi} \Delta x \right| \\
&\quad + \frac{2C_{M,l}\|G\|_\infty}{a}(2+a\Delta x)e^{-aN\Delta x}.
\end{aligned}
\tag{13.38}
$$

The error term in the third line follows from the estimate in Lemma 13.9.

To complete the estimate, we replace the function, h, with a partial sum, h_p, defined in (13.27), where p is chosen so that $N\Delta x = lp$. Using the estimate in (13.28) we see that

$$
\begin{aligned}
|\widetilde{e}_1(\xi)| \leq &\left| \int_{-N\Delta x}^{N\Delta x} h_p(x)\,\mathrm{sgn}(x) e^{-2\pi i x\xi} dx - \sum_{n=-N}^{N-1} h_p(n\Delta x) e^{-2\pi i l \Delta x \xi}\,\mathrm{sgn}(x)\Delta x \right| \\
&+ \frac{4C_{M,l}\|G\|_\infty}{a}(2+a\Delta x + lp)\left(\frac{\|G\|_1}{M}\right)^p.
\end{aligned}
\tag{13.39}
$$

We estimate this error for the sample points $\xi \in \{m\Delta k : -N < m < N\}$. So that we can accurately represent a function supported in $[-lp, lp]$, we need to require that

$$
\Delta k = \frac{1}{2lp},
\tag{13.40}
$$

which is consistent with $\Delta x = lp/N$.

The function, $\mathrm{sgn}(x)$, is not continuous at $x = 0$, but $h_p(x)\,\mathrm{sgn}(x)$ is the sum of two functions, one supported in $[-lp,\ 0]$, and the other $[0, lp]$. These functions vanish at 0 to an order determined by the smoothness of G. If $G \in \mathscr{C}_P^q([-l,0])$, i.e., q-times differentiable as an l-periodic function, then $G^{*j} \in \mathscr{C}_P^{jq}([-jl,0])$; hence, $h_p \in \mathscr{C}_P^q([-lp,lp])$.

To estimate $|\widetilde{e}_1(m\Delta k)|$ all that remains is an estimate for the error incurred in using the uniformly spaced Riemann sum on the right-hand side of (13.39) as an approximation for the Fourier series coefficients of the periodic function, $h_p(x)\operatorname{sgn}(x)$. The results from Epstein (2005) show that this depends only on the smoothness of $h_p \operatorname{sgn}(x)$ as a $2lp$-periodic function. As noted, if we assume that $g \in \mathscr{C}^q(\mathbb{R})$, then, after appropriately rescaling the results in Epstein (2005) we obtain that

$$|\widetilde{e}_1(m\Delta k)| \le 2(6lp)^{q+1}\frac{\omega^q_{h_p}\left(\frac{\Delta x}{\pi}\right)}{(2\pi N)^q} + \frac{4C_{M,l}\|G\|_\infty}{a}(2+a\Delta x+lp)\left(\frac{\|G\|_1}{M}\right)^p$$
$$\text{for } -N \le m \le N-1. \tag{13.41}$$

Here, $\omega^q_{h_p}$ is the modulus of continuity of $\partial_x^q h_p$, which is largely determined by that of $\partial_x^q G$ itself. Notice that the estimate has a factor of $(lp)^{q+1}$; l is minimized by choosing c in (13.13) so that it lies exactly at the boundary of $\operatorname{supp} g$.

A final correction is required as the values $\{h(n\Delta x) : -N \le l \le N\}$ are not actually available, only the approximations $\{h_n : -N \le l \le N\}$, which produces one further error term

$$\begin{aligned}\widetilde{e}_3(\xi) &= \sum_{n=-N}^{N-1}[h_n - h(n\Delta x)]e^{-2\pi i n\Delta x\xi}\operatorname{sgn}(x)\Delta x\\ &= \sum_{n=-N}^{N-1}[e_1(n\Delta x)+e_2(n\Delta x)]e^{-2\pi i n\Delta x\xi}\operatorname{sgn}(x)\Delta x.\end{aligned} \tag{13.42}$$

Using (13.33) and (13.31), we see that

$$|\widetilde{e}_3(\xi)| \le 2C_{M,l}\|G\|_\infty\frac{e^{\frac{-a}{2\Delta k}}}{a(1-e^{-a\Delta k^{-1}})} + \frac{2N\Delta x}{M}\int\limits_{N\Delta k<|\xi|}|\widehat{g}(\xi)|d\xi. \tag{13.43}$$

Putting together all these results we see that

$$\begin{aligned}|\mathscr{H}_a(\widehat{h})(m\Delta k) - \mathscr{H}(\widehat{h})(m\Delta k)| \le{}& \left(\frac{\|G\|_1}{M}\right)^p\frac{C_{M,l}\|G\|_\infty}{a}[13+a\Delta x+lp]\\ &+12lp\omega^q_{h_p}\left(\frac{\Delta x}{\pi}\right)\left(\frac{3\Delta x}{\pi}\right)^q + \frac{lp}{M}\int\limits_{|\xi|>\frac{1}{2\Delta x}}|\widehat{G}(\xi)|d\xi\end{aligned} \tag{13.44}$$

While these estimates are rather pessimistic, they show that good results can be expected if $\|G\|_1/M$ is small, $g(x)$ is smooth, Δk is small, but

$(2N\Delta k)^{-1} = \Delta x$ is also small. If $G \in \mathscr{C}_P^q([-l,0])$ for a $1 < q$, then the last term can be estimated by

$$\frac{lp}{M} \int_{|\xi| > \frac{1}{2\Delta x}} |\widehat{G}(\xi)| d\xi \leq \frac{N\|G^{[q]}\|_1}{2(q-1)} \left(\frac{\Delta x}{\pi}\right)^q, \tag{13.45}$$

and therefore,

$$|\mathscr{H}_a(\widehat{h})(m\Delta k) - \mathscr{H}(\widehat{h})(m\Delta k)| \leq \left(\frac{\|G\|_1}{M}\right)^p \frac{C_{M,l}\|G\|_\infty}{a}[13 + a\Delta x + lp] + \left(\frac{\Delta x}{\pi}\right)^q \left[12lp3^q \omega_{h_p}^q \left(\frac{\Delta x}{\pi}\right) + \frac{\|G^{[q]}\|_1}{2p(q-1)\Delta x}\right]. \tag{13.46}$$

Recalling that $N\Delta x = lp$, we see that if p is chosen to make the first term as small as needed, then, provided that $1 < q$, we can choose N (or Δx) to make the second term small as well. These estimates indicate that it will usually be necessary to take $2 \leq p$ showing that some degree of "oversampling" is needed to make this approach to image reconstruction practicable. These estimates also suggest that the smoother the data is, the better the HHT can be expected to perform, in sharp contrast to the iterative algorithms considered above, which perform better on piecewise constant data. We next examine how these algorithms perform on the types of data considered above for iterative algorithms.

13.4.2 Numerical Experiments

We begin our numerical experiments in the 1D case, with an image described by a function $g(x)$, which is normalized to have support in an interval of length 1. For the 1D examples we use a δ-spike, which has support in a single pixel. This facilitates the comparison between the experimental results and the analysis in Section 13.4.1. After examining how the reconstruction error in 1D depends on various parameters, we consider several 2D examples.

The error estimate in (13.46) shows that the relative error in using finitely sampled data to approximately compute $\mathscr{H}(\widehat{h})(m\Delta k)$ depends principally on two quantities

$$\left(\frac{\|G\|_1}{M}\right)^p \text{ and } \frac{1}{p}\left(\frac{\Delta x}{\pi}\right)^{q-1}. \tag{13.47}$$

Here, q is a measure of the smoothness of G, the translate of g. The "adjustable" experimental parameters are M, the height of the δ-spike; p, the extent of oversampling (or the choice of Δk); and Δx, which, once Δk is fixed, is equivalent to the number of frequencies sampled. The extent of oversampling is clearly a critical choice. As we see below, with too little oversampling, the calculation in the Fourier domain is not resolved and the resultant reconstruction has fairly serious artifacts. On the other hand, too much oversampling leads quickly to very large computational problems, especially if $d > 1$.

Example 13.11 In our first numerical experiment we explore the effect of oversampling on the reconstruction of a reasonably smooth image ($k = 3$). The height of the spike is fixed at $M = 4\|G\|_1$ and $\Delta x = 2^{-7}$. We assess the outcome of using the HHT method with $p = 2, 4$, and 8. The results are shown in Figures 13.15(a–c).

The figures in the left and center panels of Figure 13.15 all show the same extent of "physical" space; the right panels show the same range of "physical" spatial frequencies. The DFT computations in (b) and (c) are almost identical, while those in (a) differ by orders of magnitude over much of the frequency range. This shows that the twice oversampled computation is, in fact, inadequately resolved, which explains the much larger errors evident throughout the reconstruction. Note also the rapid convergence with p: from about 4 digits with $p = 4$ to about 11 digits with $p = 8$.

Another experimental parameter, c, is the location of the δ-spike. This choice is reflected in the estimate (13.46) by the value of l, which is a measure of the size of the $\operatorname{supp} G$. From the estimates it does not seem likely that this parameter choice is critical. Experimentally we have found that it should be close to the boundary of the support of G, but not too close; having the spike a distance $2\Delta x(1 - \log_2 \Delta x)$ from the $\operatorname{supp} G$ is a good choice, but a somewhat larger distance produces very similar results. Note that the errors are consistently smallest near to the location of the spike.

For our second experiment we assess the effects of the choice of $\Delta x = p/N$ and M on the reconstruction errors for three different levels of oversampling $p = 2, 4, 8$. For the most part, the errors are monotonically decreasing as a function of all three parameters. When $p = 8$ the reconstruction appears to reach machine precision very quickly, and then the errors slowly creep up as M increases or Δx decreases. Overall, these results are consistent with the error estimate in (13.46).

Example 13.12 In this experiment we use the same image considered in the previous example, but explore the effects of the choice of $\Delta x, M$ and p on the overall ℓ^2-error. We consider three levels of oversampling with $p = 2, 4,$

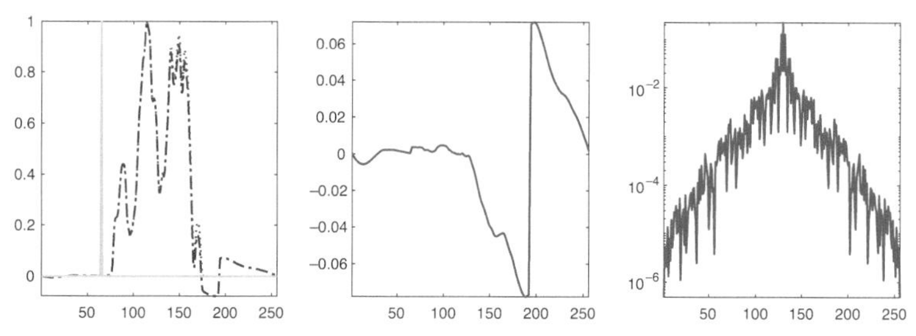

(a) Oversampling with $p = 2$. Note that the maximum error is a little more than 0.06 and the relative ℓ_2-error is 8.9×10^{-2}.

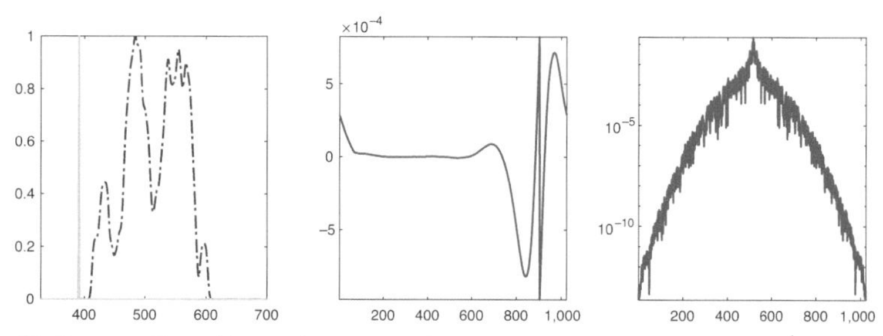

(b) Oversampling with $p = 4$. Note that the maximum error is about 9×10^{-4} and the relative ℓ_2-error is 1.05×10^{-3}.

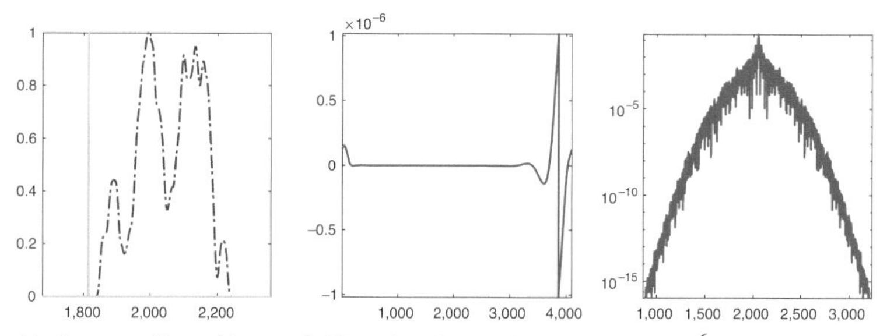

(c) Oversampling with $p = 8$. Note that the maximum error is 10^{-6} and the relative ℓ_2-error is 8.8×10^{-7}.

Figure 13.15 These plots illustrate the effects of the oversampling rate on the reconstruction error using the HHT method with a 1-pixel spike. The figures in the left panels are cropped to show just the reconstructed object; the original image is shown in pink dots and the reconstruction as a blue dot-dash curve. The δ-spike is shown as a cyan curve. The differences between the reference image and the reconstructions are shown in the center panels. The right panels shows the numerical approximation for $\log \left|1 + \frac{\widehat{h}(\xi)}{M}\right|$ used in the computation; all three plots show the same range of physical spatial frequencies.

and 8 and we look at $M \in \{2^m : m \in [2:10]\}$ and $\Delta x \in \{2^{-n} : n \in [4:12]\}$. The results of these experiments are shown in Figure 13.16. If $\widetilde{G}$ denotes the reconstructions, then the left panels show the errors $\log_{10} \|G - \widetilde{G}\|_2 / \|G\|_2$ as a function of $\log_2 M$. The different curves correspond to different values of Δx; the legends in these plots are labeled by $\log_2 \Delta x$. The right panels show the errors $\log_{10} \|G - \widetilde{G}\|_2 / \|G\|_2$ but this time as a function of $-\log_2 \Delta x$. The different curves correspond to different values of M; the legends in these plots are labeled by $\log_2 M$.

The plots in Figure 13.16(a) indicate that when $p = 2$, the principal determinant of the error is the term $\left(\frac{\|G\|_1}{M}\right)^2$, with the effect of reducing Δx quickly saturating. The only exception is when $\Delta x = 2^{-4}$, indicating that at this sample spacing the calculation is highly under resolved. Note that the minimum error attainable is $10^{-3.5}$. If $M = 2^{10}$, then this value is attained for all values of Δx except 2^{-4} and 2^{-5}. When $p = 4$, the errors depend, over their full ranges, on both Δx and M, though, once again, the dependence on M is clearly stronger. In this case the minimum error is about 10^{-8}, which is attained for $\Delta x \in \{2^{-8}, 2^{-9}, 2^{-10}, 2^{-11}, 2^{-12}\}$, provided that $M = 2^{10}$.

Finally, with $p = 8$, the dependence on M is again more pronounced. If $M \in \{2^7, 2^8, 2^9, 2^{10}\}$, then the minimum error of 10^{-12} is attained irrespective of Δx. Indeed, as Δx decreases, the errors slowly *increase* in these cases, indicating that the calculation quickly reaches the best available precision. The loss of three or four digits over machine precision can be taken as an indication of the "conditioning" of this method. At least for sufficiently smooth data, once Δx is small enough, the choice of p and M seem to be the crucial determinants of the accuracy of the reconstruction.

We conclude this section with the results of several 2D experiments. We first show the results of using a δ-spike, as was done in the 1D case, and then give examples where the δ-spike is replaced by more realizable models.

Example 13.13 The HHT approach to image reconstruction is not especially sensitive to the smoothness of the data. In this experiment we show reconstructions of the same image at three levels of smoothness, $k = 0, 3, 6$. The data are samples of the Fourier magnitude data: $\log\left|e^{2\pi i \langle \boldsymbol{\xi}, \boldsymbol{c}\rangle} + \frac{\widehat{g}_k(\boldsymbol{\xi})}{M}\right|$ for $k = 0, 3, 6$. Here, $\boldsymbol{c}$, the location of the δ-spike, is just beyond the upper left corner of the smallest rectangle containing the support of the image.

We set $M = 4\|g_k\|_1$, and we double oversample the data (relative to the support of g_k), collecting 256 samples altogether. In Figure 13.17 we show the reconstructed images in the left panels, the normalized difference images in

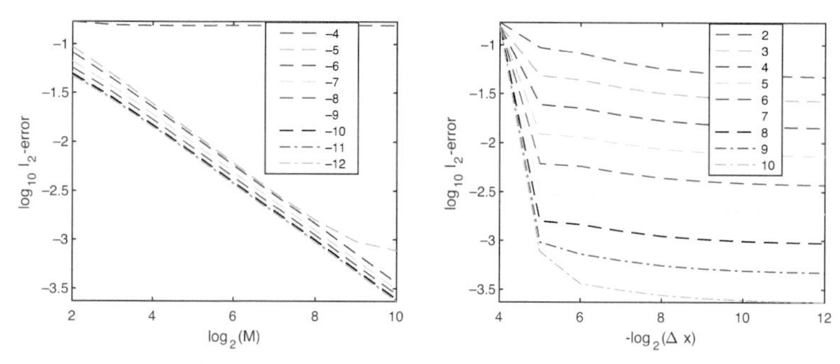

(a) Oversampling with $p = 2$; the plot on the left shows the $\log_{10} - \ell_2$-errors as a function of $\log_2(M)$, the plot on the right as a function of $\log_2(\Delta x)$.

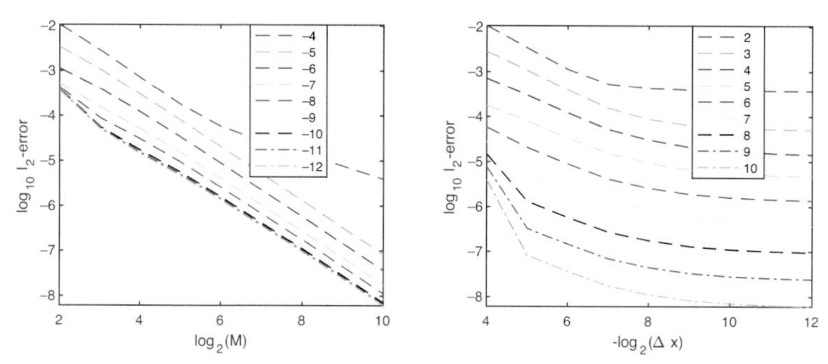

(b) Oversampling with $p = 4$; the plot on the left shows the $\log_{10} - \ell_2$-errors as a function of $\log_2(M)$, the plot on the right as a function of $\log_2(\Delta x)$.

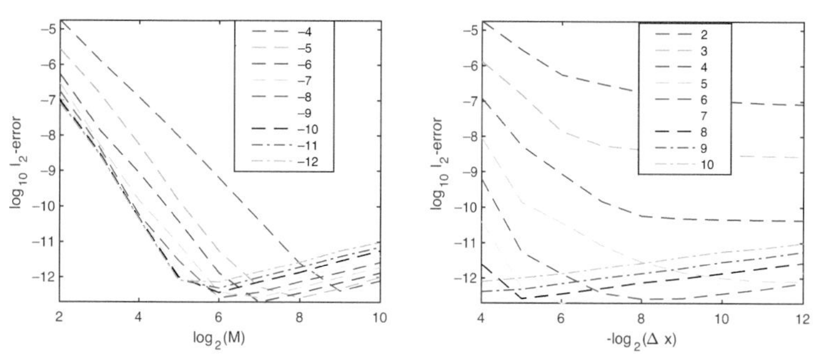

(c) Oversampling with $p = 8$; the plot on the left shows the $\log_{10} - \ell_2$-errors as a function of $\log_2(M)$, the plot on the right as a function of $\log_2(\Delta x)$.

Figure 13.16 These plots examine the accuracy of the reconstruction attained using the HHT as functions of the spatial resolution, Δx, and the height of the δ-spike, M. The plots on the left show $\log_{10} \|G - \widetilde{G}\|_2 / \|G\|_2$ as a function of $\log_2 M$, the legends are labeled by $\log_2 \Delta x$. The plots on the right show $\log_{10} \|G - \widetilde{G}\|_2 / \|G\|_2$ as a function of $\log_2 \Delta x$, the legends are labeled by $\log_2 M$.

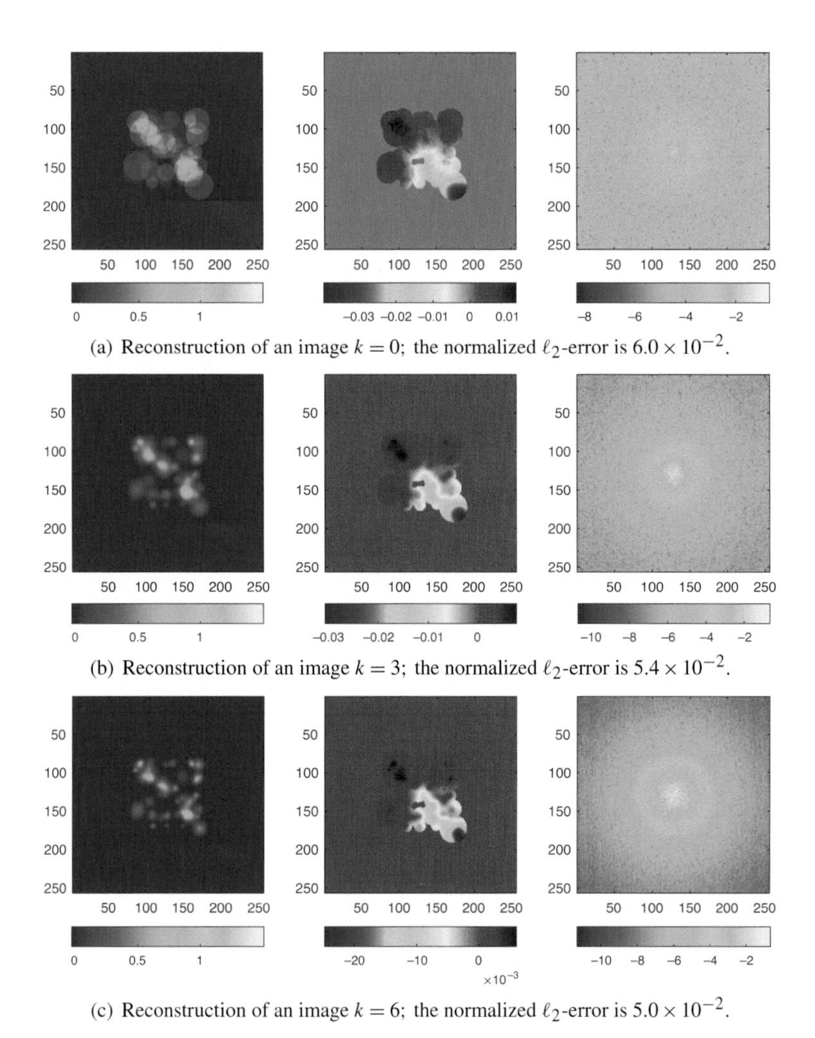

(a) Reconstruction of an image $k = 0$; the normalized ℓ_2-error is 6.0×10^{-2}.

(b) Reconstruction of an image $k = 3$; the normalized ℓ_2-error is 5.4×10^{-2}.

(c) Reconstruction of an image $k = 6$; the normalized ℓ_2-error is 5.0×10^{-2}.

Figure 13.17 The left panels show the reconstructed images; the middle panels show the error images; and the right panels show the $\log_{10}$ of the magnitude DFT data.

the center, and the $\log_{10}\left|e^{2\pi i\langle \boldsymbol{\xi}, \boldsymbol{c}\rangle} + \frac{\widehat{g}_k}{M}\right|$ in the right panels. The errors are all comparable, slightly decreasing with increasing of smoothness of the image. Note that the errors are smallest near to the δ-spike. From the right panels we see that the Fourier data decays more quickly (as expected) with increasing smoothness. In the $k = 0$ case, it has only decayed to about 10^{-4} at the boundary of the Fourier sample domain, which has not prevented the algorithm from reconstructing a fairly high-resolution image.

In the next experiment we examine the dependence of the error in the reconstruction on the parameters M and Δx for double and quadruple oversampling. This is the 2D version of Example 13.12. For this experiment we use the $k = 3$ version of the image used in the previous example, which is shown in Figure 13.17(b).

Example 13.14 In this experiment we explore the dependence of the error in the HHT reconstruction of a 2D image on the values of M, Δx, and the degree of oversampling. These results, shown in Figure 13.18, are quite similar to those in Figure 13.16. For the 2D experiment we take $p = 2, 4$:

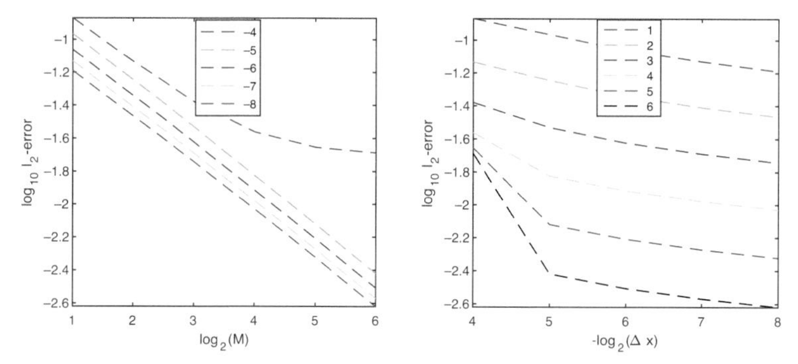

(a) Oversampling with $p = 2$; in the left panel the error is a function of $\log_2(M)$, in the right panel a function of $\log_2(\Delta x)$.

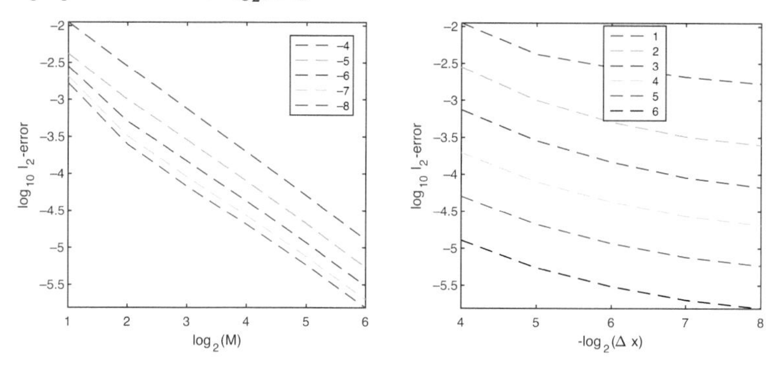

(b) Oversampling with $p = 4$; in the left panel the error is a function of $\log_2(M)$, in the right panel a function of $\log_2(\Delta x)$.

Figure 13.18 These plots show the dependence of the error in the HHT reconstruction of a 2D image on M and Δx. The plots on the left show $\log_{10} \|g - \widetilde{g}\|_2/\|g\|_2$ as a function of $\log_2 M$, the legends are labeled by $\log_2 \Delta x$. The plots on the right show $\log_{10} \|g - \widetilde{g}\|_2/\|g\|_2$ as a function of $\log_2 \Delta x$, the legends are labeled by $\log_2 M$.

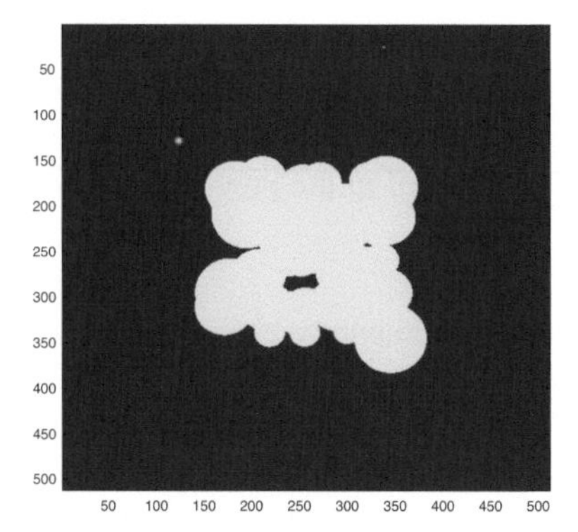

Figure 13.19 The support of the object used in the experiment below along with the Gaussian spike (above and to the left of the object).

$$M \in \{2,\ 2^2,\ 2^3,\ 2^4,\ 2^5,\ 2^6\} \text{ and } \Delta x \in \{2^{-4},\ 2^{-5},\ 2^{-6},\ 2^{-7},\ 2^{-8}\}.$$

As before, the dependence on M is stronger than that on Δx. Greater oversampling produces a better reconstruction, but in 2D, oversampling beyond $p = 4$ is computationally demanding.

For our final experiments in this section, we consider the effect of replacing a δ-spike, which is exactly supported at 1-pixel, with more realistic functions.

Example 13.15 We first use a narrow Gaussian,

$$\varphi(\boldsymbol{x}) = \frac{e^{-\frac{|x-c|^2}{\sigma^2}}}{\pi\sigma^2}. \tag{13.48}$$

Here, φ is normalized to the have total mass 1. The data is then samples of

$$|M\widehat{\varphi}(\boldsymbol{\xi}) + \widehat{g}(\boldsymbol{\xi})|. \tag{13.49}$$

It is assumed that we know $M|\widehat{\varphi}(\boldsymbol{\xi})|$, so that the measurements allow us to determine

$$\log\left|1 + \frac{\widehat{g}(\boldsymbol{\xi})}{M\widehat{\varphi}(\boldsymbol{\xi})}\right|. \tag{13.50}$$

Figure 13.19 shows the support of the object used in the experiment below along with the Gaussian, which is in the upper left portion of the image.

The Gaussian is not compactly supported, so this data cannot strictly satisfy the requirements placed on φ in Chapter 7.4. Moreover, for most reasonable functions, g, the ratio $|\widehat{g}(\boldsymbol{\xi})/\widehat{\varphi}(\boldsymbol{\xi})|$ tends to infinity as $|\boldsymbol{\xi}| \to \infty$. Nonetheless, we find that, with appropriate care, this data can be used with the Hilbert transform to obtain very good reconstructions.

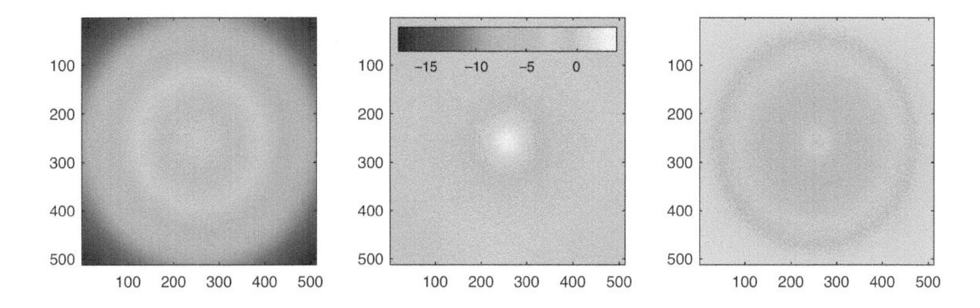

Figure 13.20 The left panel shows the $\log_{10}|\widehat{g}(\boldsymbol{\xi})|$, the middle panel shows $\log_{10}|\widehat{\varphi}(\boldsymbol{\xi})|$, and the right panel shows $\log_{10}\left|1+\frac{\widehat{g}(\boldsymbol{\xi})}{M\widehat{\varphi}(\boldsymbol{\xi})}\right|$. A single color map, shown in the middle panel, is shared by all three images. Note that in the right panel the values near the boundary are greater than 1.

In Figure 13.20 we show $\log_{10}|\widehat{g}(\boldsymbol{\xi})|$ in the left panel, $\log_{10}|\widehat{\varphi}(\boldsymbol{\xi})|$ in the center panel, and $\log_{10}\left|1+\frac{\widehat{g}(\boldsymbol{\xi})}{M\widehat{\varphi}(\boldsymbol{\xi})}\right|$ in the right. All plots are on the same color scale, shown by the color bar in the middle panel. Comparing the left panel to the middle panel it is quite clear that, eventually, $|\widehat{\varphi}(\boldsymbol{\xi})|$ has decayed much more than $|\widehat{g}(\boldsymbol{\xi})|$ so that the ratio is quite large at high frequencies, as is evident in the right panel. It seems unlikely that using all of this data to reconstruct an image will do a good job.

In fact, Figure 13.21 shows the results of using use the HHT with the full 512×512 data set. We see that essentially nothing is reconstructed correctly. In Figure 13.22 we reconstruct the image using the 256×256 portion of the data centered on $\boldsymbol{k} = \boldsymbol{0}$. This subset of the data produces a very good reconstruction.

As a final example, we consider the diffraction pattern produced by a small, hard, homogeneous ball of radius R centered at $\boldsymbol{x}_0$. Assuming that such an object diffracts a plane wave in proportion to its thickness along the direction of propagation of the wave, we use the following simple model for a 2D far-field imaging experiment:

$$\varphi(\boldsymbol{x}) = C\sqrt{R^2 - |\boldsymbol{x}-\boldsymbol{x}_0|^2}\,\chi_{[0,R]}(|\boldsymbol{x}-\boldsymbol{x}_0|). \tag{13.51}$$

The 2D Fourier transform of this function is given by

$$\widehat{\varphi}(\boldsymbol{\xi}) = C_R\frac{J_{\frac{3}{2}}(R|\boldsymbol{\xi}|)}{(R|\boldsymbol{\xi}|)^{\frac{3}{2}}} = C_R'\frac{1}{(R|\boldsymbol{\xi}|)^2}\left[\frac{\sin(R|\boldsymbol{\xi}|)}{R|\boldsymbol{\xi}|} - \cos(R|\boldsymbol{\xi}|)\right]. \tag{13.52}$$

In light of this formula, we call this spike a “Bessel spike.” The function $\sin(x) - x\cos(x)$ has its first positive zero at $x \approx 4.9341$, so care must be taken in the evaluation of $\widehat{g}(\boldsymbol{\xi})/\widehat{\varphi}(\boldsymbol{\xi})$. To implement this model, we use the Fourier

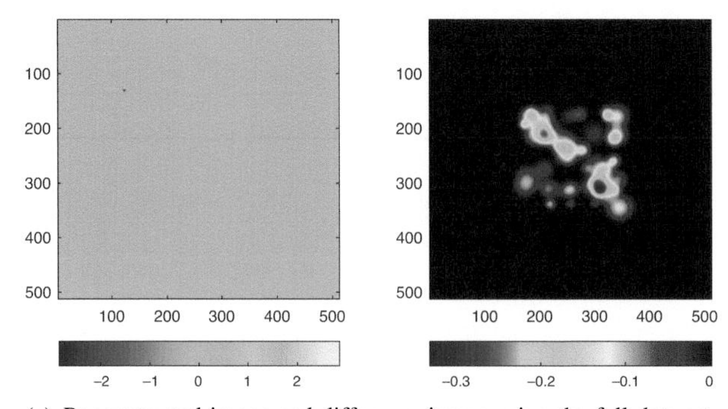

(a) Reconstructed image and difference image using the full data set. The normalized ℓ_2-error is 0.34.

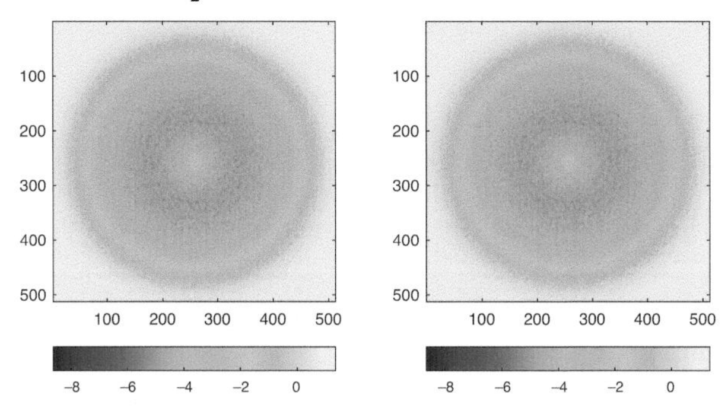

(b) The left panel shows the full data set, and the right panel, the portion used to reconstruct the image in (a), which in this case, is the full data set.

Figure 13.21 The left panel in Figure 13.21(a) shows the image reconstructed using the HHT with a Gaussian spike, without subsampling; the right panel shows the error image. The $\log_{10}$ of magnitude Fourier data, $\log_{10}\left|1+\frac{\widehat{g}(\boldsymbol{\xi})}{M\widehat{\varphi}(\boldsymbol{\xi})}\right|$, is shown in Figure 13.21(b).

space description directly, i.e., $\widehat{\varphi}$, in order to get a good approximation for the sampled data.

Example 13.16 For this experiment, we use the the Bessel spike as our approximate δ-spike, whose Fourier transform is given in (13.52). It is computed as the inverse DFT of the Fourier data sampled on a 1024×1024 grid centered at $\mathbf{0}$, with $\Delta x = 2^{-9}$, and a (nominal) spike width of about 10 pixels. The left panel of Figure 13.23 shows several sections of the spike along planes close to the peak and the right panel is a heat map. We use the same $k = 3$ object of interest for reconstruction as in the previous experiment.

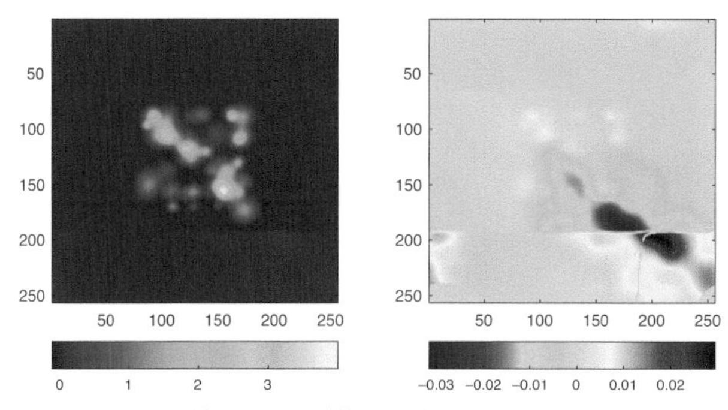

(a) Reconstructed image and difference image using the central 256×256 subset of the full data set. The normalized ℓ_2-error is 1.4×10^{-2}.

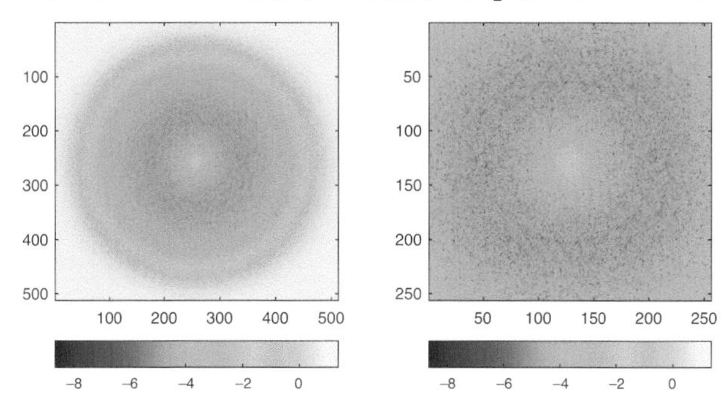

(b) The left panel shows the full data set and the right panel shows the portion used to reconstruct the image, which in this case, is the central 256×256 portion.

Figure 13.22 The left panel in Figure 13.22(a) shows the image reconstructed using the HHT with a Gaussian spike, with subsampling; the right panel shows the error image. The $\log_{10}$ of magnitude Fourier data, $\log_{10}\left|1 + \frac{\widehat{g}(\xi)}{M\widehat{\varphi}(\xi)}\right|$, is shown in Figure 13.22(b).

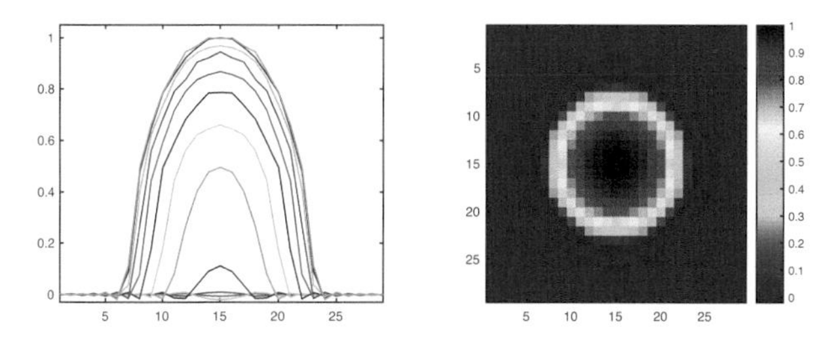

Figure 13.23 The left panel shows several 2D slices through the graph of the spike near to its peak and the right panel is a 2D heat map.

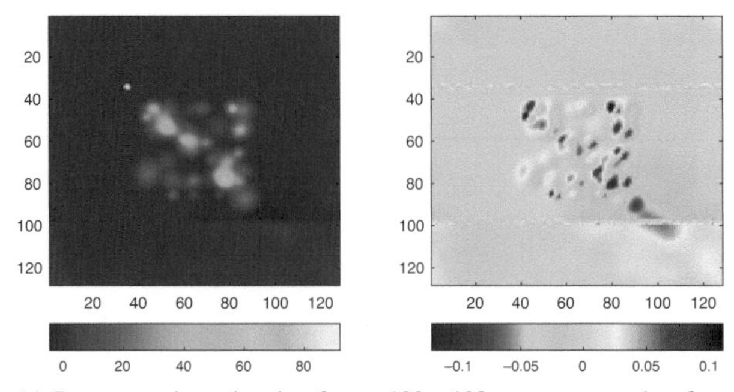

(a) Reconstruction using data from a 128×128 square centered on $\boldsymbol{k} = \mathbf{0}$. The normalized ℓ_2-error is 2.3×10^{-3}.

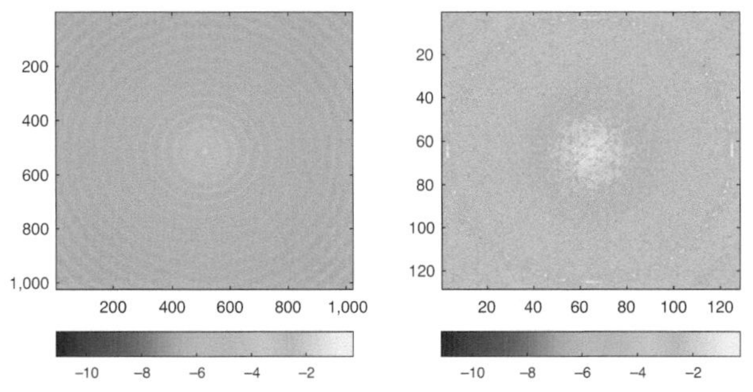

(b) The left panel shows the full data set and the right panel, the portion used to reconstruct the image in (a), which in this case, is the central 128×128 portion.

Figure 13.24 Reconstruction using a Bessel spike with $M = 2$ and $\Delta x = 2^{-9}$, using data from a 128×128 square centered on $\boldsymbol{k} = \mathbf{0}$.

In this application of the HHT, we set $M = 2$, $\Delta x = 2^{-9}$ and use double oversampling. The results are shown in Figure 13.24–13.25. The DFT magnitude images in the lower left panels have several bright circles indicating the locations of the zeros of $\sin x - x \cos x$. The reconstruction in Figure 13.24(a) uses only the central 128×128 portion of the $\boldsymbol{k}$-space data, which includes the circle produced by the first zero. The l_2-error is about 2.3×10^{-3}, and the difference image resembles those produced by iterative algorithms, which appear in earlier chapters.

In Figure 13.25(a) we show the results if we instead use the central 256×256 portion of the $\boldsymbol{k}$-space data. From the right panel in Figure 13.25(b) it is clear that this data includes samples near to a circle where $\widehat{\varphi}(\boldsymbol{\xi})$ vanishes. The l_2-error is about 1.21×10^{-2}, which is a bit larger than the previous case, but the difference image indicates that the reconstruction is

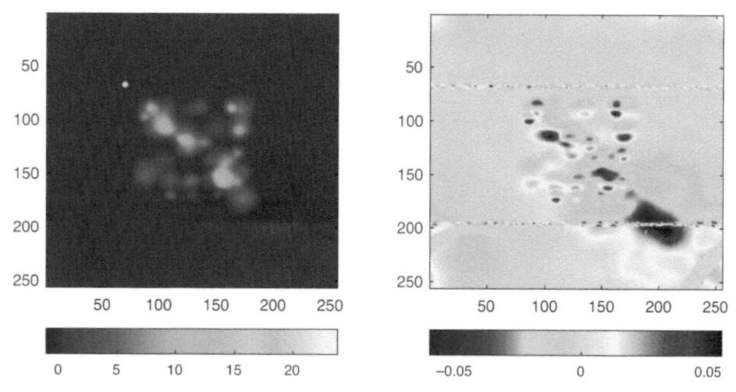

(a) Reconstruction using data from a 256×256 square centered on $\boldsymbol{k} = \mathbf{0}$. The normalized ℓ_2-error is 1.4×10^{-2}.

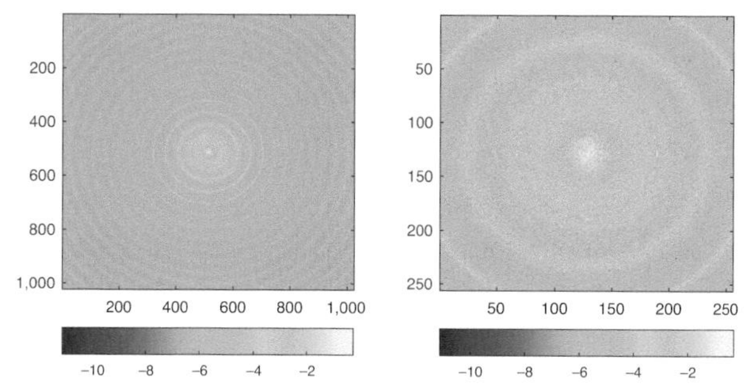

(b) The left panel shows the full data set and the right panel, the portion used to reconstruct the image in (a), which in this case, is the central 256×256 portion.

Figure 13.25 Reconstruction using a Bessel spike with $M = 2$ and $\Delta x = 2^{-9}$, using data from a 256×256 square centered on $\boldsymbol{k} = \mathbf{0}$.

quite accurate, except along two horizontal lines near the edge of the support of the image.

Taken together, these examples suggest that the HHT could be a powerful approach for recovering phase information and reconstructing images, so long as a strongly scattering object (modeled by our small, hard, homogeneous ball) is physically realizable.

13.A Appendix: Proof of Theorem 13.6

Here we give a proof of the Theorem 13.6:

Theorem 13.6 *Let $A \subset \mathbb{R}^n$ be an affine subspace, and $B_+ \subset \mathbb{R}^n$ the nonnegative orthant. Either $A \cap B_+ \neq \emptyset$, or the* $\operatorname{dist}(A, B_+)$ *is positive and is attained at a pair of points $\boldsymbol{v}_0 \in A$, $\boldsymbol{x}_0 \in B_+$.*

Proof The affine subspace $A = V + \boldsymbol{c}$, where $V \subset \mathbb{R}^n$ is a vector subspace, and $\boldsymbol{c}$ is a fixed vector. The proof of this theorem makes use of several elementary functions:

(i) For $x \in \mathbb{R}$ let $x_- = \min\{0, x\}$ and let $d(x) = (x_-)^2$.
(ii) For $\boldsymbol{v} \in \mathbb{R}^n$ let $D(\boldsymbol{v}) = \sum_{i=1}^n d(v_i)$. The is the square of the distance from $\boldsymbol{v}$ to B_+.
(iii) For $\boldsymbol{v} \in \mathbb{R}^n$ let $m(\boldsymbol{v}) = \min\{v_i : i = 1, \ldots, n\}$.

The functions d and D are convex and differentiable. The function m is continuous. The proof is divided into cases depending upon the set $W_+ = V \cap B_+$. The first observation is that if $V \cap \operatorname{int} B_+ \neq \emptyset$, then there does not exist any vector $\boldsymbol{c}$ such that $(V + \boldsymbol{c}) \cap B_+ = \emptyset$. For, in this case, there exists a vector $\boldsymbol{v} \in V$ with strictly positive coordinates. Hence, for any $\boldsymbol{c}$ there exists a positive number λ so that $\lambda \boldsymbol{v} + \boldsymbol{c} \in B_+$.

Case 1 Let $\dim W_+ = 0$. This case is relatively easy. Any vector $\boldsymbol{v} \in V$ has coordinates of both signs, and therefore, $m(\boldsymbol{v}) < 0$ for all $\boldsymbol{v} \in V \setminus \{0\}$. In particular, there is an $\alpha < 0$ so that $m(\boldsymbol{v}) \leq \alpha$ for all $\boldsymbol{v} \in V$ of unit length. This easily implies that, for any $\boldsymbol{c} \in \mathbb{R}^n$, the function $D(\boldsymbol{v} + \boldsymbol{c})$ tends to infinity as $\|\boldsymbol{v}\|_2 \to \infty$. If $\boldsymbol{c}$ is a vector, such that $(V + \boldsymbol{c}) \cap B_+ = \emptyset$, then, as a positive, proper convex function, $D(\boldsymbol{v} + \boldsymbol{c})$ assumes a positive lower bound for some $\boldsymbol{v}_0 \in V$.

Case 2 Let $\dim W_+ = \dim V$. In this case $\operatorname{int} W_+$ is a proper, open cone with vertex at $\boldsymbol{0}$ lying in V. We can, therefore, choose a basis $\{\boldsymbol{v}^1, \ldots, \boldsymbol{v}^m\}$ for V consisting of vectors with nonnegative coordinates. The indices can be reordered so that $\boldsymbol{v}_i^j = 0$ for $i = p+1, \ldots, n$, and for each $1 \leq i \leq p$, there is a j such that $\boldsymbol{v}_i^j > 0$. The function $D(\boldsymbol{v} + \boldsymbol{c})$ takes the form

$$D\left(\sum_{j=1}^m \lambda_j \boldsymbol{v}^j + \boldsymbol{c}\right) = \sum_{i=1}^p d\left(\sum_{j=1}^m \lambda_j \boldsymbol{v}_i^j + \boldsymbol{c}_i\right) + \sum_{i=p+1}^n d(c_i). \quad (13.53)$$

Clearly, the second sum on the right-hand side of (13.53) is independent of the vector in V and is a lower bound for $\operatorname{dist}(V + \boldsymbol{c}, B_+)^2$. For any $\boldsymbol{c}$ there exist $(\lambda_1, \ldots, \lambda_m)$ so that

$$\left(\sum_{j=1}^m \lambda_j \boldsymbol{v}_i^j + \boldsymbol{c}_i\right) > 0 \quad \text{for } 1 \leq i \leq p. \quad (13.54)$$

Thus, the lower bound is attained and equals the square of the distance of $V + \boldsymbol{c}$ from B_+.

Case 3 Let $0 < \dim W_+ < \dim V$. In this case there is a basis for V of the form $\{\boldsymbol{v}^1, \dots, \boldsymbol{v}^q\} \cup \{\boldsymbol{w}^1, \dots, \boldsymbol{w}^r\}$, where the vectors in $\{\boldsymbol{v}^i\}$ have nonnegative coordinates and span W_+. We reorder the indices as in Case 2. Suppose that

$$\boldsymbol{v} = \sum_{j=1}^{q} \lambda_j \boldsymbol{v}^j + \sum_{k=1}^{r} \mu_k \boldsymbol{w}^k. \tag{13.55}$$

With our choice of basis it follows that, so long as $(\mu_1, \dots, \mu_r) \neq \mathbf{0}$, the set

$$\left\{ \sum_{k=1}^{r} \mu_k \boldsymbol{w}_i^k : i = p+1, \dots, n \right\} \tag{13.56}$$

contains both positive and negative numbers. If not, then we can assume that these numbers are all nonnegative and choose $(\lambda_1, \dots, \lambda_q)$ so that

$$\sum_{j=1}^{q} \lambda_j \boldsymbol{v}^j + \sum_{k=1}^{r} \mu_k \boldsymbol{w}^k \in W_+. \tag{13.57}$$

But this contradicts the choice of $\{\boldsymbol{v}^1, \dots, \boldsymbol{v}^q\}$as a spanning set for W_+.

Hence, there is an $\alpha < 0$ so that

$$m\left(\sum_{k=1}^{r} \mu_k \boldsymbol{w}_{p+1}^k, \dots, \sum_{k=1}^{r} \mu_k \boldsymbol{w}_n^k \right) \leq \alpha \tag{13.58}$$

for $(\mu_1, \dots, \mu_r)$ of norm 1. As before we see that

$$D(\boldsymbol{v}) = \sum_{i=1}^{p} d\left(\sum_{j=1}^{q} \lambda_j \boldsymbol{v}_i^j + \sum_{k=1}^{r} \mu_k \boldsymbol{w}_i^k + \boldsymbol{c}_i \right) + \sum_{i=p+1}^{n} d\left(\sum_{k=1}^{r} \mu_k \boldsymbol{w}_i^k + \boldsymbol{c}_i \right). \tag{13.59}$$

From (13.58) it follows that the second sum defines a proper, convex function of $(\mu_1, \dots, \mu_r)$ and it is a lower bound for $D(\boldsymbol{v})$. There exists a vector $(\mu_1^0, \dots, m\mu_r^0)$ so that this sum attains its minimum value. Once again we can choose $(\lambda_1, \dots, \lambda_q)$ so that

$$\sum_{i=1}^{p} d\left(\sum_{j=1}^{q} \lambda_j \boldsymbol{v}_i^j + \sum_{k=1}^{r} \mu_k^0 \boldsymbol{w}_i^k + \boldsymbol{c}_i \right) = 0. \tag{13.60}$$

Hence, $\sum_{i=p+1}^{n} d\left(\sum_{k=1}^{r} \mu_k^0 \boldsymbol{w}_i^k + \boldsymbol{c}_i\right)$ is the squared distance of this vector in $V + \boldsymbol{c}$ to B_+, which must therefore be positive, and also gives $\operatorname{dist}(V + \boldsymbol{c}, B_+)^2$.

This completes the proof of the theorem. □

14
Concluding Remarks

In this text, we consider the classical, phase retrieval problem – where one seeks to recover an unknown function, $\rho(\boldsymbol{x})$, from measurements of the magnitude of its Fourier transform. In order to make the analysis more tractable, we focus on a discrete version of the problem, where we measure the magnitude of the discrete Fourier transform (DFT) of a function sampled on a uniform grid. Without additional information about $\rho(\boldsymbol{x})$ the problem is underdetermined. The bulk of our analysis focuses on the case where the auxiliary information is either knowledge of the support of $\rho(\boldsymbol{x})$, or the assumption that it is real, and nonnegative, along with bounds on the support of $\rho \star \rho$.

Before considering specific algorithms, we first analyze the tangent bundle of the magnitude torus, $\mathbb{A}_{\boldsymbol{a}}$, obtaining very explicit descriptions, in both the Fourier and image domains, of the fibers of the tangent and normal bundles. We then show that, for an image $\boldsymbol{f}_0$, with small support, $S_{\boldsymbol{f}_0} \subset S$, the intersections $\mathbb{A}_{\boldsymbol{a}} \cap B_S$ are usually not transversal, that is, $T_{\boldsymbol{f}_0}\mathbb{A}_{\boldsymbol{a}} \cap B_S$ is often positive dimensional. Our description of the fiber of the normal bundle provides the basis for our analysis of the transversality of the intersections of a magnitude torus with the nonnegative orthant, and the boundary of an ℓ_1-ball.

We then address the well-posedness of the phase retrieval problem when the auxiliary information is an estimate, $S \subset J$, for the support of the object. As the problem is nonlinear and the solution is usually not unique, it is necessary to discuss *local* uniqueness near to a point in $\mathbb{A}_{\boldsymbol{a}} \cap B_S$. We show that the local inverse map is Lipschitz continuous if and only if this intersection is transversal. If $\boldsymbol{f}_0 \in \mathbb{A}_{\boldsymbol{a}} \cap B_S$ is a point where the intersection is nontransversal, then the inverse has very anisotropic continuity properties: it satisfies Lipschitz estimates in many directions, but is, at best, Hölder continuous in the directions belonging to $T_{\boldsymbol{f}_0}\mathbb{A}_{\boldsymbol{a}} \cap B_S$. We also define ϵ-nonuniqueness and considered various mechanisms that lead to it.

In Chapter 4, we analyze the phase retrieval problem when the auxiliary information is that the unknown object is nonnegative, and that its autocorrelation image, see (1.21), has small support. We first show that, generically, this auxiliary information uniquely determines a solution to the phase retrieval problem, up to trivial associates. Using our description, in image space, of the tangent and normal bundles to the magnitude torus, we then analyze the transversality of the intersections, $\mathbb{A}_{\boldsymbol{a}} \cap B_+$ and $\mathbb{A}_{\boldsymbol{a}} \cap B^1_{r_1}$, which, generically, are transversal. Finally, we show that the convolution of one image with an inversion symmetric image, which is done implicitly if data is apodized using a Gaussian, inevitably leads, *in essentially all cases,* to nontransversal intersections.

In Chapters 7–8 we use our geometric analysis to understand the practical behavior of the widely used hybrid iterative maps, see Section 7.2, when the auxiliary information is a support constraint. The analysis in Section 7.3 shows that the linearization of this map at a fixed point resulting from a non-transversal intersection typically has two types of directions in which it is neutrally stable: center manifold directions and those arising from the intersection of the fiber of the tangent bundle to the torus and the linear subspace B_S. This fact strongly suggests that the iterates of such a map should stagnate, often at a substantial distance from the center manifold of a true intersection point, and this expectation is borne out in a large collection of numerical experiments. However, the extent to which nontransversality leads to an ill-posed reconstruction problem depends, to some extent, on the details of the original image.

While the fixed-point sets of hybrid iterative maps are center manifolds defined by intersection points, the intersection points themselves are not attracting. When the iterates of such a map converge, it is to points on the center manifold distant from the intersection that defines it. While small angles between the tangent space to the magnitude torus and the auxiliary set, B, do cause very slow convergence, or stagnation, even at points very distant from the intersection point itself, the effects of nontransversality are harder to assess. In some examples with a nontransversal intersection, a hybrid iterative algorithm is able reconstruct the image, almost to machine precision. Even when the intersection is nontransversal, there may exist points on the center manifold that are strongly attracting. The analysis of hybrid iterative maps along the center manifold is not yet complete, and a better understanding could lead to modifications of these algorithms with better convergence properties.

When the unknown image is nonnegative, the behavior of hybrid iterative maps was considered in Chapter 9, using both the nonnegativity and the ℓ_1-norm constraints. We show that the failure of transversality of the intersection of a magnitude torus with the nonnegative orthant is the same as the failure of transversality of the intersection with the appropriate ℓ_1-ball. This suggests that algorithms using these two types of maps will perform similarly, and they do. As the boundaries of these sets are "more convex" than a linear subspace, one also expects that these algorithms will work somewhat better than algorithms using only the support constraint, and this too is borne out in numerical experiments.

The last part of the book begins with a statistical analysis of the results of many runs, with random initial conditions, of a hybrid iterative algorithm on a single data set. There are several important lessons:

(i) The runs of the algorithm can be clustered according to their final residual, as this statistic seems to fall into a multimodal distribution. Only the runs in the cluster with the smallest residuals provide reliable results.
(ii) The logs of the residuals and true errors are usually correlated, with different clusters exhibiting different correlation coefficients.
(iii) One can compute the variances in the reconstructed phases, frequency by frequency. This data gives a reliable indicator of which phases can be accurately determined by hybrid iterative algorithms, and which cannot.

In the final chapter of this text (Chapter 13), we consider several proposals for modifying the experimental protocol that could lead to better reconstructions. In all cases, the goals are to break the infinitesimal symmetries that cause the intersection of $\mathbb{A}_{\boldsymbol{a}}$ with B to be nontransversal, and remove the directions in which the magnitude torus meets the constraint set at very small angles. The first approach involves cutting a soft object along a sharp edge. As long as the shape of this edge is known with high precision, this often allows an algorithm based on iterating a hybrid iterative map, to run to convergence with the full precision available in the data.

For the second approach, we add a known, hard object to the exterior of the object that we are trying to image. The size, placement, and hardness of this object all play a role in the success of this approach, though the most important consideration is again to have an accurate estimate of its shape. We call this second approach "external holography." There are a variety of approaches with equal claim to be called external holography, which take advantage of the interference in the scattered field provided by a *known* object exterior to the object of interest. In Jacobsen (2019) it is explained how, if the external object is small, and placed quite far from the object of interest, then the direct Fourier inversion of the measured data contains a slightly smeared copy of the object of interest. In our approach, the external object is comparable in size to the object of interest and not too far away. With a precise knowledge of the size and shape of the external object, a hybrid iterative map can be used to obtain an accurate reconstruction.

In Chapter 7.4 we introduce a new approach to phase retrieval, which is noniterative and employs the classical Hilbert transform. The underlying idea is similar to that used in external holography, but requires a different experimental setup, wherein a small, highly diffracting object is placed near to, but outside of, the object that we want to image. With a knowledge of the diffraction pattern produced by the external object, and the magnitude Fourier data of the composite object, we show that one can use the Hilbert transform, line by line in $\boldsymbol{k}$-space, to directly compute the unknown phase information for the Fourier transform of the unknown object. In numerical experiments in Section 13.4, we see that the a priori mathematical requirements on the external object can, in practice, be substantially relaxed. Indeed, the method works quite well with an external object modeled as a solid, round ball. Moreover, this approach shows promise for working almost equally well if the magnitude DFT data is replaced by samples of the magnitude of the continuum Fourier transform. Other applications of the Hilbert transform in phase retrieval can be found in Yasir and Ivan (2016), Nakajima and Asakura (1985, 1986), Nakajima (1995), and Huang et al. (2016).

There has been significant work done on continuous optimization methods for phase retrieval, but the analysis of their convergence behavior is largely outside the scope of this book. The geometric understanding of the problem developed here may help in that task. We refer the reader to a few examples from the literature on such methods, some of which have been mentioned earlier. Conjugate gradient methods were considered by Lane (1991), and Wirtinger flows (a gradient descent-like scheme) are discussed by Candès et al. (2015b). Marchesini (2007) accelerated the convergence of hybrid maps through lower-dimensional subspace saddle-point optimization, Osherovich (2011) developed a Newton-type method using a quasi-hessian for a defining function of $\mathbb{A}_a \cap B$, and Pham et al. (2019) investigated the use of generalized proximal smoothing methods. They found them to be more effective than hybrid maps for biological specimens.

We explore a different sort of Newton method: The underlying idea of Newton's method is to replace a nonlinear problem with a sequence of linear problems. The nonlinear problem at hand is to find points in the intersection, $\mathbb{A}_a \cap B$. For a point, $f \in \mathbb{A}_a$, the fiber of the tangent bundle, $T_f\mathbb{A}_a$, is the best linear approximation to $\mathbb{A}_a$ near to f. If f is near to a point, $f_0 \in \mathbb{A}_a \cap B$, then it seems reasonable to replace finding this exact intersection with the problem of finding the intersection $T_f\mathbb{A}_a \cap B$. If $B = B_S$, then this is a linear problem, and if $B = B_+$, then it is the problem of finding the intersection of two convex sets. In examples, we have shown that this algorithm can give very good results, though, as with similar "nonlinear" methods, it requires a good initial guess to be effective. It works well as a second step, after a hybrid iterative map is used to find an approximate reconstruction and can substantially increases the accuracy of the approximation with a small number of iterates.

In summary, classical, phase retrieval provides a rich set of mathematical questions for further analysis. Results in this field have the potential to significantly impact the practice of coherent diffraction imaging (CDI), a powerful imaging modality with a great deal of unrealized potential. Because of the inherent ill-posedness of the problem, robust inversion schemes will likely require higher quality auxiliary information, or a modification of the experimental protocol. It is our belief that the material collected here will provide a useful introduction to, and conceptual framework for, mathematical scientists interested in exploring this fascinating topic. These tools should also prove useful for the analysis and design of new experimental protocols, and more successful reconstruction methods for solving phase retrieval problems.

15
Notational Conventions

This book consistently uses a variety of notational conventions that are intended to make the text more readable. As some are not entirely standard or self explanatory, we review them here.

- Italic letters $a, b, \ldots, x, y, z$, etc., are used to denote real or complex numbers.
- Boldface letters $\boldsymbol{a}$, $\boldsymbol{f}$, $\boldsymbol{g}$, $\boldsymbol{j}$, $\boldsymbol{k}$, $\boldsymbol{n}$, $\boldsymbol{t}$, $\boldsymbol{x}$, $\boldsymbol{y}$, $\boldsymbol{\tau}$, $\boldsymbol{v}$, $\boldsymbol{\varphi}$, etc., are used to denote points in a vector space, with the following conventions:

  (i) The boldface letter $\boldsymbol{a}$ usually refers to a vector of magnitude DFT data.
  (ii) The boldface letters $\boldsymbol{f}, \boldsymbol{g}$ refer to sampled images and $\widehat{\boldsymbol{f}}, \widehat{\boldsymbol{g}}$ their DFTs.
  (iii) The boldface letters $\boldsymbol{x}, \boldsymbol{y}$ are used to denote points in a vector space, which may or may not be related to images.
  (iv) The boldface letters $\boldsymbol{j}, \boldsymbol{k}$ are used as indices and belong to an integer lattice, $\mathbb{Z}^d$.

- If V is a vector space of dimension N, with $\boldsymbol{v} = (v_1, \ldots, v_N)$ a point in V, then the ℓ_p–norms, $\|\boldsymbol{v}\|_p$, are defined for $p = 1, 2, \infty$ by

  (i) $p = 1$, $\|\boldsymbol{v}\|_1 = \sum_{j=1}^{N} |v_j|$,
  (ii) $p = 2$, $\|\boldsymbol{v}\|_2 = \left[\sum_{j=1}^{N} |v_j|^2\right]^{\frac{1}{2}}$, and
  (iii) $p = \infty$, $\|\boldsymbol{v}\|_\infty = \max\{|v_1|, \ldots, |v_N|\}$.

- If V is a vector space of dimension, then for $r > 0$ and $\boldsymbol{v} \in V$, the open ℓ_2–ball of radius r, centered at $\boldsymbol{v}$ is denoted by

$$B_r(\boldsymbol{v}) = \{\boldsymbol{x} \in V : \|\boldsymbol{x} - \boldsymbol{v}\|_2 < r\}. \tag{15.1}$$

- Let X be a set and let Y be a subset of X; the characteristic function of Y, χ_Y, is defined for $x \in X$ by

$$\chi_Y(x) = \begin{cases} 1 & \text{if } x \in Y \\ 0 & \text{if } x \notin Y. \end{cases} \tag{15.2}$$

- If X is a set, then $\mathrm{Id} : X \to X$ is the identity map, $\mathrm{Id}(x) = x$, for all $x \in X$.
- J denotes a finite, rectangular subset of $\mathbb{Z}^d$. Points $\boldsymbol{j} \in J$ are often used to index an image in order to retain the information of its underlying dimensionality. If $\boldsymbol{f} \in \mathbb{R}^J$, then its $\boldsymbol{j}$th coordinate is denoted $f_{\boldsymbol{j}}$.
- An image represented by a point $\boldsymbol{f} \in \mathbb{R}^J$ (or $\mathbb{C}^J$) is usually extended periodically to all of $\mathbb{R}^{\mathbb{N}^d}$, (or $\mathbb{C}^{\mathbb{N}^d}$) with its values in J representing a single period. Such an image is called J-periodic.
- If $\boldsymbol{v} \in \mathbb{N}^d$ and $\boldsymbol{f} \in \mathbb{R}^J$ is periodically extended, then the translate of $\boldsymbol{f}$ by $\boldsymbol{v}$ is denoted $\boldsymbol{f}^{(\boldsymbol{v})}$ with

$$f_{\boldsymbol{j}}^{(\boldsymbol{v})} = f_{\boldsymbol{j}-\boldsymbol{v}}. \tag{15.3}$$

- The support of the image $\boldsymbol{f}$ is defined to be the set

$$S_{\boldsymbol{f}} = \{\boldsymbol{j} \in J : |f_{\boldsymbol{j}}| > 0\}. \tag{15.4}$$

- If $\epsilon > 0$ $\boldsymbol{f}$ is an image, then the set $S_{\boldsymbol{f}}^{\epsilon}$ is defined to be

$$S_{\boldsymbol{f}}^{\epsilon} = \{\boldsymbol{j} \in J : |f_{\boldsymbol{j}}| > \epsilon \|\boldsymbol{f}\|\}. \tag{15.5}$$

  In numerical experiments, unless otherwise stated, the support constraint is *defined* by the set $S_{\boldsymbol{f}}^{10^{-14}}$, which, in this context, may be denoted by $S_{\boldsymbol{f}}$.
- If $S \subset J \subset \mathbb{N}^d$, then the p-pixel neighborhood of S, denoted S_p, is

$$S_p = \{\boldsymbol{j} \in J : \text{ there exists } \boldsymbol{k} \in S \text{ with } \|\boldsymbol{j} - \boldsymbol{k}\|_\infty \le p\}. \tag{15.6}$$

- The DFT mapping vectors in $\mathbb{R}^J$ to itself, as defined in (1.7), is denoted by $\mathscr{F}$. The underlying dimension, d, is clear from the context. The vector, $\mathscr{F}(\boldsymbol{x})$, is also indicated by addition of a "hat": $\hat{\boldsymbol{x}} = \mathscr{F}(\boldsymbol{x})$.
- The Hilbert transform, denoted by $\mathscr{H}: L^2(\mathbb{R}) \to L^2(\mathbb{R})$, is defined in (7.60).
- If $\boldsymbol{a} \in \mathbb{R}_+^J$, then $\mathbb{A}_{\boldsymbol{a}}$ is the magnitude torus defined by

$$\begin{aligned} \mathbb{A}_{\boldsymbol{a}} &= \{\boldsymbol{f} \in \mathbb{R}^J : |\widehat{f_{\boldsymbol{k}}}| = a_{\boldsymbol{k}} \quad \text{for all } \boldsymbol{k} \in J\} \text{ for real images,} \\ \mathbb{A}_{\boldsymbol{a}} &= \{\boldsymbol{f} \in \mathbb{C}^J : |\widehat{f_{\boldsymbol{k}}}| = a_{\boldsymbol{k}} \quad \text{for all } \boldsymbol{k} \in J\} \text{ for complex images.} \end{aligned} \tag{15.7}$$

  The DFT representation of this torus is denoted by $\widehat{\mathbb{A}}_{\boldsymbol{a}}$; it equals the set $\{\widehat{\boldsymbol{f}} : \boldsymbol{f} \in \mathbb{A}_{\boldsymbol{a}}\}$.

- If $\boldsymbol{f} \in \mathbb{C}^J$, then $\boldsymbol{a}_f$ is the vector of moduli of $\widehat{\boldsymbol{f}}$,

$$a_{f\boldsymbol{k}} = |\widehat{f}_{\boldsymbol{k}}|. \tag{15.8}$$

- If $M \subset \mathbb{R}^N$ is a submanifold and $\boldsymbol{x} \in M$, then $T_{\boldsymbol{x}}M$ denotes the fiber of the tangent bundle to M at $\boldsymbol{x}$, thought of as an affine subspace of $\mathbb{R}^N$ passing through $\boldsymbol{x}$. The underlying vector space is denoted by $T^0_{\boldsymbol{x}}M$.
- If $M \subset \mathbb{R}^N$ is a submanifold and $\boldsymbol{x} \in M$, then $N_{\boldsymbol{x}}M$ denotes the fiber of the normal bundle to M at $\boldsymbol{x}$, thought of as an affine subspace of $\mathbb{R}^N$ passing through $\boldsymbol{x}$. The underlying vector space is denoted by $N^0_{\boldsymbol{x}}M$.
- If $W \subset \mathbb{R}^N$, then $P_W : \mathbb{R}^N \to W$ is the *nearest point map*, that is,

$$P_W(\boldsymbol{x}) = \underset{\boldsymbol{y}\in W}{\arg\min}\, \|\boldsymbol{x} - \boldsymbol{y}\|_2. \tag{15.9}$$

  It is defined on the complement of a set of dimension at most $N-1$.
- The reflection, R_W, around the set $W \subset \mathbb{R}^N$ is defined by

$$R_W(\boldsymbol{x}) = 2P_W(\boldsymbol{x}) - \boldsymbol{x}. \tag{15.10}$$

- If $A, B \subset \mathbb{R}^N$, then

$$D_{AB}(\boldsymbol{x}) = \boldsymbol{x} + P_A \circ R_B(\boldsymbol{x}) - P_B(\boldsymbol{x}). \tag{15.11}$$

- If $A, B \subset \mathbb{R}^N$, then $d_{AB} : A \times B \to \mathbb{R}_+$ is defined by

$$d_{AB}(\boldsymbol{x}, \boldsymbol{y}) = \|\boldsymbol{x} - \boldsymbol{y}\|_2. \tag{15.12}$$

References

Alaifari, R., Daubechies, I., Grohs, P., and Yin, R. 2019. Stable phase retrieval in infinite dimensions. *Found. Comput. Math.*, **19**, 896–900. https://link.springer.com/article/10.1007%2Fs10208-018-9399-7

Barmherzig, D. A., Sun, J., Candès, E. J., Lane, T. J., and Li, P.-N. 2019a. Dual-reference design for holographic phase retrieval. *13th International Conference on Sampling Theory and Applications (SampTA)*, pp. 1–4.

Barmherzig, D. A., Sun, J., Candès, E. J., Lane, T. J., and Li, P-N. 2019b. Holographic phase retrieval and optimal reference design. *Inverse Probl.*, **35**(9), 094001.

Barnett, A., Epstein, C. L., Greengard, L., and Magland, J. 2020. Geometry of the Phase Retrieval Problem. *Inverse Probl.*, **36**(9), 094003.

Bauschke, H. H. and Borwein, J. M. 1993. On the convergence of von Neumann's alternating projection algorithm for two sets. *Set-Valued Anal.*, **1**(2), 185–212.

Bauschke, H. H., and Borwein, J. M. 1996. On projection algorithms for solving convex feasibility problems. *SIAM Rev.*, **38**(3), 367–426.

Bauschke, H. H., Combettes, P. L., and Russell, L. D. 2002. Phase retrieval, error reduction algorithm, and Fienup variants: A view from convex optimization. *J. Opt. Soc. Am. A*, **19**, 1334–1345.

Beinert, R. 2017. Non-negativity constraints in the one-dimensional discrete-time phase retrieval problem. *Information and Inference: A Journal of the IMA*, **6**, 213–224.

Borwein, J. M. 2012. Maximum entropy and feasibility methods for convex and nonconvex inverse problems. *Optimization*, **61**(1), 1–33.

Borwein, J. M., and Sims, B. 2011. The Douglas–Rachford algorithm in the absence of convexity. In H. Bauschke, R. Burachik, P. Combettes, et al. (eds.) *Fixed-Point Algorithms for Inverse Problems in Science and Engineering*, pp. 93–109. Springer Optimization and Its Applications, Vol. 49. New York: Springer.

Bruck, Yu. M., and Sodin, L. G. 1979. On the ambiguity of the image reconstruction problem. *Opt. Commun.*, **30**, 304–308.

Cahill, J., Casazza, P. G., and Daubechies, I. 2016. Phase retrieval in infinite-dimensional Hilbert spaces. *Trans. Amer. Math. Soc., Series B*, **3**, 63–76.

Candès, E. J., Eldar, Y. C., Strohmer, T., and Voroninski, V. 2015a. Phase retrieval via matrix completion. *SIAM Rev*, **57**, 225–251.

Candès, E. J., Li, X., and Soltanolkotabi, M. 2015b. Phase retrieval via Wirtinger flow. *IEEE Trans. Information Theory*, **61**(4), 1985–2007.

Chapman, H. N., Barty, A., Marchesini, S., et al. 2006. High-resolution *ab initio* three-dimensional x-ray diffraction microscopy. *J. Opt. Soc. Am. A*, **23**, 1179–1200.

Conca, A., Edidin, D., Hering, M., and Vinzant, C. 2015. An algebraic characterization of injectivity in phase retrieval. *Appl. Comput. Harmon. A.*, **38**(2), 346–356.

Crimmins, T. R. and Fienup, J. R. 1983. Uniqueness of phase retrieval for functions with sufficiently disconnected support. *J. Opt. Soc. Am.*, **73**, 218–221.

Deutsch, F. 1985. Rate of convergence of the method of alternating projections. In B. Brosowski and F. Deutsch (eds) *Parametric Optimization and Approximation (Oberwolfach, 1983)*, pp. 96–107. Internat. Schriftenreihe Numer. Math., Vol. 72. Basel: Birkhäuser.

Dierolf, M., Menzel, A., Thibault, P., et al. 2010. Ptychographic X-ray computed tomography at the nanoscale. *Nature*, **467**, 436–439.

Elser, V. 2003. Phase retrieval by iterated projections. *JOSA A*, **20**(1), 40–55.

Elser, V., Lan, T-Y., and Bendory, T. 2018. Benchmark problems for phase retrieval. *SIAM J. Imaging Sci.*, **11**(4), 2429–2455.

Elser, V., Rankenburg, I., and Thibault, P. 2007. Searching with iterated maps. *P. Natl. Acad. Sci. USA*, **104**(2), 418–423.

Epstein, C. L. 2005. How well does the finite Fourier transform approximate the Fourier transform? *Comm. Pure and App. Math.*, **58**, 1421–1435.

Epstein, C. L. and Schotland, J. 2008. The bad truth about Laplace's transform. *SIAM Review*, **50**(3), 504–520.

Fienup, J. R. 1978. Reconstruction of an object from the modulus of its Fourier transform. *Opt. Lett.*, **3**, 27–29.

Fienup, J. R. 1982. Phase retrieval algorithms: A comparison. *Appl. Opt.*, **21**, 2758–2769.

Fienup, J. R. 1987. Reconstruction of a complex-valued object from the modulus of its Fourier transform using a support constraint. *J. Opt. Soc. Am. A*, **4**, 118–123.

Fienup, J. R. and Kowalczyk, A. M. 1990. Phase retrieval for a complex-valued object by using a low-resolution image. *J. Opt. Soc. Am. A*, **7**(3), 450–458.

Fienup, J. R. and Wackerman, C. C. 1986. Phase-retrieval stagnation problems and solutions. *J. Opt. Soc. Am. A*, **3**, 1897–1907.

Fienup, J. R., Crimmins, T. R. and Holsztynski, W. 1982. Reconstruction of the support of an object from the support of its autocorrelation. *J. Opt. Soc. Am.*, **72**, 610–624.

Fienup, J. R., Crimmins, T. R. and Thelen, B. J. 1990. Improved bounds on object support from autocorrelation support and application to phase retrieval. *J. Opt. Soc. Am. A*, **7**(1), 3–13.

Gerchberg, R. W. and Saxton, W. O. 1972. A practical algorithm for the determination of the phase from image and diffraction plane pictures. *Optik*, **35**(2), 237–246.

Gonsalves, R. A. 1976. Phase retrieval from modulus data. *J. Opt. Soc. Am.*, **66**(9), 961–964.

Gravel, S. and Elser, V. 2008. Divide and concur: A general approach to constraint satisfaction. *Phys. Rev. E*, **78**(3), 036706–5.

Griffiths, P. and Harris, J. 1978. *Principles of Algebraic Geometry*. New York: Wiley-Interscience.

Guizar-Sicairos, M. and Fienup, J. R. 2007. Holography with extended reference by autocorrelation linear differential operation. *Opt. Express*, **15**(26), 17592–17612.

Guizar-Sicairos, M. and Fienup, J. R. 2008. Direct image reconstruction from a Fourier intensity pattern using HERALDO. *Opt. Lett.*, **33**(22), 2668–2670.

Guizar-Sicairos, M., Johnson, I., Diaz, A., et al. 2014. High-throughput ptychography using Eiger: Scanning X-ray nano-imaging of extended regions. *Opt. Express*, **22**(12), 14859–14870.

Hayes, M. H. 1982. The reconstruction of a multidimensional sequence from the phase or magnitude of its Fourier transform. *IEEE Trans. on Acoustics, Speech and Sig. Proc.*, **30**, 140–153.

Hesse, R. and Luke, D. R. 2013. Nonconvex notions of regularity and convergence of fundamental algorithms for feasibility problems. *SIAM J. Optimiz.*, **23**(4), 2397–2419.

Huang, K., and Eldar, Y. C. 2017. Phase retrieval using a conjugate symmetric reference. *International Conference on Sampling Theory and Applications (SampTA)*, Vol. 134, pp. 331–335

Huang, K., Eldar, Y. C., and Sidiropoulos, N. D. 2016. Phase retrieval from 1D Fourier measurements: Convexity, uniqueness, and algorithms. *IEEE T. Signal Proces.*, **64**(23), 6105–6117.

Jacobsen, C. 2019. *X-ray Microscopy*, 1st ed. Advances in Microscopy and Microanalysis. Cambridge: Cambridge University Press.

John, F. 1960. Continuous dependence on data for solutions of partial differential equations with a prescribed bound. *Commun. Pur. App. Math.*, **13**(4), 551–585.

Katznelson, Y. 1968. *An Introduction to Harmonic Analysis*. New York-London-Sydney: John Wiley & Sons.

Lane, R. G. 1991. Phase retrieval using conjugate gradient minimization. *J. Mod. Optic.*, **38**(9), 1797–1813.

Lee, J. M. 2018. *Introduction to Riemannian Manifolds*, 2nd ed. Graduate Texts in Mathematics, Vol. 176. New York: Springer.

Leshem, B., Raz, O., Jaffe, A., and Nadler, B. 2018. The discrete sign problem: Uniqueness, recovery algorithms and phase retrieval applications. *Appl. Comput. Harmon. A.*, **45**(3), 463–485.

Levin, E. and Bendory, T. 2019. A note on Douglas-Rachford, subgradients, and phase retrieval. arXiv: abs/1911.13179

Li, G. and Pong, T. K. 2016. Douglas–Rachford splitting for nonconvex optimization with application to nonconvex feasibility problems. *Math. Program.*, **159**(1), 371–401.

Lindstrom, S. B. and Sims, B. 2021. Survey: Sixty years of Douglas–Rachford. *J. Aust. Math. Soc.*, **110**(3), 333–370.

Luke, D. R. 2005. Relaxed averaged alternating reflections for diffraction imaging. *Inverse Probl.*, **21**(1), 37–50.

Luke, D. R. and Martins, A. L. 2020. Convergence analysis of iterative algorithms for phase retrieval. In T. Salditt, A. Egner, and D. R. Luke (eds.) *Nanoscale Photonic Imaging* 583–601. Topics in Applied Physics, Vol. 134. Cham: Springer.

Marchesini, S. 2007. Phase retrieval and saddle-point optimization. *J. Opt. Soc. Am. A*, **24**(10), 3289–3296.

Marchesini, S., Tu, Y.-C., and Wu, H.-T. 2016. Alternating projection, ptychographic imaging and phase synchronization. *Appl. Comput. Harmon. A.*, **41**(3), 815–851.

Maretzke, S. and Hohage, T. 2016. *Pinhole-CDI: Unique and Deterministic Phase Retrieval via Beam-Confinement.* http://ip.math.uni-goettingen.de/data-smaretzke/poster_coherence_2016-06_smaretzke.pdf.

Maretzke, S. and Hohage, T. 2017. Stability estimates for linearized near-field phase retrieval in X-ray phase constrast imaging. *SIAM J. Appl. Math.*, **77**, 384–408.

Milnor, J. 1963. *Morse Theory.* Based on lecture notes by M. Spivak and R. Wells. Annals of Mathematics Studies, No. 51. Princeton, NJ: Princeton University Press.

Nakajima, N. 1995. Phase Retrieval Using the Properties of Entire Functions. Advances in Imaging and Electron Physics, Vol. 93. New York: Elsevier.

Nakajima, N. and Asakura, T. 1985. A new approach to two-dimensional phase retrieval. *Optica Acta: International Journal of Optics*, **32**(6), 647–658.

Nakajima, N. and Asakura, T. 1986. Two-dimensional phase retrieval using the logarithmic Hilbert transform and the estimation technique of zero information. *J. Phys. D: Appl. Phys.*, **19**(3), 319–331.

Osherovich, E. 2011. *Numerical methods for phase retrieval.* arXiv:1203.4756v1 [physics.optics]. PhD Thesis, Technion.

Pfeiffer, F. 2018. X-ray ptychography. *Nat. Photonics*, **12**, 9–17.

Pham, M., Yin, P., Rana, A., Osher, S., and Miao, J. 2019. Generalized proximal smoothing for phase retrieval. *Microsc. Microanal.*, **25**(S2), 118119.

Phan, H. M. 2016. Linear convergence of the Douglas-Rachford method for two closed sets. *Optimization*, **65**(2), 369–385.

Podorov, S. G., Pavlov, K. M., and Paganin, D. M. 2007. A non-iterative reconstruction method for direct and unambiguous coherent diffractive imaging. *Opt. Express*, **15**(16), 9954–9962.

Rodenburg, J. M., Hurst, A. C., Cullis, A. G., et al. 2007. Hard-X-ray lensless imaging of extended objects. *Phys. Rev. Lett.*, **98**(Jan), 034801.

Sanz, J. L. C. 1985. Mathematical considerations for the problem of Fourier transform phase retrieval from magnitude. *SIAM J. Appl. Math.*, **45**, 651–664.

Seldin, J. H. and Fienup, J. R. 1990. Numerical investigation of the uniqueness of phase retrieval. *J. Opt. Soc. Am. A*, **7**(3), 412–427.

Shechtman, Y., Eldar, Y. C., Szameit, A., and Segev, M. 2011. Sparsity based sub-wavelength imaging with partially incoherent light via quadratic compressed sensing. *Opt. Express*, **19**(16), 14807–14822.

Sidorenko, P., Kfir, O., Shechtman, Y., et al. 2015. Sparsity-based super-resolved coherent diffraction imaging of one-dimensional objects. *Nat. Commun.*, **6**(1), 8209.

Spivak, M. 1965. *Calculus on Manifolds. A Modern Approach to Classical Theorems of Advanced Calculus*. New York and Amsterdam: W. A. Benjamin, Inc.

Spivak, M. 1979. *A Comprehensive Introduction to Differential Geometry. Vol. I.* 2nd ed. Wilmington, CA: Publish or Perish, Inc.

Stein, E. M. and Shakarchi, R. 2003a. *Complex Analysis*. Princeton Lectures in Analysis, Vol. 2. Princeton, NJ: Princeton University Press.

Stein, E. M. and Shakarchi, R. 2003b. *Fourier Analysis*. Princeton Lectures in Analysis, Vol. 1. Princeton, NJ: Princeton University Press.

Taylor, M. E. 1996. *Partial Differential Equations. II*. Applied Mathematical Sciences, Vol. 116. New York: Springer-Verlag.

Thibault, P., Dierolf, M., Menzel, A., et al. 2008. High-resolution scanning X-ray diffraction microscopy. *Science*, **321**, 379–382.

von Neumann, J. 1950. *Functional Operators, Vol. II*. Reprint of 1933 lecture notes. Princeton, NJ: Princeton University Press.

Watson, G. N. 1922. *A Treatise on the Theory of Bessel Functions*. Cambridge: Cambridge University Press.

Weinberger, S. 2004. On the topological social choice model. *J. Econom. Theory*, **115**(2), 377–384.

Wu, S.-P., Boyd, S. P., and Vandenberghe, L. 1996. FIR filter design via semidefinite programming and spectral factorization. *IEEE Conference on Decision and Control*, **1**, 271–276.

Yasir, P. A. A. and Ivan, J. S. 2016. Phase estimation using phase gradients obtained through Hilbert transform. *J. Opt. Soc. Am. A*, **33**(10), 2010–2019.

Index

Printed in the United States
by Baker & Taylor Publisher Services